Linear Algebra and Its Applications with R

Textbooks in Mathematics

Series editors:
Al Boggess, Kenneth H. Rosen

Functional Linear Algebra
Hannah Robbins

Introduction to Financial Mathematics
With Computer Applications
Donald R. Chambers, Qin Lu

Linear Algebra
An Inquiry-Based Approach
Jeff Suzuki

The Geometry of Special Relativity
Tevian Dray

Mathematical Modeling in the Age of the Pandemic
William P. Fox

Games, Gambling, and Probability
An Introduction to Mathematics
David G. Taylor

Financial Mathematics
A Comprehensive Treatment in Discrete Time
Giuseppe Campolieti, Roman N. Makarov

Maple™ Projects of Differential Equations
Robert P. Gilbert, George C. Hsiao, Robert J. Ronkese

Practical Linear Algebra
A Geometry Toolbox, Fourth Edition
Gerald Farin, Dianne Hansford

An Introduction to Analysis, Third Edition
James R. Kirkwood

Student Solutions Manual for Gallian's Contemporary Abstract Algebra, Tenth Edition
Joseph A. Gallian

Elementary Number Theory
Gove Effinger, Gary L. Mullen

Philosophy of Mathematics
Classic and Contemporary Studies
Ahmet Cevik

Linear Algebra and Its Applications with R
Ruriko Yoshida

https://www.routledge.com/Textbooks-in-Mathematics/book-series/CANDHTEXBOOMTH

Linear Algebra and Its Applications with R

Ruriko Yoshida

Naval Postgraduate School

CRC Press
Taylor & Francis Group
Boca Raton London New York

CRC Press is an imprint of the
Taylor & Francis Group, an **informa** business

A CHAPMAN & HALL BOOK

First edition published 2021
by CRC Press
6000 Broken Sound Parkway NW, Suite 300, Boca Raton, FL 33487-2742

and by CRC Press
2 Park Square, Milton Park, Abingdon, Oxon, OX14 4RN

© 2021 Taylor & Francis Group, LLC

CRC Press is an imprint of Taylor & Francis Group, LLC

Library of Congress Cataloging-in-Publication Data

Names: Yoshida, Ruriko, author.
Title: Linear algebra and its applications with R / Ruriko Yoshida, Naval
    Postgraduate School.
Description: First edition. | Boca Raton : Chapman & Hall/CRC Press, 2021.
    | Series: Textbooks in mathematics | Includes bibliographical references
    and index.
Identifiers: LCCN 2021000953 (print) | LCCN 2021000954 (ebook) | ISBN
    9780367486846 (hardback) | ISBN 9781003042259 (ebook)
Subjects: LCSH: Algebras, Linear--Data processing. | R (Computer program
    language)
Classification: LCC QA185.D37 Y67 2021 (print) | LCC QA185.D37 (ebook) |
    DDC 512/.5--dc23
LC record available at https://lccn.loc.gov/2021000953
LC ebook record available at https://lccn.loc.gov/2021000954

ISBN: 978-0-367-48684-6 (hbk)
ISBN: 978-1-032-02052-5 (pbk)
ISBN: 978-1-003-04225-9 (ebk)

Typeset in CMR10
by KnowledgeWorks Global Ltd.

DOI: 10.1201/9781003042259

To my husband, Pete.

Contents

Preface   xiii

1 Systems of Linear Equations and Matrices   1
  1.1 Introductory Example from Statistics   1
  1.2 What Is a Matrix? What Is a Vector?   2
     1.2.1 Task Completion Checklist   2
     1.2.2 Working Examples   3
     1.2.3 Matrix and Vector   4
     1.2.4 Checkmarks   15
     1.2.5 Conceptual Quizzes   16
     1.2.6 Regular Exercises   20
     1.2.7 Lab Exercises   22
     1.2.8 Practical Applications   23
     1.2.9 Supplements with python Code   26
  1.3 Systems of Linear Equations   27
     1.3.1 Task Completion Checklist   27
     1.3.2 Working Examples   28
     1.3.3 System of Linear Equations   29
     1.3.4 Checkmarks   46
     1.3.5 Conceptual Quizzes   46
     1.3.6 Regular Exercises   48
     1.3.7 Lab Exercises   53
     1.3.8 Practical Applications   54
     1.3.9 Supplements with python Code   56
  1.4 Echelon Form   57
     1.4.1 Task Completion Checklist   58
     1.4.2 Working Examples   58
     1.4.3 Echelon Form   59
     1.4.4 Checkmarks   72
     1.4.5 Conceptual Quizzes   72
     1.4.6 Regular Exercises   75
     1.4.7 Lab Exercises   78
     1.4.8 Practical Applications   80
     1.4.9 Supplements with python Code   81
  1.5 Discussion   82

2 Matrix Arithmetic   85

2.1   Introductory Example from Statistics   85

2.2   Matrix Operations   87

     2.2.1   Task Completion Checklist   87

     2.2.2   Working Examples   88

     2.2.3   Matrix Operations   88

     2.2.4   Checkmarks   98

     2.2.5   Conceptual Quizzes   99

     2.2.6   Regular Exercises   102

     2.2.7   Lab Exercises   105

     2.2.8   Practical Applications   105

     2.2.9   Supplements with `python` Code   107

2.3   Properties of Matrix Operations and Matrix Inverse   108

     2.3.1   Task Completion Checklist   109

     2.3.2   Working Examples   109

     2.3.3   Properties of Matrix Operations and Matrix Inverse   .   110

     2.3.4   Checkmarks   119

     2.3.5   Conceptual Quizzes   120

     2.3.6   Regular Exercises   121

     2.3.7   Lab Exercises   123

     2.3.8   Practical Applications   124

     2.3.9   Supplements with `python` Code   130

2.4   Elementary Matrices   131

     2.4.1   Task Completion Checklist   131

     2.4.2   Working Examples   131

     2.4.3   Elementary Matrices   132

     2.4.4   Checkmarks   136

     2.4.5   Conceptual Quizzes   137

     2.4.6   Regular Exercises   137

     2.4.7   Lab Exercises   139

     2.4.8   Practical Applications   142

     2.4.9   Supplements with `python` Code   145

2.5   Discussion   146

3 Determinants   149

3.1   Introductory Example from Astronomy   149

3.2   Determinants   151

     3.2.1   Task Completion Checklist   152

     3.2.2   Working Examples   152

3.3   Introduction of Determinants   153

     3.3.1   Checkmarks   157

     3.3.2   Conceptual Quizzes   157

     3.3.3   Regular Exercises   158

     3.3.4   Lab Exercises   161

     3.3.5   Practical Applications   164

| | | 3.3.6 | Supplements with `python` Code | 165 |

3.4	Properties of Determinants	166
	3.4.1	Task Completion Checklist	166
	3.4.2	Working Examples	166
	3.4.3	Properties of Determinants	167
	3.4.4	Checkmarks	174
	3.4.5	Conceptual Quizzes	174
	3.4.6	Regular Quizzes	177
	3.4.7	Lab Exercises	179
	3.4.8	Practical Applications	180
	3.4.9	Supplements with `python` Code	181
3.5	Cramer's Rule	182
	3.5.1	Task Completion Checklist	182
	3.5.2	Working Examples	182
	3.5.3	Cramer's Rule	184
	3.5.4	Checkmarks	187
	3.5.5	Conceptual Quizzes	188
	3.5.6	Regular Exercises	188
	3.5.7	Lab Exercises	192
	3.5.8	Practical Applications	193
	3.5.9	Supplements with `python` Code	194
3.6	Discussion	194

4 Vector Spaces **199**
4.1	Introductory Example from Data Science	199
4.2	Vector Spaces and Subspaces	201
	4.2.1	Task Completion Checklist	201
	4.2.2	Working Examples	202
	4.2.3	Vector Spaces and Vector Subspaces	203
	4.2.4	Checkmarks	209
	4.2.5	Conceptual Quizzes	209
	4.2.6	Regular Exercises	209
	4.2.7	Lab Exercises	210
	4.2.8	Practical Applications	210
	4.2.9	Supplements with `python` Code	212
4.3	Null Space, Column Space, and Row Space	213
	4.3.1	Task Completion Checklist	213
	4.3.2	Working Examples	214
	4.3.3	Null Space, Column Space, and Row Space	215
	4.3.4	Checkmarks	222
	4.3.5	Conceptual Quizzes	222
	4.3.6	Regular Exercises	223
	4.3.7	Lab Exercises	225
	4.3.8	Practical Applications	226
	4.3.9	Supplements with `python` Code	229

4.4   Spanning Sets and Bases   230
      4.4.1   Task Completion Checklist   230
      4.4.2   Working Examples   231
      4.4.3   Spanning Sets and Bases   232
      4.4.4   Checkmarks   239
      4.4.5   Conceptual Quizzes   240
      4.4.6   Regular Exercises   242
      4.4.7   Lab Exercises   246
      4.4.8   Practical Applications   247
      4.4.9   Supplements with python Code   255
4.5   Coordinates Systems and Change of Basis   255
      4.5.1   Task Completion Checklist   255
      4.5.2   Working Examples   256
      4.5.3   Coordinate Systems and Change of Basis   257
      4.5.4   Checkmarks   262
      4.5.5   Conceptual Quizzes   263
      4.5.6   Regular Exercises   264
      4.5.7   Lab Exercises   266
      4.5.8   Practical Applications   266
      4.5.9   Supplements with python Code   272
4.6   Discussion   272

5   Inner Product Space   277
5.1   Introductory Example from Statistics   277
5.2   Inner Products   277
      5.2.1   Task Completion Checklist   278
      5.2.2   Working Examples   279
      5.2.3   Inner Products   279
      5.2.4   Checkmarks   285
      5.2.5   Conceptual Quizzes   286
      5.2.6   Regular Exercises   288
      5.2.7   Lab Exercises   289
      5.2.8   Practical Applications   290
      5.2.9   Supplements with python Code   291
5.3   Angles and Orthogonality   294
      5.3.1   Task Completion Checklist   294
      5.3.2   Working Examples   295
      5.3.3   Angles and Orthogonality   296
      5.3.4   Checkmarks   303
      5.3.5   Conceptual Quizzes   303
      5.3.6   Regular Exercises   305
      5.3.7   Lab Exercises   307
      5.3.8   Practical Applications   309
      5.3.9   Supplements with python Code   310
5.4   Discussion   310

6 Eigen Values and Eigen Vectors **313**
6.1 Introductory Example from Data Science: Image Compression 313
6.2 Eigen Values and Eigen Vectors 314
    6.2.1 Task Completion Checklist 315
    6.2.2 Working Examples 315
    6.2.3 Eigen Values and Eigen Vectors 317
    6.2.4 Checkmarks 322
    6.2.5 Conceptual Quizzes 323
    6.2.6 Regular Exercises 325
    6.2.7 Lab Exercises 326
    6.2.8 Practical Applications 327
    6.2.9 Supplements with `python` Code 328
6.3 Diagonalization 330
    6.3.1 Task Completion Checklist 330
    6.3.2 Working Examples 330
    6.3.3 Diagonalization 332
    6.3.4 Checkmarks 336
    6.3.5 Conceptual Quizzes 336
    6.3.6 Regular Exercises 338
    6.3.7 Lab Exercises 338
    6.3.8 Practical Applications 339
    6.3.9 Supplements with `python` Code 341
6.4 Discussion 341

7 Linear Regression **345**
7.1 Introductory Example from Statistics 345
7.2 Simple Linear Regression 346
    7.2.1 Task Completion Checklist 346
    7.2.2 Basic Terminology 347
    7.2.3 Simple Linear Regression 347
    7.2.4 Multiple Linear Regression 351
    7.2.5 Checkmarks 355
    7.2.6 Practical Applications 356
    7.2.7 Supplements with `python` Code 357

8 Linear Programming **359**
8.1 Introductory Example from Optimization 359
8.2 Linear Programming 359
    8.2.1 Task Completion Checklist 360
    8.2.2 Basic Terminology 361
    8.2.3 Linear Programming 364
    8.2.4 Checkmarks 366
    8.2.5 Practical Applications 366
    8.2.6 Supplements with `python` Code 368

9 Network Analysis   **371**
  9.1 Introductory Example from Network Analysis 371
  9.2 Graphs and Networks 373
      9.2.1 Task Completion Checklist 373
      9.2.2 Basic Terminology 374
      9.2.3 Properties of Laplacian Matrices 381
      9.2.4 Checkmarks 384
      9.2.5 Practical Applications 384
      9.2.6 Supplements with `python` Code 386
  9.3 Discussion 387

Appendices   **389**

A Introduction to `RStudio` via Amazon Web Services (AWS)   **391**
  A.1 Setting Up AWS 391
  A.2 Basics in `RStudio` 392

B Introduction to `R`   **395**
  B.1 Display in `R` 395
  B.2 Setting Up the `R` Programming Environment 395
  B.3 Getting Started with Objects in `R` 396
      B.3.1 Assigning an Object (Variable) 397
      B.3.2 Boolean Operators 398
  B.4 Saving a Session and Data 399

Bibliography   **401**

Index   **405**

Preface

Distinctive Features

Linear Algebra from the View of Its Applications

Linear algebra is the foundation for many other fields, such as data science, statistics, computer science, bioinformatics, physics, economics, and many more. For example, linear algebra is a subject essential to learning since linear models based on linear algebra are the base models in statistical learning. However, from our teaching experience, students in data science and bioinformatics programs tend not to realize the importance of materials from linear algebra courses. We think this is because classical linear algebra courses tend to cover only mathematical contexts but not the computational aspect of linear algebra or its applications to data science and bioinformatics. In order to appreciate tools from linear algebra, students need to do computations with relatively large simulated and/or empirical data sets. Therefore, we cover lectures as well as lab exercises on simulated and empirical data sets, which we cover extensively in this book. In this book, each section will consist of three elements: (1) view from an application using working examples, (2) basic lecture materials, and (3) lab exercises with R and python. The goal of this textbook is to provide students a theoretical basis via lecture materials which can then be applied to the practical R and python problems, providing the tools needed for real-world applications.

Emphasis on Computational Linear Algebra with R

Today, because of the advances in computers, we are able to generate large-scale data sets. Even though we are able to conduct large computations with software such as R and python, we can use many tricks from linear algebra, such as singular value decomposition, to make a computation faster. These tricks often provide the key to making a large-scale computation feasible. Therefore, the lab exercises with R and supplements with python in each section demonstrate how tools from linear algebra can help solve problems in applied science. These exercises include computational experiments with empirical and simulated data sets.

Lab Exercises with R and python on Simulated and Empirical Data

Each section starts with working examples to demonstrate the tools from linear algebra. We go through the section with computational examples with R, so that students learn how to get started with R. Then, each section has lab exercises with R. These exercises start from easy computations, such as computing determinants of matrices, to practical applications on simulated and empirical data sets with R. Supplemental materials with python follow the lab exercises. These supplements with python include a translation of computational examples in the section on python code, allowing students to use their choice of R or python to work on the problems.

Note that through this book, commands preceded by $ are at the command line, > are in an R console, and >>> are in a python console. Basics and an introduction to R and Amazon Web Services (AWS) can be found in Appendices B and A.

Goals

There are three primary goals of this book for students: the importance of linear algebra in many science fields; computational experiences with R and python; and building computational skills with them.

Importance of Linear Algebra

Tools from linear algebra are relevant and useful to solve problems, especially computational problems, in academic and non-academic disciplines. From our teaching experience, linear algebra is often taught without adequate emphasis on practical applications, which causes students to gloss over its importance. This book is designed from first principles to demonstrate the importance of linear algebra through working computational examples with R and python. All working examples with R and python are available on the author's website so that it is easy for students to test out these examples and to actually see how these tricks work in practical applications.

Computational Experience with R and python

For students to see how important tools from linear algebra are in solving computational problems, they have to actually **program**. Demonstrations and theory are not enough for students to actually learn how to compute. S. A. Ambrose et al. said in [4] that goal-directed practice with targeted feedback is critical to learning for students. Therefore, each section has computational exercises with R from simple problems to practical applications focused on the materials in the section after defining its materials and computational examples. One of the great things about computational exercises is that students will see the results from R as immediate feedback. We include R tutorials which include how to install R in the Appendix. So if a student has never seen R, they can get started without any additional help.

Computational Skills with R and python

To encourage students to work on computations, we devote a subsection to R computational labs, which include easy computations to practical applications. In each practical application, we walk through applications of materials in the section. In the author's course, we spend half of the class time working computational exercises in labs, which is extremely useful. We start from very easy exercises following computational examples, so that less computationally oriented students can follow the exercises. We end with practical applications with simulated and empirical data sets so that such students will be comfortable with their computational skills at the end of the course. While we use R primarily for computational examples, code supplements are available for students who feel more comfortable with python. We choose R mainly because the R data structure is based on matrix operations and is similar to matlab. We supplement code with python since python is one of the most popular languages in data science, optimization, and computer science. One of the goals in this book is to provide practical computational skills to students in a classroom.

How this Book Is Organized

Each chapter uses the same structure. Introductory examples and problems show students why we care about the materials in the chapter. Then we introduce the goals of the chapter so that students will know exactly what they can expect to see in the chapter. After introducing sections of the chapter, we add regular exercises as supplements. These questions are word problems which are combinations of the materials throughout the chapter. Then we end with supplementary lab exercises with R and python for computational problems covering all sections in the chapter.

Each section consists of the following:

1. Goals in the section

2. Introductory working examples

3. Materials and definitions in the section

4. Checkmarks for the materials in the section

5. Conceptual quizzes

6. Regular exercises

7. Lab exercises with R

8. Supplements with python

Goals are stated in the beginning of each section so that students will know exactly what to expect in the following pages. Then, introductory examples are introduced. These examples are used throughout the section, and computational examples with R and its outputs follow. After definitions and properties of the materials in the section, a list of checkmarks is posted so that students can verify that they understand the materials introduced in the section. Conceptual quizzes are for students to test themselves to understand concepts of the materials in simple true/false and multiple-choice questions. Regular exercises test whether students understand definitions and mathematical properties of the materials; and lab exercises with R build students' computational skills in R. The python supplement provides alternative python code for all examples in the section which use R.

Course Design

We want students to learn materials in each course and we want them to be able to solve problems after finishing a course on linear algebra. We teach practical ways to solve problems with **confidence and competence** by learning along with students. The outline of each lecture is shown in Table 0.1.

Time management	We divide each one-hour class into four components: • 10 minutes of discussion/review, • 20 minutes of lecture, • 20 minutes of solving problems, and • 10 minutes of additional discussion.
Guided practice	In addition to weekly homework and exams, during the 20 minutes of solving problems together with students, we give them questions and give them time to solve these in groups. Then we discuss how each group solved the problems and discuss what they did correctly and where they made mistakes. This will give immediate feedback to students. This creates the environment where students feel comfortable asking questions. While they are working on problems, we walk through the classroom so that we can see someone who might be struggling to solve problems and we can help weaker students. Also, this makes clear what students can do and they cannot do for each material in the course.
Immediate feedback	This period of solving problems during each class also gives students immediate feedback on their performance. This makes clear to me how and where students make mistakes and also we understand the students' thought processes about solving the problem. This is also immediate feedback to my teaching as well, which we can use to improve in the future.
Learning along with students	We focus on the learning center approach. This means that we learn along with all students in the classroom. We create an environment in which we are all learning together, so that they can ask anything if they are not sure. We tell students clearly that we are learning two ways: we teach students how exciting it is to discover new materials in mathematical courses, and also students teach me how to teach materials clearer and in a better way.
Developing confidence and competence	We teach class materials for my students' future. Immediately, this means they can be capable for the next courses they are taking, but also that they can solve problems by themselves after they graduate. For example, we teach them how to use Excel, R, and RStudio, and how to use Google for solving problems. Then, during exams, we let them use these tools to solve problems in exams, since in real-world applications we use all of these tools to solve problems.

TABLE 0.1
Outline of course development

1

Systems of Linear Equations and Matrices

Solving a system of linear equations appears in many problems in academic and non-academic disciplines since this is one of the simplest and most effective models to solve practical problems in many areas. The working example is a two-way contingency table from categorical data analysis, a classic form of analysis from statistics. The walking example in this chapter is from statistics. Categorical data analysis is one of classical statistical analysis and the example is called a *two-way contingency table* from categorical data analysis.

1.1 Introductory Example from Statistics

	Jan	Feb	March	April	May	June	July	Aug	Sep	Oct	Nov	Dec
Jan	1	0	0	0	1	2	0	0	1	0	1	0
Feb	1	0	0	1	0	0	0	0	0	1	0	2
March	1	0	0	0	2	1	0	0	0	0	0	1
April	3	0	2	0	0	0	1	0	1	3	1	1
May	2	1	1	1	1	1	1	1	1	1	1	0
June	2	0	0	0	1	0	0	0	0	0	0	0
July	2	0	2	1	0	0	0	0	1	1	1	2
Aug	0	0	0	3	0	0	1	0	0	1	0	2
Sep	0	0	0	1	1	0	0	0	0	0	1	0
Oct	1	1	0	2	0	0	1	0	0	1	1	0
Nov	0	1	1	1	2	0	0	2	0	1	1	0
Dec	0	1	1	0	0	0	1	0	0	0	0	0

TABLE 1.1
Relationship between birth-month and death-month: data gathered to test the hypothesis of association between the birth day and the death day. Columns represent the month of the birth day and rows represent the month of the death day.

Table 1.1 is from [16] and it records the month of birth and death for 82 descendants of Queen Victoria. This data is used to analyze an association, i.e., relation, between birth-month and death-month. It used to be thought that there were some relations between people's birth-month and death-month. Tables like Table 1.1 are called *two-way contingency tables* and particularly we call Table 1.1 a 12 × 12 contingency table. Also notice that Table 1.1 has many zeros. Even though some numbers in the table are non-zero, these numbers are very small integers, such as 1, 2, or 3. Such tables are called

DOI: 10.1201/9781003042259-1

sparse contingency tables and give scientists headaches. Big Data is a popular term in scientific fields and if we are dealing with "Big Data", dealing with sparse contingency tables is a real problem.

One application of a system of linear equations is to test the data security problem using Table 1.1 to demonstrate. While this is an old data set, more recent data set summaries can sometimes be exploited to reveal personal information. In a large survey-such as the US Census-it is sometimes possible to identify an individual respondent's answers. Before the release of a summary, we can test whether such a possibility exists. In order to make sure if it is safe to release, we need to set up a system of linear equations and inequalities [38].

We will show how these problems can be formulated in terms of systems of linear equations (and linear inequalities) with smaller working examples in the following sections and we will return to this introductory example.

1.2 What Is a Matrix? What Is a Vector?

1.2.1 Task Completion Checklist

- Prior to the Lecture:

  1. The preface to see how this chapter is organized.
  2. Install R in your computer if needed.
  3. The Appendix for basics in R programming.

- During the Lecture:

  1. Read the definitions of a matrix and a vector.
  2. Create examples of a matrix and a vector.
  3. Try some examples without the help of a computer.
  4. With R, learn how to perform computational examples in this section.

- After the Lecture:

  1. Take conceptual quizzes to make sure you understand the materials in this section.
  2. Do some regular exercises.
  3. Conduct lab exercises with R.
  4. Conduct practical applications with R.
  5. If you are interested in python, read the supplement in this section and conduct lab exercises and practical applications with python.

TABLE 1.2
Information on variables of "Smarket" data set.

Year	This is the year which an observation was recorded.
Lag1	This is the percentage returned for previous day.
Lag2	This is the percentage returned for the second day previous.
Lag3	This is the percentage returned for the third day previous.
Lag4	This is the percentage returned for the fourth day previous.
Lag5	This is the percentage returned for the fifth day previous.
Volume	This is the volume of shares traded, i.e., the number of daily shares traded in billions of dollars.
Today	This is the percentage returned for today.
Direction	This is the categorical measurement in "Up" or "Down." This indicates whether the market had a positive (Up) or negative (Down) return on the day.

1.2.2 Working Examples

In this section we play with the data set from R package ISLR: "Smarket" [24]. This data contains information of the S&P 500 stock index between 2001 and 2005. The "Smarket" data set from the ISLR package contains four variables: "Year", "Lag1", "Lag2", "Lag3", "Lag4", "Lag5", "Volume", "Today", and "Direction." Information on each variable is listed in Table 1.2.

To add the ISLR package you type:

```
library(ISLR)
```

In order to upload the "Smarket" data set from the ISLR package, you need to type the following in R:

```
data(Smarket)
```

The "Smarket" data set has nine different measurements, and the sample size, i.e., the number of people in the data set, is 1250. If you type:

```
Smarket
```

then you can see a 2-dimensional array with 1250 observations and 9 variables. If you type

```
head(Smarket)
```

you can see the first six observations of the data set. The output is shown as the following:

```
> head(Smarket)
  Year  Lag1   Lag2   Lag3   Lag4  Lag5 Volume  Today Direction
1 2001 0.381 -0.192 -2.624 -1.055 5.010 1.1913  0.959        Up
```

```
2 2001  0.959  0.381 -0.192 -2.624 -1.055 1.2965  1.032        Up
3 2001  1.032  0.959  0.381 -0.192 -2.624 1.4112 -0.623        Down
4 2001 -0.623  1.032  0.959  0.381 -0.192 1.2760  0.614        Up
5 2001  0.614 -0.623  1.032  0.959  0.381 1.2057  0.213        Up
6 2001  0.213  0.614 -0.623  1.032  0.959 1.3491  1.392        Up
```

In the output you can ignore the first column because they are just indices for observations.

This data set is used to predict whether the market will be down or up based on the percentages returned for previous few days. This can be seen as a 1250×9 **matrix**. You will see the definition of a **matrix** in the following section.

1.2.3   Matrix and Vector

In this section we discuss matrices and vectors, and then we discuss how we can create a matrix and a vector in the R programming environment.

Definition 1 *For positive integers n, an n-**dimensional** vector is a 1-dimensional n array.*

Example 1 *The following 1-dimensional array*

$$\begin{bmatrix} 907 \\ 220 \\ 625 \\ 502 \end{bmatrix}$$

is a 4-dimensional vector. In R, we can use c() function to create a vector. For example,

```
v <- c(907,220,625,502)
```

*The notation of <- is equivalent to an = sign. We are assigning **v** as the 4-dimensional vector above.*

*When you type **v** in R, then you will see*

```
> v
[1] 907 220 625 502
```

Example 2 *The following 1-dimensional array*

$$\begin{bmatrix} 2 \\ 3 \\ 4 \\ 5 \\ 6 \end{bmatrix}$$

is a 5-dimensional vector. In R, *using ":" you can easily create a vector of a sequence of numbers. In this example, you can type*

```
v <- 2:6
```

to create the vector. If you type v *in* R *it returns:*

```
> v
[1] 2 3 4 5 6
```

This colon ":" in R *is very useful for programming.*

Definition 2 *For positive integers m, n, an $m \times n$* **matrix** *is a 2-dimensional $m \times n$ array.*

Example 3 *"Smarket" data set is a 1250×9 matrix.*

Example 4 *Suppose we have*

$$\begin{bmatrix} 1 & 1 & 1 & 1 \\ 1 & 2 & 3 & 4 \end{bmatrix}.$$

This is a 2×4 matrix.

Example 5 *Suppose we have the following 2-dimensional array:*

$$\begin{bmatrix} 1 & 1 & 0 & 0 \\ 0 & 0 & 1 & 1 \\ 1 & 0 & 1 & 0 \\ 0 & 1 & 0 & 1 \end{bmatrix}.$$

This is a 4×4 matrix.

Definition 3 *Suppose we have an $m \times n$ matrix such that*

$$\begin{bmatrix} x_{1,1} & x_{1,2} & \cdots & x_{1,n} \\ x_{2,1} & x_{2,2} & \cdots & x_{2,n} \\ \vdots & \vdots & \vdots & \vdots \\ x_{m,1} & x_{m,2} & \cdots & x_{m,n} \end{bmatrix}.$$

Then $m \times 1$ matrices

$$\begin{bmatrix} x_{1,1} \\ x_{2,1} \\ \vdots \\ x_{m,1} \end{bmatrix}, \begin{bmatrix} x_{1,2} \\ x_{2,2} \\ \vdots \\ x_{m,2} \end{bmatrix}, \cdots, \begin{bmatrix} x_{1,n} \\ x_{2,n} \\ \vdots \\ x_{m,n} \end{bmatrix}$$

are called **column vectors** *of the matrix. Also* $1 \times n$ *matrices such that*

$$\begin{bmatrix} x_{1,1} & x_{1,2} & \cdots & x_{1,n} \end{bmatrix},$$
$$\begin{bmatrix} x_{2,1} & x_{2,2} & \cdots & x_{2,n} \end{bmatrix},$$
$$\vdots$$
$$\begin{bmatrix} x_{m,1} & x_{m,2} & \cdots & x_{m,n} \end{bmatrix}$$

are called **row vectors** *of the matrix.*

Example 6 *Here, we show column vectors and row vectors of the matrix from Example 4. Suppose we have a 2×4 matrix such that*

$$\begin{bmatrix} 1 & 1 & 1 & 1 \\ 1 & 2 & 3 & 4 \end{bmatrix}.$$

Then its column vectors are

$$\begin{bmatrix} 1 \\ 1 \end{bmatrix}, \begin{bmatrix} 1 \\ 2 \end{bmatrix}, \begin{bmatrix} 1 \\ 3 \end{bmatrix}, \begin{bmatrix} 1 \\ 4 \end{bmatrix},$$

and its row vectors are

$$\begin{bmatrix} 1 & 1 & 1 & 1 \end{bmatrix}, \begin{bmatrix} 1 & 2 & 3 & 4 \end{bmatrix}.$$

Example 7 *In this example we show column vectors and row vectors of the matrix from Example 5. Recall we have a 4×4 matrix:*

$$\begin{bmatrix} 1 & 1 & 0 & 0 \\ 0 & 0 & 1 & 1 \\ 1 & 0 & 1 & 0 \\ 0 & 1 & 0 & 1 \end{bmatrix}.$$

Then its column vectors are

$$\begin{bmatrix} 1 \\ 0 \\ 1 \\ 0 \end{bmatrix}, \begin{bmatrix} 1 \\ 0 \\ 0 \\ 1 \end{bmatrix}, \begin{bmatrix} 0 \\ 1 \\ 1 \\ 0 \end{bmatrix}, \begin{bmatrix} 0 \\ 1 \\ 0 \\ 1 \end{bmatrix},$$

and its row vectors are

$$\begin{bmatrix} 1 & 1 & 0 & 0 \end{bmatrix}, \begin{bmatrix} 0 & 0 & 1 & 1 \end{bmatrix}, \begin{bmatrix} 1 & 0 & 1 & 0 \end{bmatrix}, \begin{bmatrix} 0 & 1 & 0 & 1 \end{bmatrix}.$$

Example 8 *This is again from Example 4. Now we show how to create a matrix in* R *using a 2×4 matrix*

$$\begin{bmatrix} 1 & 1 & 1 & 1 \\ 1 & 2 & 3 & 4 \end{bmatrix}.$$

In R *we can create this matrix by calling the matrix() function.*

```
M <- matrix(c(1,1,1,2,1,3,1,4),nrow=2,ncol=4)
```

Here the c() function stores elements in the matrix. The order of the elements in the c() function matters. As the default setting, it starts from the first element of the first column vector to the last element of the first column vector. Then it goes from the first element of the second column vector to the last element of the second column vector, and so on. "nrow" defines the number of row vectors and "ncol" defines the number of column vectors. For this example the number of row vectors is 2 and the number of column vectors is 4, so we set "nrow=2" and "ncol=4." As before <- means that we assign this matrix as a variable M.

If you type M in R, *then you will see the output as follows:*

```
> M
     [,1] [,2] [,3] [,4]
[1,]   1    1    1    1
[2,]   1    2    3    4
```

Now suppose we want to have just the first column vector of this matrix M, then we type:

```
M[1,]
```

Then the output looks like

```
> M[1,]
[1] 1 1 1 1
```

Similarly, if we want to have the second column of the matrix M, then we type

```
M[,2]
```

The output looks like this:

```
> M[,2]
[1] 1 2
```

Now we show another way to make a matrix from combining row vectors or column vectors. First, we show how to create a matrix by combining vectors with the function rbind(). Let us define r1 and r2 as the first row vector and the second row vector of the matrix. In R *we can define as*

```
r1 <- c(1, 1, 1, 1)
r2 <- c(1, 2, 3, 4)
```

Then we use the rbind() function to create a matrix M as

```
M <- rbind(r1, r2)
```

If you type M in R, *then the output looks*

```
> M
   [,1] [,2] [,3] [,4]
r1   1    1    1    1
r2   1    2    3    4
```

Here you can ignore r1 and r2 in the output as they are left over from the rbind(). In statistical learning or data science, this information can be useful, but for now you can just ignore this.

To create a matrix from column vectors in R we can use the function cbind(). For this example we first create four column vectors c1, c2, c3, and c4 with a function c().

```
c1 <- c(1, 1)
c2 <- c(1, 2)
c3 <- c(1, 3)
c4 <- c(1, 4)
```

Then you use the cbind() function to create a matrix as follows:

```
M <- cbind(c1,c2,c3,c4)
```

If you type M, then you can see how M looks in R*:*

```
> M
     c1 c2 c3 c4
[1,]  1  1  1  1
[2,]  1  2  3  4
```

Again, for now we ignore c1, c2, c3, and c4 in the output.
If you want to extract the element in the 2th row and the 3th column, you can do:

```
M[2,3]
```

Example 9 *We look at the working example in this section the "Smarket" data set from* ISLR *package. Using the head() function you can see the first six observations:*

```
> head(Smarket)
  Year   Lag1   Lag2   Lag3   Lag4   Lag5 Volume  Today Direction
1 2001  0.381 -0.192 -2.624 -1.055  5.010 1.1913  0.959        Up
2 2001  0.959  0.381 -0.192 -2.624 -1.055 1.2965  1.032        Up
3 2001  1.032  0.959  0.381 -0.192 -2.624 1.4112 -0.623      Down
4 2001 -0.623  1.032  0.959  0.381 -0.192 1.2760  0.614        Up
5 2001  0.614 -0.623  1.032  0.959  0.381 1.2057  0.213        Up
6 2001  0.213  0.614 -0.623  1.032  0.959 1.3491  1.392        Up
```

If you want to pick up the second observation, i.e., the second row vector, you just type:

```
Smarket[2,]
```

Then the output looks like this:

```
> Smarket[2,]
  Year Lag1  Lag2   Lag3   Lag4   Lag5 Volume Today Direction
2 2001 0.959 0.381 -0.192 -2.624 -1.055 1.2965 1.032        Up
```

If you want to have the third column of the matrix, we can type Smarket[,3] and see all 1250 entries in the column.

Using the rbind() function you can also create a new matrix. For example, let us pick the third, 4th, 7th, 9th and 10th rows of "Smarket" matrix. Then let us make a new matrix called M2.

```
M2 <- rbind(Smarket[3,], Smarket[4,], Smarket[7,],
Smarket[9,], Smarket[10,])
```

Then, if you type M2, you will see the following output:

```
> M2
   Year   Lag1   Lag2   Lag3   Lag4   Lag5 Volume  Today Direction
3  2001  1.032  0.959  0.381 -0.192 -2.624 1.4112 -0.623      Down
4  2001 -0.623  1.032  0.959  0.381 -0.192 1.2760  0.614        Up
7  2001  1.392  0.213  0.614 -0.623  1.032 1.4450 -0.403      Down
9  2001  0.027 -0.403  1.392  0.213  0.614 1.1640  1.303        Up
10 2001  1.303  0.027 -0.403  1.392  0.213 1.2326  0.287        Up
```

If we want to know the number of row vectors and the number of column vectors of a matrix, you can use the dim() function in R. *In a real-world problem, sometimes it is not obvious how many row or column vectors are there. With the "Smarket" data set here, if we just type "dim(Smarket)" then* R *returns the following output:*

```
> dim(Smarket)
[1] 1250    9
```

The first output of the dim() function is the number of row vectors and the second output of the dim() function is the number of column vectors. In this example, the number of row vectors is 1250 and the number of column vectors is 9.

There are special vectors and matrices, some of which we will cover now. More special matrices will be introduced in Section 2.2, on "Matrix Operations."

Definition 4 *A* **zero vector** *or a* **null vector** *is an n-dimensional vector with all zeros as its elements.*

Example 10 *For n = 5, the 5-dimensional zero vector is*

$$\begin{bmatrix} 0 \\ 0 \\ 0 \\ 0 \\ 0 \end{bmatrix}.$$

In R *you can create the 5-dimensional zero vector using the rep() function.*

```
rep(0, 5)
```

The first argument is the value you want to assign as its element and the second argument is the dimension of the vector. The output from R *is as follows:*

```
> rep(0, 5)
[1] 0 0 0 0 0
```

If you want to create a 10-dimensional vector with all 1s as its elements then you type rep(1, 10). Then the output in R *looks like this:*

```
> rep(1,10)
 [1] 1 1 1 1 1 1 1 1 1 1
```

This rep() function can be very helpful in R *programming.*

Definition 5 *Suppose we have an m × n matrix. If m = n, then we call this matrix a* **square matrix**.

Example 11 *From Example 5. A 4 × 4 matrix:*

$$\begin{bmatrix} 1 & 1 & 0 & 0 \\ 0 & 0 & 1 & 1 \\ 1 & 0 & 1 & 0 \\ 0 & 1 & 0 & 1 \end{bmatrix}$$

is a square matrix since the number of row vectors and the number of column vectors are equal.

Definition 6 *The* **identity matrix**, I_n, *of size n is an n × n square matrix such that all elements in the ith row and ith column equal 1 for all i from 1 to n, and otherwise all 0.*

Example 12 *For $n = 3$, an identity matrix of size 3 is*

$$I_3 = \begin{bmatrix} 1 & 0 & 0 \\ 0 & 1 & 0 \\ 0 & 0 & 1 \end{bmatrix},$$

and for $n = 4$, an identity matrix of size 4 is

$$I_4 = \begin{bmatrix} 1 & 0 & 0 & 0 \\ 0 & 1 & 0 & 0 \\ 0 & 0 & 1 & 0 \\ 0 & 0 & 0 & 1 \end{bmatrix}.$$

To create the identity matrix of size n in R you can use the diag() function. If you type diag(n), then you can create the identity matrix of size n.

Example 13 *We create here the identity matrix of size 3 and the identity matrix of size 4 in R with the diag() function. If you type*

```
diag(3)
```

then you will see the output as follows:

```
> diag(3)
     [,1] [,2] [,3]
[1,]   1    0    0
[2,]   0    1    0
[3,]   0    0    1
```

Similarly, if you type diag(4), then you will see the following output in R:

```
> diag(4)
     [,1] [,2] [,3] [,4]
[1,]   1    0    0    0
[2,]   0    1    0    0
[3,]   0    0    1    0
[4,]   0    0    0    1
```

There are several special types of matrices. Each type of matrix has special properties which we will discuss in later chapters. These properties can be very useful when we conduct computations. Therefore, here, we define some of the special types of matrices.

Notation 1.1 *Suppose we have a matrix A. a_{ij} is an (i, j)th entry of the matrix A.*

Example 14 *Suppose we have a matrix*

$$A = \begin{bmatrix} 1 & 2 \\ 3 & 4 \end{bmatrix}.$$

Then $a_{11} = 1$, $a_{12} = 2$, $a_{21} = 3$, $a_{22} = 4$.

Example 15 *Suppose we have a matrix*

$$A = \begin{bmatrix} 1 & 2 & 3 \\ 4 & 5 & 6 \end{bmatrix}.$$

Then $a_{11} = 1$, $a_{12} = 2$, $a_{13} = 3$, $a_{21} = 4$, $a_{22} = 5$, $a_{23} = 6$.

Notation 1.2 *Suppose we have an* $m \times n$ *matrix* A *for any positive integers* m *and* n. *Then we notate* A *as*

$$A = \begin{bmatrix} a_{11} & a_{12} & \cdots & a_{1n} \\ a_{21} & a_{22} & \cdots & a_{2n} \\ \vdots & \vdots & \vdots & \vdots \\ a_{m1} & a_{m2} & \cdots & a_{mn} \end{bmatrix}.$$

Example 16 *Suppose* $m = 2$ *and* $n = 2$. *Then we have*

$$A = \begin{bmatrix} a_{11} & a_{12} \\ a_{21} & a_{22} \end{bmatrix}.$$

Example 17 *Suppose* $m = 2$ *and* $n = 3$. *Then we have*

$$A = \begin{bmatrix} a_{11} & a_{12} & a_{13} \\ a_{21} & a_{22} & a_{23} \end{bmatrix}.$$

Definition 7 *Suppose we have an* $m \times n$ *matrix* A *such that*

$$A = \begin{bmatrix} a_{11} & a_{12} & \cdots & a_{1n} \\ a_{21} & a_{22} & \cdots & a_{2n} \\ \vdots & \vdots & \vdots & \vdots \\ a_{m1} & a_{m2} & \cdots & a_{mn} \end{bmatrix}.$$

The **diagonal entries** *of* A *are entries of* A *such that*

$$a_{ij} \text{ for all } j = i.$$

Definition 8 *Suppose we have an* $m \times n$ *matrix* A *such that*

$$A = \begin{bmatrix} a_{11} & a_{12} & \cdots & a_{1n} \\ a_{21} & a_{22} & \cdots & a_{2n} \\ \vdots & \vdots & \vdots & \vdots \\ a_{m1} & a_{m2} & \cdots & a_{mn} \end{bmatrix}.$$

If $a_{ij} = 0$ *for all* $j \neq i$, *then we call* A *a* **diagonal matrix**.

Example 18 *The identity matrix of size* n *for any* $n \geq 1$ *is a diagonal matrix.*

Example 19 *Suppose*

$$A = \begin{bmatrix} -1 & 0 & 0 & 0 \\ 0 & 0 & 0 & 0 \\ 0 & 0 & 2 & 0 \\ 0 & 0 & 0 & 10 \end{bmatrix}.$$

Then A is a diagonal matrix.

Definition 9 *Suppose we have a square matrix A such that*

$$A = \begin{bmatrix} a_{11} & a_{12} & \cdots & a_{1m} \\ a_{21} & a_{22} & \cdots & a_{2m} \\ \vdots & \vdots & \vdots & \vdots \\ a_{m1} & a_{m2} & \cdots & a_{mm} \end{bmatrix}.$$

If $a_{ij} = a_{ji}$ for all $i = 1, \ldots, m$ and $j = 1, \ldots, m$, then we say A is a **symmetric matrix**.

Example 20 *Any square diagonal matrices are symmetric matrices.*

Example 21 *Suppose we have*

$$A = \begin{bmatrix} -1 & 2 \\ 2 & 0 \end{bmatrix}.$$

Then A is a symmetric matrix.

Example 22 *Suppose we have*

$$A = \begin{bmatrix} -1 & 2 & 0 & 5 \\ 2 & 0 & 0 & 1 \\ 0 & 0 & 12 & -1 \\ 5 & 1 & -1 & 9 \end{bmatrix}.$$

Then A is a symmetric matrix.

Definition 10 *Suppose we have an $m \times n$ matrix A such that*

$$A = \begin{bmatrix} a_{11} & a_{12} & \cdots & a_{1n} \\ a_{21} & a_{22} & \cdots & a_{2n} \\ \vdots & \vdots & \vdots & \vdots \\ a_{m1} & a_{m2} & \cdots & a_{mn} \end{bmatrix}.$$

If $a_{ij} = 0$ for all $i > j$, then we say A is an **upper triangular matrix**.

Example 23 *Any square diagonal matrices are upper triangular matrices.*

Example 24 *Suppose we have*

$$A = \begin{bmatrix} -1 & 2 \\ 0 & 0 \end{bmatrix}.$$

Then A is an upper triangular matrix.

Example 25 *Suppose we have*

$$A = \begin{bmatrix} -1 & 2 & 0 & 5 \\ 0 & 0 & 0 & 1 \\ 0 & 0 & 12 & -1 \\ 0 & 0 & 0 & 9 \end{bmatrix}.$$

Then A is an upper triangular matrix.

Example 26 *Suppose we have*

$$A = \begin{bmatrix} -1 & 2 & 0 & 5 & -1 \\ 0 & 0 & 0 & 1 & 2 \\ 0 & 0 & 12 & -1 & -3 \\ 0 & 0 & 0 & 9 & -4 \end{bmatrix}.$$

Then A is an upper triangular matrix.

Example 27 *Suppose we have*

$$A = \begin{bmatrix} -1 & 2 & 0 & 5 \\ 0 & 0 & 0 & 1 \\ 0 & 0 & 12 & -1 \\ 0 & 0 & 0 & 9 \\ 0 & 0 & 0 & 0 \end{bmatrix}.$$

Then A is an upper triangular matrix.

Definition 11 *Suppose we have an $m \times n$ matrix A such that*

$$A = \begin{bmatrix} a_{11} & a_{12} & \ldots & a_{1n} \\ a_{21} & a_{22} & \ldots & a_{2n} \\ \vdots & \vdots & \vdots & \vdots \\ a_{m1} & a_{m2} & \ldots & a_{mn} \end{bmatrix}.$$

If $a_{ij} = 0$ for all $i < j$, then we say A is a **lower triangular matrix**.

Example 28 *Any square diagonal matrices are lower triangular matrices.*

Example 29 *Suppose we have*

$$A = \begin{bmatrix} -1 & 0 \\ 2 & 0 \end{bmatrix}.$$

Then A is a lower triangular matrix.

Example 30 *Suppose we have*

$$A = \begin{bmatrix} -1 & 0 & 0 & 0 \\ 2 & 0 & 0 & 0 \\ 0 & 0 & 12 & 0 \\ 5 & 1 & -1 & 9 \end{bmatrix}.$$

Then A is a lower triangular matrix.

Example 31 *Suppose we have*

$$A = \begin{bmatrix} -1 & 0 & 0 & 0 & 0 \\ 2 & 0 & 0 & 0 & 0 \\ 0 & 0 & 12 & 0 & 0 \\ 5 & 1 & -1 & 9 & 0 \end{bmatrix}.$$

Then A is a lower triangular matrix.

Example 32 *Suppose we have*

$$A = \begin{bmatrix} -1 & 0 & 0 & 0 \\ 2 & 0 & 0 & 0 \\ 0 & 0 & 12 & 0 \\ 5 & 1 & -1 & 9 \\ 1 & 0 & 1 & 0 \end{bmatrix}.$$

Then A is a lower triangular matrix.

1.2.4 Checkmarks

- The definition of an n-dimensional vector.
- You can create a sequence of increasing number using a colon ":."
- The definition of an $m \times n$ matrix.
- The definition of a row vector of a matrix.
- The definition of a column vector of a matrix.
- The c() function in R.
- The matrix() function in R.
- You can extract a column vector from a matrix in R.
- You can extract a row vector from a matrix in R.
- You can create a matrix with the rbind() function in R.
- You can create a matrix with the cbind() function in R.

- You can extract the element in the ith row and the jth column of the matrix in R.

- You can extract the number of row vectors and the number of column vectors with the dim() function.

- The definition of the n-dimensional zero vector.

- You can create the n-dimensional zero vector with the function rep() in R.

- The definition of a square matrix.

- The definition of a symmetric matrix.

- The definition of an upper triangular matrix.

- The definition of a lower triangular matrix.

- The definition of the identity matrix of size n.

- You can create the identity matrix of size n with the diag() function in R.

- The definition of a diagonal matrix.

1.2.5   Conceptual Quizzes

Quiz 1 True or False: *The matrix*

$$\begin{bmatrix} 1 & 2 & 3 & 4 & 5 \\ 0 & 1 & 0 & 0 & 1 \\ 0 & 0 & 0 & 1 & 0 \end{bmatrix}$$

is a 5×3 matrix.

Quiz 2 True or False: *The matrix*

$$\begin{bmatrix} 1 & 0 & 0 & 0 & 1 \\ 0 & 1 & 0 & 0 & 0 \\ 0 & 0 & 1 & 0 & 1 \\ 0 & 0 & 0 & 1 & 0 \end{bmatrix}$$

is a square matrix.

Quiz 3 True or False: *The matrix*

$$\begin{bmatrix} 1 & 0 & 0 & 0 & 0 \\ 0 & 1 & 0 & 0 & 0 \\ 0 & 0 & 1 & 0 & 0 \\ 0 & 0 & 0 & 1 & 0 \end{bmatrix}$$

is the identity matrix of size 5.

Quiz 4 True or False: *The matrix*

$$\begin{bmatrix} 1 & 0 & 0 & 0 & 1 \\ 0 & 1 & 0 & 0 & 0 \\ 0 & 0 & 1 & 0 & 1 \\ 0 & 0 & 0 & 1 & 0 \end{bmatrix}$$

is a diagonal matrix.

Quiz 5 True or False: *The matrix*

$$\begin{bmatrix} 1 & 0 & 0 & 0 & 1 \\ 0 & 1 & 0 & 0 & 0 \\ 0 & 0 & 1 & 0 & 1 \\ 0 & 0 & 0 & 1 & 0 \end{bmatrix}$$

is an upper triangular matrix.

Quiz 6 True or False: *The matrix*

$$\begin{bmatrix} 1 & 0 & 0 & 0 & 1 \\ 0 & 1 & 0 & 0 & 0 \\ 0 & 0 & 1 & 0 & 1 \\ 0 & 0 & 0 & 1 & 0 \end{bmatrix}$$

is a lower triangular matrix.

Quiz 7 True or False: *A row vector of an $m \times n$ matrix is a $1 \times n$ matrix.*

Quiz 8 True or False: *A row vector of an $m \times n$ matrix is an $n \times 1$ matrix.*

Quiz 9 True or False: *A row vector of an $m \times n$ matrix is a $1 \times m$ matrix.*

Quiz 10 True or False: *A column vector of an $m \times n$ matrix is a $1 \times n$ matrix.*

Quiz 11 True or False: *A column vector of an $m \times n$ matrix is an $m \times 1$ matrix.*

Quiz 12 True or False: *A column vector of an $m \times n$ matrix is a $1 \times m$ matrix.*

Quiz 13 Multiple Choice: *Suppose we have a 2 matrix*

$$A = \begin{bmatrix} 1 & 2 & 1 \\ 2 & 0 & 3 \end{bmatrix}.$$

$a_{21} =$

    (i) 1

(ii) 2

(iii) 0

(iv) 3

Quiz 14 Multiple Choice*: Suppose we have a 2 matrix*

$$A = \begin{bmatrix} 1 & 2 & 1 \\ 2 & 0 & 3 \end{bmatrix}.$$

$a_{23} =$

(i) 1

(ii) 2

(iii) 0

(iv) 3

Quiz 15 Multiple Choice*: Suppose we have a 3 matrix*

$$A = \begin{bmatrix} 1 & 2 & 1 \\ 2 & 0 & 3 \\ 3 & 1 & 2 \end{bmatrix}.$$

$a_{31} =$

(i) 1

(ii) 2

(iii) 0

(iv) 3

Quiz 16 Multiple Choice*: Suppose we have a 3 matrix*

$$A = \begin{bmatrix} 1 & 2 & 1 \\ 2 & 0 & 3 \\ 3 & 1 & 2 \end{bmatrix}.$$

$a_{33} =$

(i) 1

(ii) 2

(iii) 0

(iv) 3

Quiz 17 Multiple Choice*: The number of row vectors of the following matrix*

$$\begin{bmatrix} 1 & 2 & 3 & 4 & 5 \\ 1 & 1 & 1 & 1 & 1 \\ 1.2 & 0.1 & 0.5 & 1.5 & 0 \\ 0.2 & 0.3 & 0.4 & 0.5 & 0.6 \end{bmatrix}$$

is

   (i) 2

   (ii) 3

   (iii) 4

   (iv) 5

Quiz 18 Multiple Choice*: The first row vector of the following matrix*

$$\begin{bmatrix} 1 & 2 & 3 & 4 & 5 \\ 1 & 1 & 1 & 1 & 1 \\ 1.2 & 0.1 & 0.5 & 1.5 & 0 \\ 0.2 & 0.3 & 0.4 & 0.5 & 0.6 \end{bmatrix}$$

is

   (i)
$$\begin{bmatrix} 1 & 2 & 3 & 4 & 5 \end{bmatrix}$$

   (ii)
$$\begin{bmatrix} 1 & 1 & 1 & 1 & 1 \end{bmatrix}$$

   (iii)
$$\begin{bmatrix} 1.2 & 0.1 & 0.5 & 1.5 & 0 \end{bmatrix}$$

   (iv)
$$\begin{bmatrix} 0.2 & 0.3 & 0.4 & 0.5 & 0.6 \end{bmatrix}$$

Quiz 19 Multiple Choice*: The element in the second row and the third column of the following matrix*

$$\begin{bmatrix} 1 & 2 & 3 & 4 & 5 \\ 1 & 1 & 1 & 1 & 1 \\ 1.2 & 0.1 & 0.5 & 1.5 & 0 \\ 0.2 & 0.3 & 0.4 & 0.5 & 0.6 \end{bmatrix}$$

is

   (i) 1

   (ii) 0.5

   (iii) 0.1

   (iv) 3

1.2.6   Regular Exercises

Exercise 1.1 *Write the identity matrix of size* 3.

Exercise 1.2 *Create a* 3 × 4 *matrix.*

Exercise 1.3 *Create a* 3 × 5 *matrix.*

Exercise 1.4 *Create a* 2 × 4 *matrix.*

Exercise 1.5 *Create a* 2 × 5 *matrix.*

Exercise 1.6 *Consider the following matrix:*

$$\begin{bmatrix} 5 & 16 & 3 & 16 \\ 6 & 19 & 9 & 5 \\ 10 & 17 & 19 & 13 \end{bmatrix}.$$

1. What is the number of row vectors?

2. What is the number of column vectors?

3. List all row vectors.

4. List all column vectors.

Exercise 1.7 *Consider the following matrix:*

$$\begin{bmatrix} 12 & 13 & 19 & 3 \\ 6 & 20 & 8 & 12 \\ 3 & 11 & 19 & 19 \end{bmatrix}.$$

1. What is the number of row vectors?

2. What is the number of column vectors?

3. List all row vectors.

4. List all column vectors.

Exercise 1.8 *Consider the following matrix:*

$$\begin{bmatrix} 3 & 11 & 10 & 4 \\ 14 & 12 & 9 & 9 \\ 8 & 20 & 2 & 13 \end{bmatrix}.$$

1. What is the number of row vectors?

2. What is the number of column vectors?

3. List all row vectors.

4. List all column vectors.

Exercise 1.9 *Consider the following matrix:*

$$\begin{bmatrix} 3 & 10 & 6 & 4 \\ 2 & 1 & 15 & 19 \\ 8 & 16 & 8 & 2 \\ 18 & 16 & 16 & 5 \\ 5 & 20 & 3 & 20 \end{bmatrix}.$$

1. What is the number of row vectors?

2. What is the number of column vectors?

3. List all row vectors.

4. List all column vectors.

Exercise 1.10 *Consider the following matrix:*

$$\begin{bmatrix} 15 & 19 & 8 & 5 & 16 \\ 11 & 5 & 14 & 20 & 12 \\ 18 & 5 & 7 & 12 & 4 \\ 19 & 9 & 4 & 12 & 13 \\ 17 & 20 & 16 & 17 & 19 \end{bmatrix}.$$

1. What is the number of row vectors?

2. What is the number of column vectors?

3. List all row vectors.

4. List all column vectors.

Exercise 1.11 *Consider the following matrix:*

$$\begin{bmatrix} 12 & 9 & 8 & 4 & 5 \\ 2 & 0 & 4 & 4 & 6 \\ 9 & 0 & 3 & 0 & 0 \\ 9 & 0 & 0 & 0 & 1 \\ 1 & 0 & 0 & 0 & 0 \end{bmatrix}.$$

Re-order the columns of the matrix to make an upper triangular matrix.

Exercise 1.12 *Consider the following matrix:*

$$\begin{bmatrix} 12 & 9 & 8 & 4 & 5 \\ 1 & 3 & 0 & 4 & 0 \\ 0 & 0 & 3 & 0 & 0 \\ 2 & 5 & 0 & 0 & 0 \\ 1 & 0 & 0 & 0 & 0 \end{bmatrix}.$$

Re-order the rows of the matrix to make a lower triangular matrix.

Exercise 1.13   *1. Construct an example of a* 3×4 *upper triangular matrix.*

2. Construct an example of a 4×3 *upper triangular matrix.*

3. Construct an example of a 3×4 *lower triangular matrix.*

4. Construct an example of a 4×3 *lower triangular matrix.*

Exercise 1.14

1.2.7   Lab Exercises

Lab Exercise 1 *In* R, *create a 3-dimensional vector v such that*

$$v = \begin{bmatrix} 12 \\ 7 \\ 20 \end{bmatrix}.$$

Lab Exercise 2 *In* R, *create a 5-dimensional vector v such that*

$$v = \begin{bmatrix} 19 \\ 2 \\ 4 \\ 9 \\ 6 \end{bmatrix}.$$

Lab Exercise 3 *In* R, *with the matrix() function, create a* 2×5 *matrix vector M such that*

$$M = \begin{bmatrix} 14 & 1 & 8 & 2 & 17 \\ 20 & 6 & 1 & 10 & 6 \end{bmatrix}.$$

Lab Exercise 4 *In* R, *with the matrix() function, create a* 3×6 *matrix vector M such that*

$$M = \begin{bmatrix} 16 & 7 & 1 & 14 & 9 & 10 \\ 5 & 10 & 2 & 16 & 11 & 16 \\ 9 & 15 & 8 & 2 & 6 & 9 \end{bmatrix}.$$

Lab Exercise 5 *In* R, *with the matrix() function and a colon "*:*", create a* 6×6 *matrix vector M such that*

$$M = \begin{bmatrix} 1 & 7 & 13 & 19 & 25 & 31 \\ 2 & 8 & 14 & 20 & 26 & 32 \\ 3 & 9 & 15 & 21 & 27 & 33 \\ 4 & 10 & 16 & 22 & 28 & 34 \\ 5 & 11 & 17 & 23 & 29 & 35 \\ 6 & 12 & 18 & 24 & 30 & 36 \end{bmatrix}.$$

Lab Exercise 6 *Create the 20-dimensional zero vector with the rep() function in* R.

Lab Exercise 7 *Create the identity matrix of size 20 with the diag() function in* R.

Lab Exercise 8 *In* R*, with the rbind() function, create a* 2×5 *matrix vector* M *such that*
$$M = \begin{bmatrix} 14 & 1 & 8 & 2 & 17 \\ 20 & 6 & 1 & 10 & 6 \end{bmatrix}.$$

Lab Exercise 9 *In* R*, with the cbind() function, create a* 2×5 *matrix vector* M *such that*
$$M = \begin{bmatrix} 14 & 1 & 8 & 2 & 17 \\ 20 & 6 & 1 & 10 & 6 \end{bmatrix}.$$

Lab Exercise 10 *In* R*, with the rbind() function, create a* 3×6 *matrix vector* M *such that*
$$M = \begin{bmatrix} 16 & 7 & 1 & 14 & 9 & 10 \\ 5 & 10 & 2 & 16 & 11 & 16 \\ 9 & 15 & 8 & 2 & 6 & 9 \end{bmatrix}.$$

Lab Exercise 11 *In* R*, with the cbind() function, create a* 3×6 *matrix vector* M *such that*
$$M = \begin{bmatrix} 16 & 7 & 1 & 14 & 9 & 10 \\ 5 & 10 & 2 & 16 & 11 & 16 \\ 9 & 15 & 8 & 2 & 6 & 9 \end{bmatrix}.$$

Lab Exercise 12 Challenge Problem *In* R*, use "help(matrix)" to figure out the following question: with the matrix() function and a colon ":", create a* 6×6 *matrix vector* M *such that*
$$M = \begin{bmatrix} 1 & 2 & 3 & 4 & 5 & 6 \\ 7 & 8 & 9 & 10 & 11 & 12 \\ 13 & 14 & 15 & 16 & 17 & 18 \\ 19 & 20 & 21 & 22 & 23 & 24 \\ 25 & 26 & 27 & 28 & 29 & 30 \\ 31 & 32 & 33 & 34 & 35 & 36 \end{bmatrix}.$$

1.2.8   Practical Applications

We will analyze the "Smarket" data set from the ISLR package in R. For this data fortunately the observations, i.e., row vectors, are ordered by time, "Year." It is very hard to see the trends of these percentages via human eyes, but we can try to extract some information from this data using what we learned in this section. Since these observations are ordered, we can see how these percentages are up or down in terms of time. For example, we want to know the trend of the percentage returned for the previous day (the second column vector of the "Smarket" matrix), and we can plot them in terms of time. In order to do so, we will use the plot() function in R.

First we will extract information on the number of row vectors (the number of observations) and the number of columns of the "Smarket" matrix.

```
dim(Smarket)
```

We know we have 1250 row vectors and 9 column vectors.

In the plot() function, the first argument is the x-axis, and the second argument is the y-axis. In this example, since all observations are ordered by their record time from oldest to newest, we will assign the index for each observation by using a colon ":."

```
x <- 1:1250
```

Then we use the plot() function. Here we assign the index of the observations as the x-axis and the percentage for the previous day (the second column of the matrix) as the y-axis.

```
plot(x, Smarket[,2],type="l")
```

Here type="l" means drawing lines instead of circle points (the default for the plot() function is circle points).

Then R will output the following plot in Figure 1.1. This is called a **line plot**.

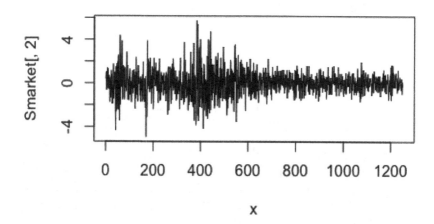

FIGURE 1.1

This is a plot to see the trend of the percentage for the previous day in the "Smarket" data set from the ISLR package in R.

From Figure 1.1, it is very hard to see the trend since it is a little crowded, so we will focus on the observations from the year 2001. The first 242 observations in the data set are from the year 2001, so we will take the first 242 observations using a colon ":" command. The command

```
Smarket[1:242,2]
```

displays percentages for the previous day from the year 2001. Also, for the index, we will take the first 242 indices.

```
x[1:242]
```

Now we will use the plot() function.

```
plot(x, Smarket[1:242,2],type="l")
```

Then R produces the plot in Figure 1.2.

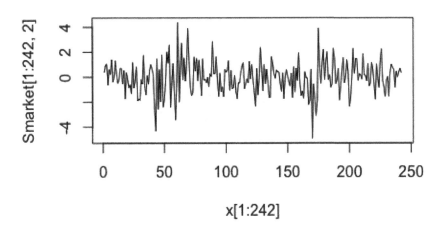

FIGURE 1.2
This is a plot to see the trend of the percentage for the previous day in the year 2001 in the "Smarket" data set from the ISLR package in R.

Lab Exercise 13 *Draw a line plot of the percentage return for 2 days previous from the "Smarket" data set from the* ISLR *package in* R. *Do the same thing for the observations in the year 2001 (i.e., the first 242 observations).*

Lab Exercise 14 *Draw a line plot of the percentage return for 3 days previous from the "Smarket" data set from the* ISLR *package in* R. *Do the same thing for the observations in the year 2001 (i.e., the first 242 observations).*

Lab Exercise 15 *Draw a line plot of the percentage return for 4 days previous from the "Smarket" data set from the* ISLR *package in* R. *Do the same thing for the observations in the year 2001 (i.e., the first* 242 *observations).*

Lab Exercise 16 *Draw a line plot of the percentage return for 5 days previous from the "Smarket" data set from the* ISLR *package in* R. *Do the same thing for the observations in the year 2001 (i.e., the first* 242 *observations).*

Lab Exercise 17 *From all line plots, what do you think? Is there any trend? And if so, how would you characterize the trend? How about in the year 2001?*

1.2.9   Supplements with python Code

This section is meant to help students who are interested in coding in python as well. We summarize what we learned in this section with python instead of R.

In python, you need the numpy package in order to create and manipulate matrices in python. In order to install the numpy package, type

```
pip install numpy
```

or

```
pip3 install numpy
```

in a terminal window.

To call the numpy package in python, use

```
import numpy as np
```

Let us consider this matrix

$$\begin{bmatrix} 1 & 1 & 1 & 1 \\ 1 & 2 & 3 & 4 \end{bmatrix}.$$

In python you just type

```
M = np.array([[1, 1, 1, 1],
[1, 2, 3, 4]])
```

This assigns M as a matrix shown above. When you type M then you will see

```
>>> M
array([[1, 1, 1, 1],
       [1, 2, 3, 4]])
```

The biggest difference between R and python is that python starts counting at zero, not one. If you like to extract the first row vector of the matrix M, type

```
M[0]
```

Note that it is not M[1]. If you type M[1], then it will return the second row vector.

If you like to extract the first column of this matrix M, you type:

```
M[:,0]
```

Again, if you type M[:,1], then it will return the second column.

If you would like to extract the element in the ith row and jth column of the matrix, just type M[i-1][j-1]. For example, if you would like to extract the element of the 2nd row and 3rd column of the example matrix M, then just type

```
M[1][2]
```

1.3 Systems of Linear Equations

Solving systems of linear equations is the foundation of linear algebra and also a problem found in many areas, such as optimization, statistics, engineering, etc. In this section, we will discuss how to define a system of linear equations and how to solve it.

1.3.1 Task Completion Checklist

- During the Lecture:

  1. Read the definitions of a linear equation and a system of linear equations.
  2. Transform a system of linear equations into a matrix.
  3. Read the definition of the elementary row reduction.
  4. Try to solve some examples without the help of a computer.
  5. With R, try to solve examples in this section.
  6. Plot the geometric view of a system of linear equations with two variables with R.
  7. Plot the geometric view of a system of linear equations with three variables with R.

- After the Lecture:

  1. Take conceptual quizzes to make sure you understand the materials in this section.

2. Do some regular exercises.

3. Conduct lab exercises with R.

4. Conduct practical applications with R.

5. If you are interested in python, read the supplement in this section and conduct lab exercises and practical applications with python.

1.3.2   Working Examples

We will use one of the most classical problems in optimization, called the "transportation problem." Per its name, we are trying to allocate transportation of goods (supplies) to other locations according to the demand for those goods. In this working example, we are using a Guinness data set [5]. This data set was collected from Guinness Ghana Ltd. and contains information on the supply of Malta Guinness from two production sites, Kaasi and Achimota, to nine key distributors geographically scattered in the regions of Ghana. They collected data twice and the data sets shown in Tables 1.3 and 1.4 was collected between July 7th 2010 to June 8th 2011. Table 1.3 shows the demand from the distributors and supplies from the production sites. Table 1.4 shows the cost of transportation from a production site to a distributor, written in thousands. The question is: how we can ship Guinness beers efficiently to distributors? Minimizing cost is an important problem for the company.

TABLE 1.3
Demands and Supplies in thousands
for Malta Guinness

Demand		Supply	
FTA	465	Achimota	1298
RICKY	605	Kaasi	1948
OBIBAJK	451		
KADOM	338		
NAATO	260		
LESK	183		
DCEE	282		
JOEMAN	127		
KBOA	535		

To solve this question we use a system of linear equations (and inequalities). This system of linear equations will define feasible ways to distribute the beer to distributors. They may not be optimal, but the solution(s) of this system of linear equations are feasible ways to distribute them. If there is no solution, there is no way to satisfy the demands and supplies in the problem and the company has to find another way to satisfy both supplies and demands.

TABLE 1.4

Cost to transport from
production sites to distributors
for Malta Guinness in thousands

	Achimota	Kaasi
FTA	39.99	145.36
RICKY	126.27	33.82
OBIBAJK	102.70	154.05
KADOM	81.68	64.19
NAATO	38.81	87.90
LESK	71.99	107.98
DCEE	31.21	65.45
JOEMAN	22.28	39.08
KBOA	321.04	167.38

1.3.3 System of Linear Equations

To define a system of equations, we will set up a smaller example of the transportation problem described in the previous section. A beer company has two production sites, A and B, and they want to transport their beers to two distributors, C and D. The demand from distributor C is 542 beers per week, and the demand from distributor D is 422 beers per week. The supply from production site A is 475 beers per week, and the supply from production site B is 489 beers per week. Can these sites produce enough beers to satisfy the demands from the distributors?

In this example, we can set up variables x_{ij}, the amount of beer shipped from a production site i to the distributor j. In the transportation problem the easiest thing is to set up the table as follows:

	C	D	total
A	x_{AC}	x_{AD}	475
B	c_{BC}	x_{BD}	489
total	542	422	964

The total supply from production site A is 475 and the total supply from production site B is 489. The total demand from distributor C is 542 and the total demand from distributor D is 422. Therefore we have the following system of equations:

$$
\begin{aligned}
x_{AC} &+ x_{AD} & & & &= 475 \\
& & x_{BC} &+ x_{BD} &= 489 \\
x_{AC} & &+ x_{BC} & &= 542 \\
& x_{AD} & &+ x_{BD} &= 422
\end{aligned}
$$

Definition 12 *A **linear equation** with the variables $x_1, \ldots, x_n$ is an equation written as*

$$
a_1 x_1 + a_2 x_2 \ldots + a_n x_n = b
$$

where $a_1, \ldots, a_n, b$ *are real numbers.* $a_1, \ldots, a_n$ *are called* **coefficients.**

Now we have the definition of a system of linear equations:

Definition 13 *A* **system of linear equations** *is a collection of one or more equations with the same set of variables, for example,* $x_1, \ldots, x_n$:

$$
\begin{array}{ccccccc}
a_{11}x_1 & + & a_{12}x_2 & \ldots & + & a_{1n}x_n & = & b_1 \\
a_{21}x_1 & + & a_{22}x_2 & \ldots & + & a_{2n}x_n & = & b_2 \\
\vdots & & & & & & & \vdots \\
a_{m1}x_1 & + & a_{m2}x_2 & \ldots & + & a_{mn}x_n & = & b_m
\end{array}
$$

where $a_{11}, \ldots, a_{1n}, \ldots, a_{mn}$ *are coefficients and* $b_1, \ldots b_m$ *are real numbers.*

Example 33 *The set of linear equations with variables* x_{AC}, x_{AD}, x_{BC} *and* x_{BD} *such that*

$$
\begin{array}{ccccccc}
3x_{AC} & + & x_{AD} & & & & = & 475 \\
& & & x_{BC} & + & x_{BD} & = & 489 \\
x_{AC} & & & + & 2x_{BC} & & = & 542 \\
& & x_{AD} & & & + & x_{BD} & = & 422
\end{array}
$$

is a system of linear equations.

In general, a system of linear equations with four variables is difficult to see geometrically. First, we show systems of linear equations with two variables so that we can demonstrate them geometrically.

Example 34 *The set of linear equations with variables* x_1 *and* x_2 *such that*

$$
\begin{array}{ccccc}
x_1 & + & x_2 & = & 4 \\
2x_1 & + & x_2 & = & 5
\end{array}
$$

is a system of linear equations.

A **solution** of a system of linear equations with the variables $x_1, \ldots, x_n$ such that

$$
\begin{array}{ccccccc}
a_{11}x_1 & + & a_{12}x_2 & \ldots & + & a_{1n}x_n & = & b_1 \\
a_{21}x_1 & + & a_{22}x_2 & \ldots & + & a_{2n}x_n & = & b_2 \\
\vdots & & & & & & & \vdots \\
a_{m1}x_1 & + & a_{m2}x_2 & \ldots & + & a_{mn}x_n & = & b_m
\end{array}
$$

is a vector $\begin{bmatrix} s_1 \\ \vdots \\ s_n \end{bmatrix}$ which satisfies all equations such that

$$
\begin{array}{ccccccc}
a_{11}s_1 & + & a_{12}s_2 & \ldots & + & a_{1n}s_n & = & b_1 \\
a_{21}s_1 & + & a_{22}s_2 & \ldots & + & a_{2n}s_n & = & b_2 \\
\vdots & & & & & & & \vdots \\
a_{m1}s_1 & + & a_{m2}s_2 & \ldots & + & a_{mn}s_n & = & b_m.
\end{array}
$$

Example 35 *Suppose we have the system of linear equations with variables x_1 and x_2 such that*

$$\begin{array}{rcrcl} x_1 & + & x_2 & = & 4 \\ 2x_1 & + & x_2 & = & 5. \end{array}$$

A vector $\begin{bmatrix} 1 \\ 3 \end{bmatrix}$ *is a solution of the system since*

$$\begin{array}{rcrcl} 1 & + & 3 & = & 4 \\ 2 \cdot 1 & + & 3 & = & 5. \end{array}$$

Geometrically, each linear equation in this system of linear equations defines a line (Figure 1.3). Since there are two linear equations in this system, there are two lines in this system. In this example, the two lines meet at the unique point, a vector $\begin{bmatrix} 1 \\ 3 \end{bmatrix}$. In the case of two variables, a solution is defined by where the linear equations in the system meet.

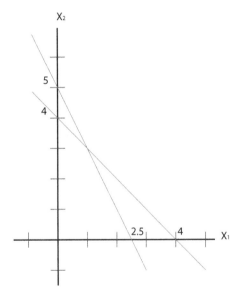

FIGURE 1.3
Geometrical view of the system of linear equations when there is a unique solution.

If these lines do not meet, then there is no solution to satisfy the system of linear equations. If they overlap, then there are multiple solutions, i.e., infinitely many solutions. For example, suppose we have a system of linear equations such that

$$\begin{array}{rcrcl} x_1 & + & x_2 & = & 4 \\ -2x_1 & - & 2x_2 & = & 5, \end{array}$$

then as Figure 1.4 shows, there is no crossing between two lines and therefore there is no solution.

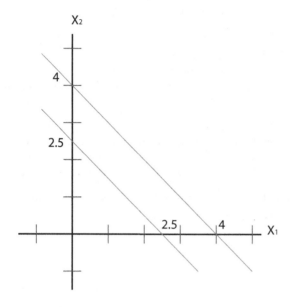

FIGURE 1.4

Geometrical view of the system of linear equations when there is no solution.

Suppose we have the following system of linear equations:

$$
\begin{aligned}
-x_1 \quad - \quad x_2 &= \quad 4 \\
x_1 \quad + \quad x_2 &= \quad -4.
\end{aligned}
$$

Then since the lines overlap to each other, there are infinitely many solutions shown in Figure 1.5.

A system of linear equations has either:

1. no solution;

2. exactly one solution;

3. infinitely many solutions.

Matrix Notation

A system of linear equations can be written as a matrix. It is easy to visualize and using **matrix operations**, we can solve a system of linear equations. We will discuss how to solve a system of linear equations in the next section.

Suppose we have the system of linear equations with variables x_1 and x_2 such that

$$
\begin{aligned}
x_1 \quad + \quad x_2 &= \quad 4 \\
2x_1 \quad + \quad x_2 &= \quad 5.
\end{aligned}
$$

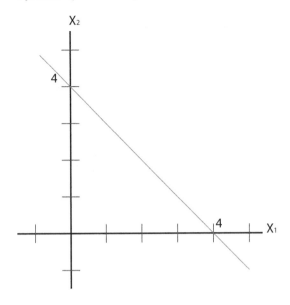

FIGURE 1.5
Geometrical view of the system of linear equations when there are infinitely many solutions.

Then we can write this system as a matrix:

$$\begin{bmatrix} 1 & 1 & 4 \\ 2 & 1 & 5 \end{bmatrix}.$$

This is called the **augmented matrix** of the system of linear equations. Also the matrix

$$\begin{bmatrix} 1 & 1 \\ 2 & 1 \end{bmatrix}$$

without the last column of the augmented matrix is called the **coefficient matrix** of the system of linear equations.

Example 36 *The system of linear equations*

$$\begin{array}{rrrrr} x_1 & + & x_2 & = & 4 \\ -2x_1 & - & 2x_2 & = & 5 \end{array}$$

has the augmented matrix

$$\begin{bmatrix} 1 & 1 & 4 \\ -2 & -2 & 5 \end{bmatrix}.$$

Example 37 *The system of linear equations*

$$\begin{array}{rrrrrrr} x_1 & - & x_2 & + & 4x_3 & = & 1 \\ 2x_1 & & & - & x_3 & = & -1.5 \end{array}$$

has the augmented matrix

$$\begin{bmatrix} 1 & -1 & 4 & 1 \\ 2 & 0 & -1 & -1.5 \end{bmatrix}.$$

The element of the second row and the second column of the matrix is 0 *because we have a coefficient of* x_2 *equal to* 0 *in the second linear equation.*

Example 38 *Go back to the first example of this section. A beer company has two production sites A and B and they want to transport their beer to two distributors C and D. The demand from distributor C is 542 beers per week, and the demand from distributor D is 422 beers per week. The supply from production site A is 475 beers per week, and the supply from production site B is 489 beers per week. We want to know if these sites produce enough beers to satisfy the demands from the distributors. This problem can set up a system of linear equations such that*

$$\begin{array}{rcrcrcl} x_{AC} & + & x_{AD} & & & & = & 475 \\ & & & & x_{BC} & + & x_{BD} & = & 489 \\ x_{AC} & & & + & x_{BC} & & & = & 542 \\ & & x_{AD} & & & + & x_{BD} & = & 422 \end{array}$$

where the variables x_{ij} *are the amounts of beer shipped from production site i to distributor j. Then we have the augmented matrix:*

$$\begin{bmatrix} 1 & 1 & 0 & 0 & 475 \\ 0 & 0 & 1 & 1 & 489 \\ 1 & 0 & 1 & 0 & 542 \\ 0 & 1 & 0 & 1 & 422 \end{bmatrix}.$$

In R, *you can type this as*

```
M <- matrix(c(1,0,1,0,1,0,0,1,0,1,1,0,0,1,0,1,475,489,542,422),
nrow=4,ncol=5)
```

Then, if you type M, you see the following:

```
> M
     [,1] [,2] [,3] [,4] [,5]
[1,]    1    1    0    0  475
[2,]    0    0    1    1  489
[3,]    1    0    1    0  542
[4,]    0    1    0    1  422
```

If you would like to have the coefficient matrix, then you can use the command "-5" to remove the fifth column of the matrix in R*:*

```
M2 <- M[,-5]
```

Then R *outputs*

```
> M2
     [,1] [,2] [,3] [,4]
[1,]   1    1    0    0
[2,]   0    0    1    1
[3,]   1    0    1    0
[4,]   0    1    0    1
```

Solving a System of Linear Equations

Adding or multiplying two equations in the system does not change a solution of the system. Also, multiplying the whole linear equation by a constant scalar would not change a solution of the system. By applying a sequence of these operations we try to find a solution of a system of linear equations. In this section you will see how to solve a system of linear equations via the **elementary row reduction** or **Gaussian Elimination**. The elementary row operations are equivalent to operations like adding or multiplying two equations in the system and multiplying a whole linear equation by a constant scalar, but they are operations on its augmented matrix. We will demonstrate such operations by the following example.

Suppose we have the following system of 3 linear equations with the variables x_1, x_2, x_3:

$$\begin{array}{rcrcrcr} x_1 & - & x_2 & + & 4x_3 & = & 1 \\ 2x_1 & & & - & x_3 & = & -1.5 \\ -x_1 & + & x_2 & & & = & 2. \end{array}$$

With this example, we show how the elementary row reduction method works. First you compute the augmented matrix of this system:

$$\begin{bmatrix} 1 & -1 & 4 & 1 \\ 2 & 0 & -1 & -1.5 \\ -1 & 1 & 0 & 2 \end{bmatrix}.$$

Here we will go through, step by step, a notion of a system of linear equations and a notion of its augmented matrix:

$$\begin{array}{rcrcrcr} x_1 & - & x_2 & + & 4x_3 & = & 1 \\ 2x_1 & & & - & x_3 & = & -1.5 \\ -x_1 & + & x_2 & & & = & 2 \end{array} \qquad \begin{bmatrix} 1 & -1 & 4 & 1 \\ 2 & 0 & -1 & -1.5 \\ -1 & 1 & 0 & 2 \end{bmatrix}.$$

1. We add the first equation to the third equation. This new equation becomes the new third equation:

$$\begin{array}{rcrcrcr} x_1 & - & x_2 & + & 4x_3 & = & 1 \\ -x_1 & + & x_2 & & & = & 2 \\ \hline & & & & 4x_3 & = & 3 \end{array}$$

Thus a new system of linear equations and its augmented matrix are:

$$
\begin{aligned}
x_1 \;-\; x_2 \;+\; 4x_3 &= 1 \\
2x_1 \qquad\qquad -\; x_3 &= -1.5 \\
4x_3 &= 3
\end{aligned}
\qquad
\begin{bmatrix}
1 & -1 & 4 & 1 \\
2 & 0 & -1 & -1.5 \\
0 & 0 & 4 & 3
\end{bmatrix}.
$$

2. Then we divide the third equation by 4. The new system of equations and its augmented matrix are:

$$
\begin{aligned}
x_1 \;-\; x_2 \;+\; 4x_3 &= 1 \\
2x_1 \qquad\qquad -\; x_3 &= -1.5 \\
x_3 &= 0.75
\end{aligned}
\qquad
\begin{bmatrix}
1 & -1 & 4 & 1 \\
2 & 0 & -1 & -1.5 \\
0 & 0 & 1 & 0.75
\end{bmatrix}.
$$

3. Now we add the third equation to the second equation. This becomes the new second equation:

$$
\begin{aligned}
2x_1 \qquad -\; x_3 &= -1.5 \\
x_3 &= 0.75 \\
\hline
2x_1 \qquad\qquad &= -0.75
\end{aligned}
$$

Thus, the new system of linear equations and its augmented matrix are:

$$
\begin{aligned}
x_1 \;-\; x_2 \;+\; 4x_3 &= 1 \\
2x_1 \qquad\qquad &= -0.75 \\
x_3 &= 0.75
\end{aligned}
\qquad
\begin{bmatrix}
1 & -1 & 4 & 1 \\
2 & 0 & 0 & -0.75 \\
0 & 0 & 1 & 0.75
\end{bmatrix}.
$$

4. Now we divide the second equation by 2. The new system of linear equations and its augmented matrix are:

$$
\begin{aligned}
x_1 \;-\; x_2 \;+\; 4x_3 &= 1 \\
x_1 \qquad\qquad &= -0.375 \\
x_3 &= 0.75
\end{aligned}
\qquad
\begin{bmatrix}
1 & -1 & 4 & 1 \\
1 & 0 & 0 & -0.375 \\
0 & 0 & 1 & 0.75
\end{bmatrix}.
$$

5. Exchange the first equation and the second equation: The new system of linear equations and its augmented matrix are:

$$
\begin{aligned}
x_1 \qquad\qquad &= -0.375 \\
x_1 \;-\; x_2 \;+\; 4x_3 &= 1 \\
x_3 &= 0.75
\end{aligned}
\qquad
\begin{bmatrix}
1 & 0 & 0 & -0.375 \\
1 & -1 & 4 & 1 \\
0 & 0 & 1 & 0.75
\end{bmatrix}.
$$

6. Now we take the second equation and subtract the first equation:

$$
\begin{aligned}
x_1 \;-\; x_2 \;+\; 4x_3 &= 1 \\
-(x_1 \qquad\qquad &= -0.375) \\
\hline
-\; x_2 \;+\; 4x_3 &= 1.375
\end{aligned}
$$

This equation becomes the new second equation. The new system of linear equations and its augmented matrix are:

$$
\begin{aligned}
x_1 \qquad\qquad &= -0.375 \\
-\; x_2 \;+\; 4x_3 &= 1.375 \\
x_3 &= 0.75
\end{aligned}
\qquad
\begin{bmatrix}
1 & 0 & 0 & -0.375 \\
0 & -1 & 4 & 1.375 \\
0 & 0 & 1 & 0.75
\end{bmatrix}.
$$

7. Similarly we take the second equation and subtract 4 times the third equation:

$$
\begin{array}{rcl}
-\ x_2\ +\ 4x_3 & = & 1.375 \\
-4(x_3 & = & 0.75) \\
\hline
-\ x_2 & = & -1.625
\end{array}
$$

This becomes the new second equation. The new system of linear equations and its augmented matrix are:

$$
\begin{array}{rcl}
x_1 & = & -0.375 \\
-\ x_2 & = & -1.625 \\
x_3 & = & 0.75
\end{array}
\qquad
\left[\begin{array}{cccc}
1 & 0 & 0 & -0.375 \\
0 & -1 & 0 & -1.625 \\
0 & 0 & 1 & 0.75
\end{array}\right].
$$

8. The last step is to multiply the second equation by -1. Then we have the solution:

$$
\begin{array}{rcl}
x_1 & = & -0.375 \\
x_2 & = & 1.625 \\
x_3 & = & 0.75
\end{array}
\qquad
\left[\begin{array}{cccc}
1 & 0 & 0 & -0.375 \\
0 & 1 & 0 & 1.625 \\
0 & 0 & 1 & 0.75
\end{array}\right].
$$

What you have seen in the example above is an elementary row reduction applied to the augmented matrix. Elementary row operations are operations on an augmented matrix. These are equivalent to the operations on a system of linear equations such that adding or multiplying two equations in the system does not change a solution of the system, nor does multiplying the whole linear equation by a constant scalar.

Definition 14 *The* **elementary row operations** *or* **Gaussian Eliminations** *are operations in a matrix defined as follows:*

1. *Replacement: Replace one row by the sum of the row itself and another row.*

2. *Interchange: Interchange two rows.*

3. *Scaling: Multiplication of a whole row by a constant.*

These operations in an augmented matrix can be translated to operations in its system of linear equations, which are shown in Table 1.5.

Uniqueness and Existence of a Solution

In solving a system of linear equations, there are two fundamental questions you have to ask:

1. Is there a solution to the system of linear equations?

2. If so, is it a unique solution?

TABLE 1.5
Operations in a system of linear equations and elementary row operations on its augmented matrix.

Elementary row operations	System of linear equations
Replacement	Replace an equation by adding itself to another equation
Interchange	Interchanging two equations
Scaling	Multiplying a whole equation by a constant

If it is not a unique solution, then the system of linear equations has infinitely many solutions. In order to find answers to these questions, we can apply a sequence of elementary row operations to the augmented matrix of a system of linear equations until the lower triangular part of its coefficient matrix is all zero. We will explain with some examples.

Example 39 *Suppose we have the system of linear equations such that:*

$$\begin{aligned}
x_1 + 2x_2 + 3x_3 &= 6 \\
-2x_1 + 3x_2 - 2x_3 &= -1 \\
-x_1 + 2x_2 + x_3 &= 2.
\end{aligned}$$

Its augmented matrix is

$$\begin{bmatrix} 1 & 2 & 3 & 6 \\ -2 & 3 & -2 & -1 \\ -1 & 2 & 1 & 2 \end{bmatrix}.$$

1. *We add the first equation to the third equation. This new equation becomes the new third equation. This is equivalent to the Replacement operation:*

$$\begin{aligned}
x_1 + 2x_2 + 3x_3 &= 6 \\
-x_1 + 2x_2 + x_3 &= 2 \\
\hline
4x_2 + 4x_3 &= 8
\end{aligned}$$

Thus the new system of linear equations and its augmented matrix are:

$$\begin{aligned}
x_1 + 2x_2 + 3x_3 &= 6 \\
-2x_1 + 3x_2 - 2x_3 &= -1 \\
4x_2 + 4x_3 &= 8
\end{aligned} \qquad \begin{bmatrix} 1 & 2 & 3 & 6 \\ -2 & 3 & -2 & -1 \\ 0 & 4 & 4 & 8 \end{bmatrix}.$$

2. *Then we switch the second equation and the third equation. This is equivalent to the Interchange operation. The new system of equations and its augmented matrix are:*

$$\begin{aligned}
x_1 + 2x_2 + 3x_3 &= 6 \\
4x_2 + 4x_3 &= 8 \\
-2x_1 + 3x_2 - 2x_3 &= -1
\end{aligned} \qquad \begin{bmatrix} 1 & 2 & 3 & 6 \\ 0 & 4 & 4 & 8 \\ -2 & 3 & -2 & -1 \end{bmatrix}.$$

3. Now we multiply the third equation by $1/2$. *This is equivalent to the Scaling operation. Thus the new system of linear equations and its augmented matrix are:*

$$\begin{array}{rrrrr} x_1 & + & 2x_2 & + & 3x_3 & = & 6 \\ & & 4x_2 & + & 4x_3 & = & 8 \\ -x_1 & + & \frac{3}{2}x_2 & - & x_3 & = & -\frac{1}{2} \end{array} \qquad \begin{bmatrix} 1 & 2 & 3 & 6 \\ 0 & 4 & 4 & 8 \\ -1 & \frac{3}{2} & -1 & -\frac{1}{2} \end{bmatrix}.$$

4. We add the first equation to the third equation. This new equation becomes the new third equation. This is equivalent to the Replacement operation:

$$\begin{array}{rrrrr} x_1 & + & 2x_2 & + & 3x_3 & = & 6 \\ -x_1 & + & \frac{3}{2}x_2 & - & x_3 & = & -\frac{1}{2} \\ \hline & & \frac{7}{2}x_2 & + & 2x_3 & = & \frac{11}{2} \end{array}$$

The new system of linear equations and its augmented matrix are:

$$\begin{array}{rrrrr} x_1 & + & 2x_2 & + & 3x_3 & = & 6 \\ & & 4x_2 & + & 4x_3 & = & 8 \\ & & \frac{7}{2}x_2 & + & 2x_3 & = & \frac{11}{2} \end{array} \qquad \begin{bmatrix} 1 & 2 & 3 & 6 \\ 0 & 4 & 4 & 8 \\ 0 & \frac{7}{2} & 2 & \frac{11}{2} \end{bmatrix}.$$

5. We multiply the third equation by $-8/7$. *This is equivalent to the Scaling operation. The new system of linear equations and its augmented matrix are:*

$$\begin{array}{rrrrr} x_1 & + & 2x_2 & + & 3x_3 & = & 6 \\ & & 4x_2 & + & 4x_3 & = & 8 \\ & & -4x_2 & - & \frac{16}{7}x_3 & = & -\frac{44}{7} \end{array} \qquad \begin{bmatrix} 1 & 2 & 3 & 6 \\ 0 & 4 & 4 & 8 \\ 0 & -4 & -\frac{16}{7} & -\frac{44}{7} \end{bmatrix}.$$

6. Now we add the second equation to the third equation:

$$\begin{array}{rrrrr} & & 4x_2 & + & 4x_3 & = & 8 \\ & & -4x_2 & - & \frac{16}{7}x_3 & = & -\frac{44}{7} \\ \hline & & & & \frac{12}{7}x_3 & = & \frac{12}{7} \end{array}$$

This equation becomes the new third equation. This operation is equivalent to the Replacement operation. The new system of linear equations and its augmented matrix are:

$$\begin{array}{rrrrr} x_1 & + & 2x_2 & + & 3x_3 & = & 6 \\ & & 4x_2 & + & 4x_3 & = & 8 \\ & & & & \frac{12}{7}x_3 & = & \frac{12}{7} \end{array} \qquad \begin{bmatrix} 1 & 2 & 3 & 6 \\ 0 & 4 & 4 & 8 \\ 0 & 0 & \frac{12}{7} & \frac{12}{7} \end{bmatrix}.$$

The coefficient matrix of the augmented matrix for the last system of linear equations is

$$\begin{bmatrix} 1 & 2 & 3 \\ 0 & 4 & 4 \\ 0 & 0 & \frac{12}{7} \end{bmatrix}.$$

Note that this augmented matrix forms an upper triangular matrix. In this case we have a unique solution of the system of linear equations. If we keep applying more elementary row operations, we have the solution $x_1 = 1$, $x_2 = 1$, $x_3 = 1$.

Example 40 *Suppose we have the following system of linear equations:*

$$
\begin{array}{rcrcrcr}
x_1 & + & 2x_2 & + & 3x_3 & = & 6 \\
-2x_1 & + & 3x_2 & - & 2x_3 & = & -1 \\
-x_1 & + & 5x_2 & + & x_3 & = & 2.
\end{array}
$$

Its augmented matrix is

$$
\begin{bmatrix}
1 & 2 & 3 & 6 \\
-2 & 3 & -2 & -1 \\
-1 & 5 & 1 & 2
\end{bmatrix}.
$$

As in the previous example, we apply elementary row operations to the augmented matrix:

1. We add the first equation to the second equation. This new equation becomes the new second equation. This is equivalent to the Replacement operation:

$$
\begin{array}{rcrcrcr}
x_1 & + & 2x_2 & + & 3x_3 & = & 6 \\
-2x_1 & + & 3x_2 & - & 2x_3 & = & -1 \\
\hline
-x_1 & + & 5x_2 & + & x_3 & = & 5
\end{array}
$$

Thus a new system of linear equations and its augmented matrix are:

$$
\begin{array}{rcrcrcr}
x_1 & + & 2x_2 & + & 3x_3 & = & 6 \\
-x_1 & + & 5x_2 & + & x_3 & = & 5 \\
-x_1 & + & 5x_2 & + & x_3 & = & 2
\end{array}
\qquad
\begin{bmatrix}
1 & 2 & 3 & 6 \\
-1 & 5 & 1 & 5 \\
-1 & 5 & 1 & 2
\end{bmatrix}.
$$

2. We multiply the third equation by -1. This is equivalent to the Scaling operation. The new system of linear equations and its augmented matrix are:

$$
\begin{array}{rcrcrcr}
x_1 & + & 2x_2 & + & 3x_3 & = & 6 \\
-x_1 & + & 5x_2 & + & x_3 & = & 5 \\
x_1 & - & 5x_2 & - & x_3 & = & -2
\end{array}
\qquad
\begin{bmatrix}
1 & 2 & 3 & 6 \\
-1 & 5 & 1 & 5 \\
1 & -5 & -1 & -2
\end{bmatrix}.
$$

3. We add the second equation to the third equation. This new equation becomes the new third equation. This is equivalent to the Replacement operation:

$$
\begin{array}{rcrcrcr}
-x_1 & + & 5x_2 & + & x_3 & = & 5 \\
x_1 & - & 5x_2 & - & x_3 & = & -2 \\
\hline
& & & & 0 & = & 3
\end{array}
$$

The new system of linear equations and its augmented matrix are:

$$
\begin{array}{rcrcrcr}
x_1 & + & 2x_2 & + & 3x_3 & = & 6 \\
& & 4x_2 & + & 4x_3 & = & 8 \\
& & & & 0 & = & 3
\end{array}
\qquad
\begin{bmatrix}
1 & 2 & 3 & 6 \\
-1 & 5 & 1 & 5 \\
0 & 0 & 0 & 3
\end{bmatrix}.
$$

The last equation does not make sense since 0 *is not equal to* 3. *So for this system of linear equations there is no solution.*

Solving a System of Linear Equations with R

Even though we can solve the system of linear equations with the solve() function, we are going to use the `matlib` package since it has nice functions to visualize what is going on geometrically. In order to install the `matlib` package, we just type:

```
install.packages("matlib", dependencies = TRUE)
```

Then, in order to upload the package, just type:

```
library(matlib)
```

Example 41 *We go back to Example 34. Recall we have the following system of linear equations such that*

$$
\begin{array}{rcrcl}
x_1 & + & x_2 & = & 4 \\
2x_1 & + & x_2 & = & 5.
\end{array}
$$

Its augmented matrix is

$$
\begin{bmatrix} 1 & 1 & 4 \\ 2 & 1 & 5 \end{bmatrix}.
$$

In R *you type:*

```
A <- matrix(c(1,2,1,1), nrow = 2, ncol = 2)
```

If you type A, then R *will return*

```
> A
     [,1] [,2]
[1,]   1    1
[2,]   2    1
```

Now we create a vector for the right-hand side of the system of linear equations:

```
b <- c(4, 5)
```

If you type b, then R *will return*

```
> b
[1] 4 5
```

Then, in order to solve the system of linear equations with the **matlib** *package, just type:*

```
Solve(A, b)
```

Then, R *will return*

```
> Solve(A, b)
x1   =  1
  x2 =  3
```

Without this package, if you failed to install matlib, *just type*

```
solve(A, b)
```

Then, R *will return*

```
> solve(A, b)
[1] 1 3
```

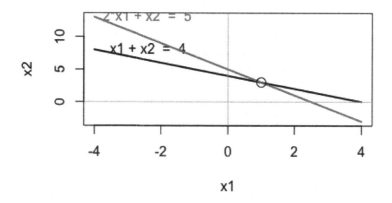

FIGURE 1.6
This is a visualization of the geometry of the system of linear equation in Example 41 from the matlib package in R.

A nice thing about this package is that you can plot the geometry of the system of linear equations. If you type

```
plotEqn(A,b)
```

Then R *outputs the image shown in Figure 1.6. In this example the two lines defined by the two equations intersect at one point* $(1, 3)$, *which is the circle in the image.*

Example 42 *This is from Example 39. The system of linear equations is:*

$$
\begin{aligned}
x_1 &+ 2x_2 + 3x_3 = 6 \\
-2x_1 &+ 3x_2 - 2x_3 = -1 \\
-x_1 &+ 2x_2 + x_3 = 2.
\end{aligned}
$$

Its augmented matrix is

$$
\begin{bmatrix}
1 & 2 & 3 & 6 \\
-2 & 3 & -2 & -1 \\
-1 & 2 & 1 & 2
\end{bmatrix}.
$$

In R, you type:

```
A <- matrix(c(1,-2,-1,2,3,2,3,-2,1), nrow = 3, ncol = 3)
```

If you type A, then R will return

```
> A
     [,1] [,2] [,3]
[1,]    1    2    3
[2,]   -2    3   -2
[3,]   -1    2    1
```

Now we create a vector for the right-hand side of the system of linear equations:

```
b <- c(6, -1, 2)
```

If you type b, then R will return

```
> b
[1]  6 -1  2
```

Then, in order to solve the system of linear equations with the `matlib` *package, just type:*

```
Solve(A, b)
```

Then, R will return

```
> Solve(A, b)
x1      = 1
  x2    = 1
    x3  = 1
```

Without this package, if you failed to install `matlib`, *just type*

```
solve(A, b)
```

Then, R *will return:*

```
> solve(A, b)
[1] 1 1 1
```

Again, this package can plot not only the dimensional geometry but also it can plot the three-dimensional geometry of the system of linear equations. If you type as follows:

```
plotEqn3d(A,b, xlim=c(0,4), ylim=c(0,4))
```

R *plots the figure shown in Figure 1.7.*

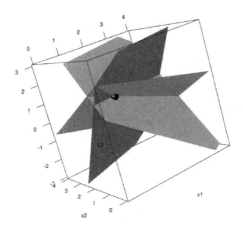

FIGURE 1.7
This is a visualization of the geometry of the system of linear equation in Example 42 from the `matlib` package in R.

In the argument of the function, "xlim=" is used to set up the range of the x-axis and "ylim" is used to set up the range of the y-axis. Then R *outputs the image shown in Figure 1.7. This is interactive. You hold the button of the mouse and rotate the image, so you can see how the linear equations intersect each other. In this example, the three linear planes defined by the three equations intersect at the solution* $(1, 1, 1)$, *which is the black dot in the image.*

We will end with the simplified example of Guinness beers in Example 38.

Example 43 *A beer company has two production sites, A and B, and they want to transport their beers to two distributors, C and D. The demand from distributor C is 542 beers per week, and the demand from distributor D is 422 beers per week. The supply from production site A is 475 beers per week, and the supply from production site B is 489 beers per week. We want to know if these production sites produce enough beers to satisfy the demands from the distributors. This problem can be set up as a system of linear equations such that*

$$
\begin{aligned}
x_{AC} \;+\; x_{AD} \quad\qquad\qquad\qquad &= 475 \\
x_{BC} \;+\; x_{BD} &= 489 \\
x_{AC} \qquad\quad +\; x_{BC} \qquad\qquad &= 542 \\
x_{AD} \qquad\qquad +\; x_{BD} &= 422
\end{aligned}
$$

where the variables x_{ij} are the number of beers shipped from a production site i to the distributor j. Then we have the augmented matrix:

$$
\begin{bmatrix}
1 & 1 & 0 & 0 & 475 \\
0 & 0 & 1 & 1 & 489 \\
1 & 0 & 1 & 0 & 542 \\
0 & 1 & 0 & 1 & 422
\end{bmatrix}.
$$

In R, *first you create the coefficient matrix A by the following:*

```
A <- matrix(c(1,0,1,0,1,0,0,1,0,1,1,0,0,1,0,1),nrow=4,ncol=4)
```

Then if you type A, R *returns*

```
> A
     [,1] [,2] [,3] [,4]
[1,]    1    1    0    0
[2,]    0    0    1    1
[3,]    1    0    1    0
[4,]    0    1    0    1
```

Now we create a vector for the right-hand side of the system of linear equations such that

```
b <- c(475, 489, 542,422)
```

Then if you type b, R *returns*

```
> b
[1] 475 489 542 422
```

Now we will use the function Solve() to solve the system of linear equations:

```
Solve(A, b)
```

Then R *returns*

```
> Solve(A, b)
x1    - 1*x4  =    53
  x2    + x4  =   422
    x3  + x4  =   489
         0 =     0
```

This means that there are infinitely many solutions since x4 is an independent variable, i.e., you can plug in any number in the variable x4.

1.3.4   Checkmarks

- The definition of a linear equation.

- The definition of a system of linear equations.

- The definition of the augmented matrix and the coefficient matrix of a system of linear equations.

- The definition of elementary row operations.

- You can solve a system of linear equations without a computer.

- You can solve a system of linear equations in R.

- You can visualize the geometry of a system of linear equations with the `matlib` package.

1.3.5   Conceptual Quizzes

Quiz 20 True or False: *Suppose we have a system of linear equations such that*

$$
\begin{array}{rcrcrcl}
-4x_1 & + & x_2 & - & 17x_3 & = & 5 \\
11x_1 & & & - & 2x_3 & = & 9 \\
x_1 & + & x_2 & + & 5x_3 & = & 3.
\end{array}
$$

Its augmented matrix is

$$
\begin{bmatrix}
-4 & 1 & -17 & 5 \\
11 & 0 & -2 & 9 \\
1 & 1 & 5 & 3
\end{bmatrix}.
$$

Quiz 21 True or False: *Suppose we have a system of linear equations such that*

$$
\begin{array}{rcrcrcl}
-x_1 & + & x_2 & + & x_3 & = & 8 \\
11x_1 & - & x_2 & - & 2x_3 & = & 2 \\
x_1 & + & x_2 & + & 5x_3 & = & 3.
\end{array}
$$

Its coefficient matrix is

$$\begin{bmatrix} -1 & 1 & 1 & 8 \\ 11 & -1 & -2 & 2 \\ 1 & 1 & 5 & 3 \end{bmatrix}.$$

Quiz 22 True or False: *Suppose we have an augmented matrix*

$$\begin{bmatrix} -1 & 1 & 1 & 8 \\ 11 & 1 & -2 & 2 \\ 1 & 1 & 5 & 3 \end{bmatrix}.$$

Switching the first row and the second row is an elementary row operation.

Quiz 23 True or False: *Suppose we have an augmented matrix*

$$\begin{bmatrix} -1 & 1 & 1 & 8 \\ 11 & 1 & -2 & 2 \\ 1 & 1 & 5 & 3 \end{bmatrix}.$$

Switching the first column and the second column is an elementary row operation.

Quiz 24 True or False: *Suppose we have an augmented matrix*

$$\begin{bmatrix} -1 & 1 & 1 & 8 \\ 11 & 1 & -2 & 2 \\ 1 & 1 & 5 & 3 \end{bmatrix}.$$

The following elementary row operation is the Scaling:

$$\begin{bmatrix} -1 & 1 & 1 & 8 \\ 11 & 1 & -2 & 2 \\ 1 & 1 & 5 & 3 \end{bmatrix} \rightarrow \begin{bmatrix} -1 & 1 & 1 & 8 \\ 1 & 1 & 5 & 3 \\ 11 & 1 & -2 & 2 \end{bmatrix}$$

Quiz 25 True or False: *Suppose we have an augmented matrix*

$$\begin{bmatrix} -1 & 1 & 1 & 8 \\ 11 & 1 & -2 & 2 \\ 1 & 1 & 5 & 3 \end{bmatrix}.$$

The following elementary row operation is the "Interchange":

$$\begin{bmatrix} -1 & 1 & 1 & 8 \\ 11 & 1 & -2 & 2 \\ 1 & 1 & 5 & 3 \end{bmatrix} \rightarrow \begin{bmatrix} -1 & 1 & 1 & 8 \\ 1 & 1 & 5 & 3 \\ 11 & 1 & -2 & 2 \end{bmatrix}$$

Quiz 26 True or False: *Suppose we have an augmented matrix*

$$\begin{bmatrix} -1 & 1 & 1 & 8 \\ 11 & 1 & -2 & 2 \\ 1 & 1 & 5 & 3 \end{bmatrix}.$$

The following elementary row operation is the "Scaling":

$$
\begin{bmatrix} -1 & 1 & 1 & 8 \\ 11 & 1 & -2 & 2 \\ 1 & 1 & 5 & 3 \end{bmatrix}
\rightarrow
\begin{bmatrix} -1 & 1 & 1 & 8 \\ 11 & 1 & -2 & 2 \\ 0 & 2 & 6 & 11 \end{bmatrix}
$$

1.3.6   Regular Exercises

Exercise 1.15 *Suppose we have a system of linear equations such that*

$$
\begin{aligned}
x_1 \;-\; 6x_2 &= 3 \\
-2x_2 &= -1.
\end{aligned}
$$

1. *Is there a solution?*

2. *If so, then is it unique? Or infinitely many?*

3. *If it is unique, then solve the system of linear equations using the elementary row reduction.*

Exercise 1.16 *Suppose we have a system of linear equations such that*

$$
\begin{aligned}
x_1 \;-\; 5x_2 &= 4 \\
3x_1 \;-\; 10x_2 &= 10.
\end{aligned}
$$

1. *Is there a solution?*

2. *If so, then is it unique? Or infinitely many?*

3. *If it is unique, then solve the system of linear equations using the elementary row reduction.*

Exercise 1.17 *Suppose we have a system of linear equations such that*

$$
\begin{aligned}
-9x_1 \;-\; 3x_2 &= 2 \\
5x_1 \;-\; 2x_2 &= 9.
\end{aligned}
$$

1. *Is there a solution?*

2. *If so, then is it unique? Or infinitely many?*

3. *If it is unique, then solve the system of linear equations using the elementary row reduction.*

Exercise 1.18 *Suppose we have a system of linear equations such that*

$$
\begin{aligned}
x_1 \;-\; 2x_2 &= -5 \\
10x_1 \;-\; x_2 &= 6.
\end{aligned}
$$

1. *Is there a solution?*

2. *If so, then is it unique? Or infinitely many?*

3. *If it is unique, then solve the system of linear equations using the elementary row reduction.*

Exercise 1.19 *Suppose we have a system of linear equations such that*

$$
\begin{aligned}
x_1 &- 2x_2 &= -5 \\
2x_1 &- 4x_2 &= 6.
\end{aligned}
$$

1. *Is there a solution?*

2. *If so, then is it unique? Or infinitely many?*

3. *If it is unique, then solve the system of linear equations using the elementary row reduction.*

Exercise 1.20 *Suppose we have a system of linear equations such that*

$$
\begin{aligned}
-6x_1 &- 10x_2 &- x_3 &= -1 \\
-9x_1 &+ 5x_2 &+ 6x_3 &= 0 \\
-3x_1 &- 7x_2 &+ x_3 &= 2.
\end{aligned}
$$

1. *Is there a solution?*

2. *If so, then is it unique? Or infinitely many?*

3. *If it is unique, then solve the system of linear equations using the elementary row reduction.*

Exercise 1.21 *Suppose we have a system of linear equations such that*

$$
\begin{aligned}
-7x_1 &+ 8x_2 &+ 2x_3 &= 7 \\
-5x_1 &- 8x_2 &+ x_3 &= 4 \\
-4x_1 &- 3x_2 &- x_3 &= -10.
\end{aligned}
$$

1. *Is there a solution?*

2. *If so, then is it unique? Or infinitely many?*

3. *If it is unique, then solve the system of linear equations using the elementary row reduction.*

Exercise 1.22 *Suppose we have a system of linear equations such that*

$$
\begin{aligned}
8x_1 &- 3x_2 &+ 2x_3 &= 3 \\
-1x_1 &- 10x_2 & &= 8 \\
-2x_1 &+ 8x_2 &- 9x_3 &= 1.
\end{aligned}
$$

1. *Is there a solution?*

2. *If so, then is it unique? Or infinitely many?*

3. *If it is unique, then solve the system of linear equations using the elementary row reduction.*

Exercise 1.23 *Suppose we have a system of linear equations such that*

$$
\begin{array}{rrrrrrr}
-3x_1 & + & 5x_2 & - & 5x_3 & = & 8 \\
-8x_1 & + & 3x_2 & + & 4x_3 & = & 6 \\
8x_1 & + & 10x_2 & + & 2x_3 & = & -9.
\end{array}
$$

1. *Is there a solution?*

2. *If so, then is it unique? Or infinitely many?*

3. *If it is unique, then solve the system of linear equations using the elementary row reduction.*

Exercise 1.24 *Suppose we have a system of linear equations such that*

$$
\begin{array}{rrrrrrr}
-5x_1 & - & 6x_2 & - & 7x_3 & = & -2 \\
5x_1 & - & 9x_2 & - & 2x_3 & = & 7 \\
x_1 & - & 8x_2 & + & 10x_3 & = & -9.
\end{array}
$$

1. *Is there a solution?*

2. *If so, then is it unique? Or infinitely many?*

3. *If it is unique, then solve the system of linear equations using the elementary row reduction.*

Exercise 1.25 *Suppose we have a system of linear equations such that*

$$
\begin{array}{rrrrrrr}
-7x_1 & + & x_2 & - & 5x_3 & = & -3 \\
 & & 5x_2 & + & 6x_3 & = & -6 \\
-4x_1 & + & 9x_2 & + & 4x_3 & = & -3.
\end{array}
$$

1. *Is there a solution?*

2. *If so, then is it unique? Or infinitely many?*

3. *If it is unique, then solve the system of linear equations using the elementary row reduction.*

Exercise 1.26 *Suppose we have a system of linear equations such that*

$$
\begin{array}{rrrrrrr}
-7x_1 & + & x_2 & - & 5x_3 & = & -3 \\
 & & 5x_2 & + & 6x_3 & = & -6 \\
-7x_1 & + & 6x_2 & + & 5x_3 & = & -9.
\end{array}
$$

1. *Is there a solution?*

2. *If so, then is it unique? Or infinitely many?*

3. *If it is unique, then solve the system of linear equations using the elementary row reduction.*

Exercise 1.27 *Suppose we have a system of linear equations such that*

$$\begin{aligned} x_1 &- 2x_2 &= -5 \\ hx_1 &- 2x_2 &= 3. \end{aligned}$$

1. *Find h so that the system of linear equations has a unique solution using the elementary row reduction.*

2. *Solve the system of linear equations.*

Exercise 1.28 *Suppose we have a system of linear equations such that*

$$\begin{aligned} -x_1 &+ x_2 &= h \\ x_1 &- 2x_2 &= 5. \end{aligned}$$

1. *Find h so that the system of linear equations has a unique solution using the elementary row reduction.*

2. *Solve the system of linear equations.*

Exercise 1.29 *Suppose we have a system of linear equations such that*

$$\begin{aligned} hx_1 &- 2x_2 &- x_3 &= -5 \\ -x_1 &- 3x_2 &- 2x_3 &= 1 \\ -x_1 &+ 3x_2 &- 5x_3 &= 2. \end{aligned}$$

1. *Find h so that the system of linear equations has a unique solution using the elementary row reduction.*

2. *Solve the system of linear equations.*

Exercise 1.30 *Suppose we have a system of linear equations such that*

$$\begin{aligned} 3x_1 &- 5x_2 &+ 2x_3 &= -4 \\ -x_1 &- x_2 &- 5x_3 &= h \\ -4x_1 &+ 3x_2 &- 5x_3 &= -3. \end{aligned}$$

1. *Find h so that the system of linear equations has a unique solution using the elementary row reduction.*

2. *Solve the system of linear equations.*

Exercise 1.31 *You are at the accounting office for a small county fair and the admission fee is \$1.50 for children and \$4.00 for adults. Suppose 2200 people came to the fair in one day and we earned \$5050. How many children and how many adults entered the fair in that day?*

Exercise 1.32 *Suppose we have 3,000 first-year undergraduate students at a university. They are required to take elementary chemistry class or linear algebra class and they cannot take both classes. 90% of students who took the chemistry class passed and 80% of students who took the linear algebra class passed in a certain year. The total of students who passed the chemistry class and students who passed the linear algebra class were 2,542. How many students took the chemistry class and how many students took the linear algebra class in the year?*

Exercise 1.33 *You and your friend are racing. Your friend is riding her horse but in order to set up the saddle, it takes her 8.5 minutes until she can start running. You can run 0.2 km every minute and her horse can run 0.5 km every minute. How far can you go before your friend and her horse catch you?*

Exercise 1.34 *You are trying to buy a cell phone plan. Company A provides a plan that costs $20 every month and 25 cents per minute. Company B provides a plan that costs $40 per month but calls cost only 8 cents per minute. When does the plan from company A and the plan from company B have the same price?*

Exercise 1.35 *Mary wants to make 5 identical candy baskets for her party. She has the budget of $61 for bouquets and wants 24 small cupcakes for each basket. For the price, chocolate cupcakes cost $0.6 each, strawberry cupcakes cost $0.4 each, and lemon cupcakes cost $0.3 each. Now, for each basket, she wants to have twice as many chocolate cupcakes as the other 2 kind of cupcakes combined. How many chocolate cupcakes, strawberry cupcakes, and lemon cupcakes are in each basket?*

Exercise 1.36 *Alex buys 4 pounds of lemon candies and 2 pounds of orange candies on one day. The total of the candies costs $10 in that day. Then he buys 2 pounds of orange candies and 3 pounds of strawberry candies on the second day and it costs him $8 in total. He then buys 3 pounds of strawberry candies and one pound of chocolates on the third day and it costs him $10 in total. On the last day he buys 2 pounds of lemon candies and 2 pounds of chocolates and it costs him $12 in total. What is the price per pound of lemon candies, orange candies, strawberry candies, and chocolates?*

Exercise 1.37 *We will end with a toy example of Guinness beers in Example 38. A beer company has two production sites A and B and they want to transport their beers to two distributors C and D. The demand from distributor C is 542 beers per week, and the demand from distributor D is 422 beers per week. The supply from production site A is 475 beers per week, and the supply from production site B is 489 beers per week. Suppose we know the number of beers shipping from production site A to distributor C is 100. Then how many beers do we have to ship from production site A to distributor D, how many beers do we have to ship from production site B to distributor C, and how many beers do we have to ship from production site B to distributor D?*

1.3.7   Lab Exercises

Lab Exercise 18 *Go back to Example 1.15. With* R, *do the following:*

1. *Solve the system of linear equations with the Solve() or the solve() function.*

2. *Plot the geometry of the system of linear equations.*

Lab Exercise 19 *Go back to Exercise 1.16. With* R, *do the following:*

1. *Solve the system of linear equations with the Solve() or the solve() function.*

2. *Plot the geometry of the system of linear equations.*

Lab Exercise 20 *Go back to Exercise 1.17. With* R, *do the following:*

1. *Solve the system of linear equations with the Solve() or the solve() function.*

2. *Plot the geometry of the system of linear equations.*

Lab Exercise 21 *Go back to Exercise 1.18. With* R, *do the following:*

1. *Solve the system of linear equations with the Solve() or the solve() function.*

2. *Plot the geometry of the system of linear equations.*

Lab Exercise 22 *Go back to Exercise 1.19. With* R, *do the following:*

1. *Solve the system of linear equations with the Solve() or the solve() function.*

2. *Plot the geometry of the system of linear equations.*

Lab Exercise 23 *Go back to Exercise 1.20. With* R, *do the following:*

1. *Solve the system of linear equations with the Solve() or the solve() function.*

2. *Plot the geometry of the system of linear equations.*

Lab Exercise 24 *Go back to Exercise 1.21. With* R, *do the following:*

1. *Solve the system of linear equations with the Solve() or the solve() function.*

2. *Plot the geometry of the system of linear equations.*

Lab Exercise 25 *Go back to Exercise 1.22. With* R, *do the following:*

1. *Solve the system of linear equations with the Solve() or the solve() function.*

2. *Plot the geometry of the system of linear equations.*

Lab Exercise 26 *Go back to Exercise 1.23. With R, do the following:*

1. *Solve the system of linear equations with the Solve() or the solve() function.*

2. *Plot the geometry of the system of linear equations.*

Lab Exercise 27 *Go back to Exercise 1.24. With R, do the following:*

1. *Solve the system of linear equations with the Solve() or the solve() function.*

2. *Plot the geometry of the system of linear equations.*

Lab Exercise 28 *Go back to Exercise 1.25. With R, do the following:*

1. *Solve the system of linear equations with the Solve() or the solve() function.*

2. *Plot the geometry of the system of linear equations.*

Lab Exercise 29 *Go back to Exercise 1.26. With R, do the following:*

1. *Solve the system of linear equations with the Solve() or the solve() function.*

2. *Plot the geometry of the system of linear equations.*

1.3.8   Practical Applications

Let us solve the working problem we showed in the beginning of this section. Recall that this data set is collected from Guinness Ghana Ltd. and it contains information on the supply of Malta Guinness from two production sites, Kaasi and Achimota, to nine key distributors geographically scattered in the regions of Ghana. They collected data sets twice and the data sets shown in Tables 1.3 and 1.4 were collected between July 7th 2010 and June 8th 2011. Table 1.3 shows the demands from the distributors and supplies from the production sites. Now we want to know if it is possible to satisfy the demand with the given supplies.

First, as we did in a smaller example at the beginning of this section, we create a 2×9 table such that

	FTA	RICKY	OBIBAJK	KADOM	NAATO	LESK	DCEE	JOEMAN	KBOA	Total
Achimota	x_{11}	x_{12}	x_{13}	x_{14}	x_{15}	x_{16}	x_{17}	x_{18}	x_{19}	1298
Kaasi	x_{21}	x_{22}	x_{23}	x_{24}	x_{25}	x_{26}	x_{27}	x_{28}	x_{29}	1948
Total	465	605	451	338	260	183	282	127	535	3246

Now we have a system of linear equations

$$
\begin{array}{rcl}
x_{11} + x_{12} + \ldots + x_{19} &=& 1298 \\
x_{21} + x_{22} + \ldots + x_{29} &=& 1948 \\
x_{11} + x_{21} &=& 465 \\
x_{12} + x_{22} &=& 605 \\
x_{13} + x_{23} &=& 451 \\
x_{14} + x_{24} &=& 338 \\
x_{15} + x_{25} &=& 260 \\
x_{16} + x_{26} &=& 183 \\
x_{17} + x_{27} &=& 282 \\
x_{18} + x_{28} &=& 127 \\
x_{19} + x_{29} &=& 535.
\end{array}
$$

The coefficient matrix for this system is

$$
\begin{bmatrix}
1 & 1 & 1 & 1 & 1 & 1 & 1 & 1 & 1 & 0 & 0 & 0 & 0 & 0 & 0 & 0 & 0 & 0 \\
0 & 0 & 0 & 0 & 0 & 0 & 0 & 0 & 0 & 1 & 1 & 1 & 1 & 1 & 1 & 1 & 1 & 1 \\
1 & 0 & 0 & 0 & 0 & 0 & 0 & 0 & 0 & 1 & 0 & 0 & 0 & 0 & 0 & 0 & 0 & 0 \\
0 & 1 & 0 & 0 & 0 & 0 & 0 & 0 & 0 & 0 & 1 & 0 & 0 & 0 & 0 & 0 & 0 & 0 \\
0 & 0 & 1 & 0 & 0 & 0 & 0 & 0 & 0 & 0 & 0 & 1 & 0 & 0 & 0 & 0 & 0 & 0 \\
0 & 0 & 0 & 1 & 0 & 0 & 0 & 0 & 0 & 0 & 0 & 0 & 1 & 0 & 0 & 0 & 0 & 0 \\
0 & 0 & 0 & 0 & 1 & 0 & 0 & 0 & 0 & 0 & 0 & 0 & 0 & 1 & 0 & 0 & 0 & 0 \\
0 & 0 & 0 & 0 & 0 & 1 & 0 & 0 & 0 & 0 & 0 & 0 & 0 & 0 & 1 & 0 & 0 & 0 \\
0 & 0 & 0 & 0 & 0 & 0 & 1 & 0 & 0 & 0 & 0 & 0 & 0 & 0 & 0 & 1 & 0 & 0 \\
0 & 0 & 0 & 0 & 0 & 0 & 0 & 1 & 0 & 0 & 0 & 0 & 0 & 0 & 0 & 0 & 1 & 0 \\
0 & 0 & 0 & 0 & 0 & 0 & 0 & 0 & 1 & 0 & 0 & 0 & 0 & 0 & 0 & 0 & 0 & 1
\end{bmatrix},
$$

which is a 11×18 matrix and the vector for the right-hand side of the system of linear equations is

$$
\begin{bmatrix}
1298 \\
1948 \\
465 \\
605 \\
451 \\
338 \\
260 \\
183 \\
282 \\
127 \\
535
\end{bmatrix},
$$

which is a 1×11 matrix.

In R, you can type

```
A <- matrix(c(1 , 1 , 1 , 1 , 1 , 1 , 1 , 1 , 1 , 0 ,
0 , 0 , 0 , 0 , 0 , 0 , 0 , 0 , 0 , 0 , 0 , 0 , 0 , 0 ,
0 , 0 , 0 , 1 , 1 , 1 , 1 , 1 , 1 , 1 , 1 , 1 , 1 , 0 ,
0 , 0 , 0 , 0 , 0 , 0 , 0 , 1 , 0 , 0 , 0 , 0 , 0 , 0 ,
0 , 0 , 0 , 1 , 0 , 0 , 0 , 0 , 0 , 0 , 0 , 0 , 1 , 0 ,
0 , 0 , 0 , 0 , 0 , 0 , 0 , 0 , 1 , 0 , 0 , 0 , 0 , 0 ,
0 , 0 , 0 , 1 , 0 , 0 , 0 , 0 , 0 , 0 , 0 , 0 , 0 , 1 ,
0 , 0 , 0 , 0 , 0 , 0 , 0 , 0 , 1 , 0 , 0 , 0 , 0 , 0 ,
0 , 0 , 0 , 0 , 1 , 0 , 0 , 0 , 0 , 0 , 0 , 0 , 0 , 1 ,
0 , 0 , 0 , 0 , 0 , 0 , 0 , 0 , 0 , 1 , 0 , 0 , 0 , 0 ,
0 , 0 , 0 , 0 , 1 , 0 , 0 , 0 , 0 , 0 , 0 , 0 , 0 , 0 ,
1 , 0 , 0 , 0 , 0 , 0 , 0 , 0 , 0 , 1 , 0 , 0 , 0 , 0 ,
0 , 0 , 0 , 0 , 0 , 1 , 0 , 0 , 0 , 0 , 0 , 0 , 0 , 0 ,
1 , 0 , 0 , 0 , 0 , 0 , 0 , 0 , 0 , 0 , 1 , 0 , 0 , 0 ,
0 , 0 , 0 , 0 , 0 , 1), nrow = 11, ncol = 18, byrow = TRUE)
```

In the argument of the matrix() function "byrow" defined the order of the entries row by row. This means the elements of the matrix are filled from the first row and to the second row and so on.

Now we write the right-hand side in R:

```
b <- c(1298, 1948, 465, 605, 451,  338,  260,  183,  282, 127, 535)
```

When you type Solve(A, b) in R, you will see:

```
> Solve(A, b)
x1              - 1*x11 - 1*x12 - 1*x13 - 1*x14 - 1*x15 - 1*x16 - 1*x17 - 1*x18  =  -1483
   x2           + x11                                                            =   605
      x3                + x12                                                     =   451
         x4                    + x13                                             =   338
            x5                        + x14                                      =   260
               x6                           + x15                               =   183
                  x7                              + x16                          =   282
                     x8                               + x17                      =   127
                        x9                               + x18  =   535
                           x10  + x11  + x12  + x13  + x14  + x15  + x16  + x17  + x18  =  1948
                                                                          0  =     0
```

This means that there are infinitely many solutions to the system of linear equations.

Remark 1.3 *This kind of problem is called the Transportation Problem and in addition to the set of linear equations, we need to add constraints that all variables must be non-negative, i.e., $x_{ij} \geq 0$ for all $i = 1, 2$ and $j = 1, \ldots 9$. All solutions of the system of linear equations are in a feasible region for the Transportation Problem, which is a well-known problem in optimization.*

1.3.9   Supplements with python Code

In python we use the package numpy to solve a system of linear equations. As in the previous section, to call the package you type:

```
import numpy as np
```

Suppose we have the system of linear equations in Example 39. Recall that the system of linear equations is:

$$
\begin{array}{rcrcrcr}
x_1 & + & 2x_2 & + & 3x_3 & = & 6 \\
-2x_1 & + & 3x_2 & - & 2x_3 & = & -1 \\
-x_1 & + & 2x_2 & + & x_3 & = & 2.
\end{array}
$$

Its augmented matrix is

$$
\begin{bmatrix}
1 & 2 & 3 & 6 \\
-2 & 3 & -2 & -1 \\
-1 & 2 & 1 & 2
\end{bmatrix}.
$$

In **python**, to create the coefficient matrix of the system of linear equations, you type

```
A = np.array([[1, 2, 3], [-2, 3, -2], [-1, 2, 1]])
```

For the right-hand side you type:

```
b = np.array([6, -1, 2])
```

To solve the system of linear equations, you can use the np.linalg.solve function. To solve this system you type

```
np.linalg.solve(A, b)
```

Then **python** outputs

```
>>> np.linalg.solve(A, b)
array([1., 1., 1.])
```

In **python** it is not very easy to visualize the geometry of a system of linear equations like we can do in R. In general R is a great tool for visualization and **python** is a great tool for large computations.

1.4   Echelon Form

In the previous section, we discussed how to apply elementary row reductions to the augmented matrix of a system of linear equations until the augmented matrix of the system becomes the **Echelon form** to solve it. In this section we officially define the Echelon form of the matrix and, from its form, determine whether the system has no solution, a unique solution, or infinitely many solutions.

1.4.1   Task Completion Checklist

- During the Lecture:

  1. Read the definition of the Echelon form.

  2. Using elementary row reductions, learn how to compute the Echelon form from the augmented matrix of a system of linear equations.

  3. Learn whether a system of linear equations has no solution, a unique solution, or infinitely many solutions from the Echelon form.

  4. Try to solve examples without the help of a computer.

  5. With R, try to solve examples in this section.

- After the Lecture:

  1. Take conceptual quizzes to make sure you understand the materials in this section.

  2. Do some regular exercises.

  3. Conduct lab exercises with R.

  4. Conduct practical applications with R.

  5. If you are interested in python, read the supplement in this section and conduct lab exercises and practical applications with python.

1.4.2   Working Examples

This data is from the 1996 General Society Survey, National Opinion Research Center.

Income	Very Dissatisfied	Little Dissatisfied	Moderately Satisfied	Very Satisfied
< 15,000	1	3	10	6
15,000 − 25,000	2	3	10	7
25,000 − 40,000	1	6	14	12
> 40,000	0	1	9	11

TABLE 1.6

Income is measured in US dollars. These questions were asked of black males in the United State.

The data in Table 1.6 are incomes in terms of US dollars and job satisfaction among black men in the US. This data is used to determine if there is a relationship between job satisfaction and income among black men in the US. As in the example in Section 1.1, this table also has small numbers. When an agency releases information, they release the data only if it does not expose personal information. For example, if this data was collected from a certain small town, then one might be able to find one specific person from this data. For example, the number of people whose income is more than $40,000 and

"Little Dissatisfied" is 1, and we might be able to find out who this person is. In order to protect personal information, we will take summaries of these data sets, such as row and column sums of this table. These summaries are known as **sufficient statistics**. These summaries are the only information needed to study these relationships.

The row and column sums from Table 1.6 are:

				Total
1	3	10	6	20
2	3	10	7	22
1	6	14	12	33
0	1	9	11	21
Total 4	13	43	36	96

In order to make sure we can protect personal information, we perturb the data. For example, if we perturb the column sums $(4, 13, 43, 63)$ to $(4.5, 14.2, 42.4, 62.9)$ and the row sums $(20, 22, 33, 21)$ to $(21.5, 20.9, 33.4, 22.1)$, then is there a 4×4 table satisfying these row and column sums? If there is no table with these row and column sums, then this perturbation does not work. This is an important problem as the agency has to be careful to perturb the information which they are going to release so that the released information will make sense in terms of statistical analysis.

1.4.3 Echelon Form

With a sequence of elementary row reductions, every matrix has a unique **echelon form** and a **reduced echelon form**. With the reduced echelon form, we can answer the fundamental questions of a system of linear equations, namely, whether there is no solution, a unique solution, or infinitely many solutions of the system.

Recall the system of linear equations from Example 39. We have the following system of linear equations:

$$
\begin{array}{rcrcrcr}
x_1 & + & 2x_2 & + & 3x_3 & = & 6 \\
-2x_1 & + & 3x_2 & - & 2x_3 & = & -1 \\
-x_1 & + & 2x_2 & + & x_3 & = & 2.
\end{array}
$$

Its augmented matrix is

$$
\begin{bmatrix}
1 & 2 & 3 & 6 \\
-2 & 3 & -2 & -1 \\
-1 & 2 & 1 & 2
\end{bmatrix}.
$$

After applying a sequence of elementary row reductions, we have

$$
\begin{bmatrix}
1 & 2 & 3 & 6 \\
0 & 4 & 4 & 8 \\
0 & 0 & \frac{12}{7} & \frac{12}{7}
\end{bmatrix}.
$$

Notice that the leading coefficient of each non-zero row is to the right of the leading entry of the row above it. This form is an example of an **echelon form**.

Definition 15 *We say a matrix is in an* **echelon form** *if the following condition is satisfied:*

1. *all nonzero rows are above rows of all zeros;*

2. *the leading entry of each non-zero row is to the right of the leading entry of the row above it; and*

3. *all entries in each column below a leading entry are all zeros.*

When apply a sequence of elementary row reductions, we can get a simpler form than an echelon form.

1. Now we have a system of linear equations and its augmented matrix such that:

$$
\begin{array}{rrrcr}
x_1 & + & 2x_2 & + & 3x_3 & = & 6 \\
 & & 4x_2 & + & 4x_3 & = & 8 \\
 & & & & \frac{12}{7}x_3 & = & \frac{12}{7}
\end{array}
\qquad
\begin{bmatrix}
1 & 2 & 3 & 6 \\
0 & 4 & 4 & 8 \\
0 & 0 & \frac{12}{7} & \frac{12}{7}
\end{bmatrix}.
$$

2. We multiply the third equation by 7/12. This is the Scaling operation. The new system of linear equations and its augmented matrix are:

$$
\begin{array}{rrrcr}
x_1 & + & 2x_2 & + & 3x_3 & = & 6 \\
 & & 4x_2 & + & 4x_3 & = & 8 \\
 & & & & x_3 & = & 1
\end{array}
\qquad
\begin{bmatrix}
1 & 2 & 3 & 6 \\
0 & 4 & 4 & 8 \\
0 & 0 & 1 & 1
\end{bmatrix}.
$$

3. We multiply the second equation by 1/4. This is the Scaling operation. The new system of linear equations and its augmented matrix are:

$$
\begin{array}{rrrcr}
x_1 & + & 2x_2 & + & 3x_3 & = & 6 \\
 & & x_2 & + & x_3 & = & 2 \\
 & & & & x_3 & = & 1
\end{array}
\qquad
\begin{bmatrix}
1 & 2 & 3 & 6 \\
0 & 1 & 1 & 2 \\
0 & 0 & 1 & 1
\end{bmatrix}.
$$

4. We subtract the third equation from the second equation. This new equation becomes the new second equation. These are the Replacement and Scaling operations:

$$
\begin{array}{rrrcr}
x_2 & + & x_3 & = & 2 \\
 & - & x_3 & = & -1 \\
\hline
x_2 & & & = & 1
\end{array}
$$

The new system of linear equations and its augmented matrix are:

$$
\begin{array}{rrrcr}
x_1 & + & 2x_2 & + & 3x_3 & = & 6 \\
 & & x_2 & & & = & 1 \\
 & & & & x_3 & = & 1
\end{array}
\qquad
\begin{bmatrix}
1 & 2 & 3 & 6 \\
0 & 1 & 0 & 1 \\
0 & 0 & 1 & 1
\end{bmatrix}.
$$

5. We subtract three times the third equation from the first equation and we also subtract two times the second equations from the first equation. The new equation becomes the new first equation. This is equivalent to a sequence of the Replacement and Scaling operations:

$$
\begin{array}{rcrcrcl}
x_1 & + & 2x_2 & + & 3x_3 & = & 6 \\
 & & - 2x_2 & & & = & -2 \\
 & & & & - 3x_3 & = & -3 \\
\hline
x_1 & & & & & = & 1
\end{array}
$$

The new system of linear equations and its augmented matrix are:

$$
\begin{array}{rcl}
x_1 & = & 1 \\
x_2 & = & 1 \\
x_3 & = & 1
\end{array}
\qquad
\left[
\begin{array}{cccc}
1 & 0 & 0 & 1 \\
0 & 1 & 0 & 1 \\
0 & 0 & 1 & 1
\end{array}
\right].
$$

The last form of the augmented matrix of the system of linear equations is called the **reduced echelon form**. Formally it is defined as follows:

Definition 16 *We say a matrix is in the* **reduced echelon form** *if it is in an echelon form with the following additional conditions:*

1. *leading entries are all 1; and*

2. *if there is a leading entry in a column, then the leading entry is the only non-zero entry in the column.*

Remark 1.4 *Every matrix has its unique reduced echelon form and we can obtain the form by a sequence of elementary row reductions.*

Gaussian Elimination

The Gaussian Elimination is an algorithm to compute the reduced echelon form of a matrix using a sequence of elementary row reductions. We have seen some examples in the previous section and one in this example of how to obtain an echelon form or the reduced echelon form of the augmented matrix for a system of linear equations in order to solve the system. Here we use the following system of linear equations to demonstrate Gaussian Elimination.

Suppose we have the following system of linear equations:

$$
\begin{array}{rcrcrcrcr}
 & & x_2 & + & 3x_3 & - & x_4 & = & 1 \\
-x_1 & + & x_2 & - & 4x_3 & & & = & 1 \\
x_1 & & & + & 2x_3 & + & 4x_4 & = & 5 \\
 & & x_2 & & & - & 4x_4 & = & -2.
\end{array}
$$

Its augmented matrix is

$$
\left[
\begin{array}{ccccc}
0 & 1 & 3 & -1 & 1 \\
-1 & 1 & -4 & 0 & 1 \\
1 & 0 & 2 & 4 & 5 \\
0 & 1 & 0 & -4 & -2
\end{array}
\right].
$$

Step 1: Order the rows of the matrix so that all nonzero rows are above rows of all zeros. In this case we do not have any zero rows, so we will not do anything for this step.

Step 2: Order the rows of the matrix so that the leading entry of each non-zero row is to the right or in the same position of the leading entry of the row above it.

In this example, since the leading non-zero coefficient of the first equation is for the variable x_2 and the second equation has the leading non-zero coefficient which is for the variable x_1, we order the equations as the following:

$$
\begin{array}{rcrcrcrcr}
-x_1 & + & x_2 & - & 4x_3 &   &   & = & 1 \\
 x_1 &   &   & + & 2x_3 & + & 4x_4 & = & 5 \\
     &   & x_2 & + & 3x_3 & - & x_4 & = & 1 \\
     &   & x_2 &   &   & - & 4x_4 & = & -2.
\end{array}
$$

Its augmented matrix is

$$
\begin{bmatrix}
-1 & 1 & -4 & 0 & 1 \\
1 & 0 & 2 & 4 & 5 \\
0 & 1 & 3 & -1 & 1 \\
0 & 1 & 0 & -4 & -2
\end{bmatrix}.
$$

Step 3: We make the leading non-zero coefficient for the second equation as the coefficient for the variable x_2.

In order to make an echelon form, we have to make a leading non-zero entry in the augmented matrix for the second row as the second column or higher. So we change the coefficient for the variable x_1 to 0 by an elementary row reduction Replacement. We add the first equation to the second equation. This new equation becomes the new second equation. This is equivalent to the Replacement operation:

$$
\begin{array}{rcrcrcrcr}
-x_1 & + & x_2 & - & 4x_3 &   &   & = & 1 \\
 x_1 &   &   & + & 2x_3 & + & 4x_4 & = & 5 \\
\hline
     &   & x_2 & - & 2x_3 & + & 4x_4 & = & 6
\end{array}
$$

Thus a new system of linear equations and its augmented matrix are:

$$
\begin{array}{rcrcrcrcr}
-x_1 & + & x_2 & - & 4x_3 &   &   & = & 1 \\
     &   & x_2 & - & 2x_3 & + & 4x_4 & = & 6 \\
     &   & x_2 & + & 3x_3 & - & x_4 & = & 1 \\
     &   & x_2 &   &   & - & 4x_4 & = & -2
\end{array}
\qquad
\begin{bmatrix}
-1 & 1 & -4 & 0 & 1 \\
0 & 1 & -2 & 4 & 6 \\
0 & 1 & 3 & -1 & 1 \\
0 & 1 & 0 & -4 & -2
\end{bmatrix}.
$$

Step 4: Similarly, we make the leading non-zero coefficient of the third equation for the variable x_3. In order to make an echelon form, we have to make a leading non-zero entry in the augmented matrix for the third row as the third column or higher. So we change the coefficients for the variables x_1 and x_2 to 0 by the elementary row reduction Replacement. We subtract the second equation from the third equation. This new equation becomes the new third equation. This is equivalent to the Replacement operation:

$$
\begin{array}{rrrrrr}
 & x_2 & - & 2x_3 & + & 4x_4 & = & 6 \\
-(& x_2 & + & 3x_3 & - & x_4 & = & 1) \\
\hline
 & & & -5x_3 & + & 5x_4 & = & 5
\end{array}
$$

Thus a new system of linear equations and its augmented matrix are:

$$
\begin{array}{rrrrrr}
-x_1 & + & x_2 & - & 4x_3 & & & = & 1 \\
 & & x_2 & - & 2x_3 & + & 4x_4 & = & 6 \\
 & & & & -5x_3 & + & 5x_4 & = & 5 \\
 & & x_2 & & & - & 4x_4 & = & -2
\end{array}
\qquad
\left[\begin{array}{ccccc}
-1 & 1 & -4 & 0 & 1 \\
0 & 1 & -2 & 4 & 6 \\
0 & 0 & -5 & 5 & 5 \\
0 & 1 & 0 & -4 & -2
\end{array}\right].
$$

Step 5: We multiply the third row by $-\frac{1}{5}$. This is a Scaling operation. Thus the new system of linear equations and its augmented matrix are:

$$
\begin{array}{rrrrrr}
-x_1 & + & x_2 & - & 4x_3 & & & = & 1 \\
 & & x_2 & - & 2x_3 & + & 4x_4 & = & 6 \\
 & & & & x_3 & - & x_4 & = & -1 \\
 & & x_2 & & & - & 4x_4 & = & -2
\end{array}
\qquad
\left[\begin{array}{ccccc}
-1 & 1 & -4 & 0 & 1 \\
0 & 1 & -2 & 4 & 6 \\
0 & 0 & 1 & -1 & -1 \\
0 & 1 & 0 & -4 & -2
\end{array}\right].
$$

Step 6: Now we make the leading non-zero coefficient for the fourth equation the coefficient for the variable x_4. We subtract the second equation from the fourth equation and make it the fourth equation. This is equivalent to the Replacement operation:

$$
\begin{array}{rrrrrr}
 & x_2 & & & - & 4x_4 & = & -2 \\
-(& x_2 & - & 2x_3 & + & 4x_4 & = & 6) \\
\hline
 & & & 2x_3 & - & 8x_4 & = & -8
\end{array}
$$

Thus the new system of linear equations and its augmented matrix are:

$$
\begin{array}{rrrrrr}
-x_1 & + & x_2 & - & 4x_3 & & & = & 1 \\
 & & x_2 & - & 2x_3 & + & 4x_4 & = & 6 \\
 & & & & x_3 & - & x_4 & = & -1 \\
 & & & & 2x_3 & - & 8x_4 & = & -8
\end{array}
\qquad
\left[\begin{array}{ccccc}
-1 & 1 & -4 & 0 & 1 \\
0 & 1 & -2 & 4 & 6 \\
0 & 0 & 1 & -1 & -1 \\
0 & 0 & 2 & -8 & -8
\end{array}\right].
$$

Then, we subtract two times the third equation from the fourth equation and make it the fourth equation. This is equivalent to the Replacement operation:

$$
\begin{array}{rrrrrr}
 & & 2x_3 & - & 8x_4 & = & -8 \\
-2(& & x_3 & - & x_4 & = & -1) \\
\hline
 & & & & -6x_4 & = & -6
\end{array}
$$

Thus the new system of linear equations and its augmented matrix are:

$$
\begin{array}{rrrrrr}
-x_1 & + & x_2 & - & 4x_3 & & & = & 1 \\
 & & x_2 & - & 2x_3 & + & 4x_4 & = & 6 \\
 & & & & x_3 & - & x_4 & = & -1 \\
 & & & & & - & 6x_4 & = & -6
\end{array}
\qquad
\left[\begin{array}{ccccc}
-1 & 1 & -4 & 0 & 1 \\
0 & 1 & -2 & 4 & 6 \\
0 & 0 & 1 & -1 & -1 \\
0 & 0 & 0 & -6 & -6
\end{array}\right].
$$

Step 7: Now we change the leading coefficient for the fourth equation to 1 by multiplying by $-\frac{1}{6}$. This is equivalent to the Scaling operation. Thus the new system of linear equations and its augmented matrix are:

$$
\begin{array}{rrrrrl}
-x_1 & + & x_2 & - & 4x_3 & & & = & 1 \\
 & & x_2 & - & 2x_3 & + & 4x_4 & = & 6 \\
 & & & & x_3 & - & x_4 & = & -1 \\
 & & & & & & x_4 & = & 1
\end{array}
\qquad
\begin{bmatrix}
-1 & 1 & -4 & 0 & 1 \\
0 & 1 & -2 & 4 & 6 \\
0 & 0 & 1 & -1 & -1 \\
0 & 0 & 0 & 1 & 1
\end{bmatrix}.
$$

The augmented matrix for the system is in an echelon form. Now we will create this augmented matrix in the reduced echelon form.

Step 8: Now we make all non-leading coefficients for the third equation 0. We add the fourth equation to the third equation, and replace the new equation with the third equation. This is equivalent to the Replacement operation:

$$
\begin{array}{rrrrl}
x_3 & - & x_4 & = & -1 \\
 & & x_4 & = & 1 \\
\hline
x_3 & & & = & 0
\end{array}
$$

Thus a new system of linear equations and its augmented matrix are:

$$
\begin{array}{rrrrrl}
-x_1 & + & x_2 & - & 4x_3 & & & = & 1 \\
 & & x_2 & - & 2x_3 & + & 4x_4 & = & 6 \\
 & & & & x_3 & & & = & 0 \\
 & & & & & & x_4 & = & 1
\end{array}
\qquad
\begin{bmatrix}
-1 & 1 & -4 & 0 & 1 \\
0 & 1 & -2 & 4 & 6 \\
0 & 0 & 1 & 0 & 0 \\
0 & 0 & 0 & 1 & 1
\end{bmatrix}.
$$

Step 9: Now we make all non-leading coefficients for the second equation 0. We add the second equation to twice the third equation, and replace the second equation. This is equivalent to the Replacement operation:

$$
\begin{array}{rrrrl}
x_2 & - & 2x_3 & + & 4x_4 & = & 6 \\
 & & 2(x_3 & & & = & 0) \\
\hline
x_2 & & & + & 4x_4 & = & 6
\end{array}
$$

Thus the new system of linear equations and its augmented matrix are:

$$
\begin{array}{rrrrrl}
-x_1 & + & x_2 & - & 4x_3 & & & = & 1 \\
 & & x_2 & & & + & 4x_4 & = & 6 \\
 & & & & x_3 & & & = & 0 \\
 & & & & & & x_4 & = & 1
\end{array}
\qquad
\begin{bmatrix}
-1 & 1 & -4 & 0 & 1 \\
0 & 1 & 0 & 4 & 6 \\
0 & 0 & 1 & 0 & 0 \\
0 & 0 & 0 & 1 & 1
\end{bmatrix}.
$$

Then, we add the second equation to the fourth equation multiplied by -4, and replace the second equation. This is equivalent to the "Replacement" operation:

$$
\begin{array}{rrrl}
x_2 & + & 4x_4 & = & 6 \\
 & & -4(x_4 & = & 1) \\
\hline
x_2 & & & = & 2
\end{array}
$$

Thus the new system of linear equations and its augmented matrix are:

$$
\begin{aligned}
-x_1 + x_2 - 4x_3 &= 1 \\
x_2 &= 2 \\
x_3 &= 0 \\
x_4 &= 1
\end{aligned}
\qquad
\begin{bmatrix}
-1 & 1 & -4 & 0 & 1 \\
0 & 1 & 0 & 0 & 2 \\
0 & 0 & 1 & 0 & 0 \\
0 & 0 & 0 & 1 & 1
\end{bmatrix}.
$$

Step 10: Now we make the leading coefficient for the first equation 1 by multiplying by -1. This is equivalent to the Scaling operation. Thus the new system of linear equations and its augmented matrix are:

$$
\begin{aligned}
x_1 - x_2 + 4x_3 &= -1 \\
x_2 &= 2 \\
x_3 &= 0 \\
x_4 &= 1
\end{aligned}
\qquad
\begin{bmatrix}
1 & -1 & 4 & 0 & -1 \\
0 & 1 & 0 & 0 & 2 \\
0 & 0 & 1 & 0 & 0 \\
0 & 0 & 0 & 1 & 1
\end{bmatrix}.
$$

Step 11: Now we make all non-leading coefficients for the first equation 0. We add the second equation to the first equation, and replace the first equation. This is equivalent to the Replacement operation:

$$
\begin{aligned}
x_1 - x_2 + 4x_3 &= -1 \\
x_2 &= 2 \\
\hline
x_1 \phantom{+ {}} 4x_3 &= 1
\end{aligned}
$$

Thus, the new system of linear equations and its augmented matrix are:

$$
\begin{aligned}
x_1 + 4x_3 &= 1 \\
x_2 &= 2 \\
x_3 &= 0 \\
x_4 &= 1
\end{aligned}
\qquad
\begin{bmatrix}
1 & 0 & 4 & 0 & 1 \\
0 & 1 & 0 & 0 & 2 \\
0 & 0 & 1 & 0 & 0 \\
0 & 0 & 0 & 1 & 1
\end{bmatrix}.
$$

Then, we subtract four times the third equation from the first equation, and replace the first equation. This is equivalent to the "Replacement" operation:

$$
\begin{aligned}
x_1 + 4x_3 &= 1 \\
-4(x_3 &= 0) \\
\hline
x_1 &= 1
\end{aligned}
$$

Thus a new system of linear equations and its augmented matrix are:

$$
\begin{aligned}
x_1 &= 1 \\
x_2 &= 2 \\
x_3 &= 0 \\
x_4 &= 1
\end{aligned}
\qquad
\begin{bmatrix}
1 & 0 & 0 & 0 & 1 \\
0 & 1 & 0 & 0 & 2 \\
0 & 0 & 1 & 0 & 0 \\
0 & 0 & 0 & 1 & 1
\end{bmatrix}.
$$

Now the augmented matrix is in the reduced echelon form. Also notice that we have the solution to the system of linear equations. If the reduced echelon form has this form, i.e., the coefficient matrix of the augmented matrix in the reduced echelon form for the system of linear equation is the identity matrix, then there is a unique solution to the system of linear equations.

Theorem 1.5 *Suppose we have a system of linear equations with the variables* $x_1, x_2, \ldots, x_n$. *If the first* $n \times n$ *matrix of the augmented matrix of the system of linear equations becomes the identity matrix after Gaussian Elimination, then the system of linear equations has a unique solution.*

Example 44 *In* R *we can also see how Gaussian Elimination works using the* matlib *package [17]. We will use the example above. Suppose we have the following system of linear equations:*

$$
\begin{array}{rcrcrcrcr}
   & & x_2 & + & 3x_3 & - & x_4 & = & 1 \\
-x_1 & + & x_2 & - & 4x_3 & & & = & 1 \\
x_1 & & & + & 2x_3 & + & 4x_4 & = & 5 \\
   & & x_2 & & & - & 4x_4 & = & -2.
\end{array}
$$

Its augmented matrix is

$$
\begin{bmatrix}
0 & 1 & 3 & -1 & 1 \\
-1 & 1 & -4 & 0 & 1 \\
1 & 0 & 2 & 4 & 5 \\
0 & 1 & 0 & -4 & -2
\end{bmatrix}.
$$

First we load the package:

```
library(matlib)
```

Then we create the coefficient matrix and the vector for the right hand side of the system of linear equations.

```
A <- matrix(c(0, -1, 1, 0, 1, 1, 0, 1, 3, -4, 2, 0, -1, 0, 4, -4), 4, 4)
b <- c(1, 1, 5, -2)
```

Here, we did not type "nrow=" and "ncol=". If it is clear we can skip typying them.
Using the showEqn() function we can see the system of linear equations.

```
showEqn(A, b)
```

Then R *shows*

```
> showEqn(A, b)
  0*x1 + 1*x2 + 3*x3 - 1*x4 =   1
 -1*x1 + 1*x2 - 4*x3 + 0*x4 =   1
  1*x1 + 0*x2 + 2*x3 + 4*x4 =   5
  0*x1 + 1*x2 + 0*x3 - 4*x4 =  -2
```

Now using the echelon() function we can see how Gaussian Elimination works. With this example, we can type in R *as:*

```
echelon(A, b, verbose=TRUE, fractions=TRUE)
```

If this option "fraction" is set as "TRUE", then it outputs in the form of rational numbers. R *outputs the following as how Gaussian Elimination works:*

```
> echelon(A, b, verbose=TRUE, fractions=TRUE)

Initial matrix:
     [,1] [,2] [,3] [,4] [,5]
[1,]  0    1    3   -1    1
[2,] -1    1   -4    0    1
[3,]  1    0    2    4    5
[4,]  0    1    0   -4   -2

row: 1

 exchange rows 1 and 2
     [,1] [,2] [,3] [,4] [,5]
[1,] -1    1   -4    0    1
[2,]  0    1    3   -1    1
[3,]  1    0    2    4    5
[4,]  0    1    0   -4   -2

 multiply row 1 by -1
     [,1] [,2] [,3] [,4] [,5]
[1,]  1   -1    4    0   -1
[2,]  0    1    3   -1    1
[3,]  1    0    2    4    5
[4,]  0    1    0   -4   -2

 subtract row 1 from row 3
     [,1] [,2] [,3] [,4] [,5]
[1,]  1   -1    4    0   -1
[2,]  0    1    3   -1    1
[3,]  0    1   -2    4    6
[4,]  0    1    0   -4   -2

row: 2

 multiply row 2 by 1 and add to row 1
     [,1] [,2] [,3] [,4] [,5]
[1,]  1    0    7   -1    0
[2,]  0    1    3   -1    1
[3,]  0    1   -2    4    6
[4,]  0    1    0   -4   -2

 subtract row 2 from row 3
     [,1] [,2] [,3] [,4] [,5]
```

```
[1,]  1    0    7   -1    0
[2,]  0    1    3   -1    1
[3,]  0    0   -5    5    5
[4,]  0    1    0   -4   -2
```

subtract row 2 from row 4

```
      [,1] [,2] [,3] [,4] [,5]
[1,]  1    0    7   -1    0
[2,]  0    1    3   -1    1
[3,]  0    0   -5    5    5
[4,]  0    0   -3   -3   -3
```

row: 3

multiply row 3 by -1/5

```
      [,1] [,2] [,3] [,4] [,5]
[1,]  1    0    7   -1    0
[2,]  0    1    3   -1    1
[3,]  0    0    1   -1   -1
[4,]  0    0   -3   -3   -3
```

multiply row 3 by 7 and subtract from row 1

```
      [,1] [,2] [,3] [,4] [,5]
[1,]  1    0    0    6    7
[2,]  0    1    3   -1    1
[3,]  0    0    1   -1   -1
[4,]  0    0   -3   -3   -3
```

multiply row 3 by 3 and subtract from row 2

```
      [,1] [,2] [,3] [,4] [,5]
[1,]  1    0    0    6    7
[2,]  0    1    0    2    4
[3,]  0    0    1   -1   -1
[4,]  0    0   -3   -3   -3
```

multiply row 3 by 3 and add to row 4

```
      [,1] [,2] [,3] [,4] [,5]
[1,]  1    0    0    6    7
[2,]  0    1    0    2    4
[3,]  0    0    1   -1   -1
[4,]  0    0    0   -6   -6
```

row: 4

multiply row 4 by -1/6

```
      [,1] [,2] [,3] [,4] [,5]
[1,]  1    0    0    6    7
[2,]  0    1    0    2    4
```

```
[3,]  0    0    1   -1   -1
[4,]  0    0    0    1    1
```

multiply row 4 by 6 and subtract from row 1

```
      [,1] [,2] [,3] [,4] [,5]
[1,]  1    0    0    0    1
[2,]  0    1    0    2    4
[3,]  0    0    1   -1   -1
[4,]  0    0    0    1    1
```

multiply row 4 by 2 and subtract from row 2

```
      [,1] [,2] [,3] [,4] [,5]
[1,]  1    0    0    0    1
[2,]  0    1    0    0    2
[3,]  0    0    1   -1   -1
[4,]  0    0    0    1    1
```

multiply row 4 by 1 and add to row 3

```
      [,1] [,2] [,3] [,4] [,5]
[1,]  1    0    0    0    1
[2,]  0    1    0    0    2
[3,]  0    0    1    0    0
[4,]  0    0    0    1    1
```

Therefore the solution for this system is $x_1 = 1$, $x_2 = 2$, $x_3 = 0$, $x_4 = 1$.

Example 45 *Now we have a system of linear equations such that*

$$
\begin{aligned}
5x_2 &+ 4x_3 &+ 7x_4 &= -9 \\
-9x_1 + 6x_2 &+ 6x_3 &+ 7x_4 &= -17 \\
4x_1 + 3x_2 & &+ 5x_4 &= -3 \\
4x_1 + 8x_2 &+ 4x_3 &+ 12x_4 &= 0.
\end{aligned}
$$

Its augmented matrix is

$$
\begin{bmatrix}
0 & 5 & 4 & 7 & -9 \\
-9 & 6 & 6 & 7 & -17 \\
4 & 3 & 0 & 5 & -3 \\
4 & 8 & 4 & 12 & 0
\end{bmatrix}.
$$

We use the `matlib` *package for this example. We create the coefficient matrix and the vector for the right hand side of the equations for the system. Then we have*

```
A <- matrix(c(0, -9, 4, 4, 5, 6, 3, 8, 4, 6, 0, 4, 7, 7, 5, 12), 4, 4)
b <- c(-9, -17, -3, 0)
```

Using the showEqn() function we can see the system of linear equations.

```
showEqn(A, b)
```

Then R *shows*

```
> showEqn(A, b)
 0*x1 + 5*x2 + 4*x3  + 7*x4  =   -9
-9*x1 + 6*x2 + 6*x3  + 7*x4  =  -17
 4*x1 + 3*x2 + 0*x3  + 5*x4  =   -3
 4*x1 + 8*x2 + 4*x3 + 12*x4  =    0
```

Now using the echelon() function, we can obtain the reduced echelon form of the augmented matrix. From this we have an echelon form of the augmented matrix as follows:

```
       [,1]     [,2]     [,3]     [,4]      [,5]
[1,]     1        0        0      1/7      -1/7
[2,]     0        1        0      31/21    67/21
[3,]     0        0        1     -2/21    -131/21
[4,]     0        0        0        0       -12
```

Therefore, the reduced echelon form of the augmented matrix is

$$\begin{bmatrix} 1 & 0 & 0 & 1/7 & -1/7 \\ 0 & 1 & 0 & 31/21 & 67/21 \\ 0 & 0 & 1 & -2/21 & -131/21 \\ 0 & 0 & 0 & 0 & 1 \end{bmatrix}.$$

The last row has all zero elements except the last entry. This row represents the following equation

$$0 = 1,$$

which does not make sense. So if the last non-zero row has the form of

$$[0, 0, \ldots, 1]$$

in the reduced echelon form of the augmented matrix of the system, then there is no solution to the system of linear equations.

Theorem 1.6 *If the reduced echelon form of the augmented matrix of a system of linear equations has the last non-zero row in the form of*

$$[0, 0, \ldots, 1]$$

then the system of linear equations does not have any solution.

Example 46 *Now we have a system of linear equations such that*

$$\begin{aligned} 5x_2 &+ 4x_3 &+ 7x_4 &= -9 \\ -9x_1 + 6x_2 &+ 6x_3 &+ 7x_4 &= -17 \\ 4x_1 + 3x_2 & &+ 5x_4 &= -3 \\ 4x_1 + 8x_2 &+ 4x_3 &+ 12x_4 &= -12. \end{aligned}$$

Its augmented matrix is

$$\begin{bmatrix} 0 & 5 & 4 & 7 & -9 \\ -9 & 6 & 6 & 7 & -17 \\ 4 & 3 & 0 & 5 & -3 \\ 4 & 8 & 4 & 12 & -12 \end{bmatrix}.$$

We use the `matlib` *package for this example. We create the coefficient matrix and the vector for the right hand side of the equations for the system. Then we have*

```
A <- matrix(c(0, -9, 4, 4, 5, 6, 3, 8, 4, 6, 0, 4, 7, 7, 5, 12), 4, 4)
b <- c(-9, -17, -3, -12)
```

Using the showEqn() function we can see the system of linear equations.

```
showEqn(A, b)
```

Then R *shows*

```
> showEqn(A, b)
 0*x1 + 5*x2 + 4*x3  + 7*x4  =   -9
-9*x1 + 6*x2 + 6*x3  + 7*x4  =  -17
 4*x1 + 3*x2 + 0*x3  + 5*x4  =   -3
 4*x1 + 8*x2 + 4*x3 + 12*x4  =  -12
```

Now using the echelon() function we obtain the reduced echelon form of the augmented matrix. From this we have the reduced echelon form of the augmented matrix as follows:

```
multiply row 3 by 7/8 and add to row 4
      [,1]   [,2]   [,3]   [,4]    [,5]
[1,]    1      0      0     1/7     5/7
[2,]    0      1      0    31/21  -41/21
[3,]    0      0      1    -2/21    4/21
[4,]    0      0      0      0       0
```

The reduced echelon form of the augmented matrix is

$$\begin{bmatrix} 1 & 0 & 0 & 1/7 & 5/7 \\ 0 & 1 & 0 & 31/21 & -41/21 \\ 0 & 0 & 1 & -2/21 & 4/21 \\ 0 & 0 & 0 & 0 & 0 \end{bmatrix}.$$

We can write the corresponding system of linear equations

$$\begin{aligned} x_1 \quad & & + \tfrac{1}{7}x_4 &= \tfrac{5}{7} \\ & x_2 & + \tfrac{31}{21}x_4 &= -\tfrac{41}{21} \\ & x_3 & - \tfrac{2}{21}x_4 &= \tfrac{4}{21} \\ & & 0 &= 0. \end{aligned}$$

Since x_4 can be any real number then there are infinitely many solutions.

Note that this reduced echelon form of the augmented matrix is not the case of Theorem 1.5 or Theorem 1.6.

Theorem 1.7 *If the reduced echelon form of the augmented matrix is not the case of Theorem 1.5 or Theorem 1.6, then the system of linear equations has infinitely many solutions.*

1.4.4 Checkmarks

- The definition of the reduced echelon form of a matrix.

- Procedure of Gaussian Elimination.

- You can apply Gaussian Elimination to obtain the reduced echelon form of the augmented matrix of a system of linear equations.

- The condition of the reduced echelon form of the augmented matrix if there is a unique solution.

- The condition of the reduced echelon form of the augmented matrix if there is no solution.

- The condition of the reduced echelon form of the augmented matrix if there are infinitely many solutions.

- You can apply Gaussian Elimination to a system of linear equations in R.

1.4.5 Conceptual Quizzes

Quiz 27 True or False: *Suppose we have a matrix*

$$\begin{bmatrix} -4 & 1 & -17 & 5 \\ 0 & 0 & -2 & 9 \\ 0 & 1 & 0 & 3 \end{bmatrix}.$$

It is an echelon form.

Quiz 28 True or False: *Suppose we have a matrix*

$$\begin{bmatrix} 0 & 1 & -17 & 5 \\ 1 & 0 & -2 & 9 \\ 0 & 0 & 5 & 3 \end{bmatrix}.$$

It is an echelon form.

Quiz 29 True or False: *Suppose we have a matrix*

$$\begin{bmatrix} 1 & 1 & -17 & 5 \\ 0 & 0 & -2 & 9 \\ 0 & 0 & 0 & 0 \end{bmatrix}.$$

It is an echelon form.

Quiz 30 True or False: *Suppose we have a matrix*

$$\begin{bmatrix} 1 & 1 & -17 & 5 \\ 0 & 0 & -2 & 9 \\ 0 & 0 & 0 & 0 \end{bmatrix}.$$

It is the reduced echelon form.

Quiz 31 True or False: *Suppose we have a matrix*

$$\begin{bmatrix} 1 & 2 & 0 & 5 \\ 0 & 0 & 1 & 9 \\ 0 & 0 & 0 & 0 \end{bmatrix}.$$

It is the reduced echelon form.

Quiz 32 True or False: *There is a matrix which may be row reduced to many matrices in reduced echelon form by different sequences of row operations.*

Quiz 33 True or False: *The echelon form of a matrix is unique.*

Quiz 34 True or False: *Gaussian Elimination can be only applied to augmented matrices for systems of linear equations.*

Quiz 35 True or False: *If we have the following augmented matrix for a system of linear equations, then the system of linear equations does not have a solution:*

$$\begin{bmatrix} 0 & 0 & 0 & 2 & 0 & 0 \end{bmatrix}.$$

Quiz 36 Multiple Choice: *Suppose we have the reduced echelon form of the augmented matrix of a system of linear equations as follows:*

$$\begin{bmatrix} 1 & 0 & 0 & 0 & 5 \\ 0 & 1 & 0 & 0 & -7 \\ 0 & 0 & 1 & 0 & 21 \\ 0 & 0 & 0 & 0 & 1 \end{bmatrix}.$$

Then choose one of the following items:

1. *There is a unique solution.*

2. *There is no solution.*

3. *There are infinitely many solutions.*

Quiz 37 Multiple Choice: *Suppose we have the reduced echelon form of the augmented matrix of a system of linear equations as follows:*

$$\begin{bmatrix} 1 & 0 & 0 & 0 & 1/7 \\ 0 & 1 & 0 & 0 & -1/2 \\ 0 & 0 & 1 & 0 & 4 \\ 0 & 0 & 0 & 1 & 0 \end{bmatrix}.$$

Then choose one of the following items:

1. *There is a unique solution.*

2. *There is no solution.*

3. *There are infinitely many solutions.*

Quiz 38 Multiple Choice*: Suppose we have the reduced echelon form of the augmented matrix of a system of linear equations as follows:*

$$\begin{bmatrix} 1 & 2 & 0 & 0 & 5/4 \\ 0 & 0 & 1 & 0 & -1 \\ 0 & 0 & 0 & 1 & 2 \\ 0 & 0 & 0 & 0 & 0 \end{bmatrix}.$$

Then choose one of the the following items:

1. *There is a unique solution.*

2. *There is no solution.*

3. *There are infinitely many solutions.*

Quiz 39 Multiple Choice*: Suppose we have the reduced echelon form of the augmented matrix of a system of linear equations as follows:*

$$\begin{bmatrix} 1 & 0 & 0 & 0 & 0 & 5 \\ 0 & 1 & 0 & 0 & 0 & -7 \\ 0 & 0 & 1 & 0 & 0 & 21 \\ 0 & 0 & 0 & 0 & 1 & 1 \\ 0 & 0 & 0 & 0 & 0 & 0 \end{bmatrix}.$$

Then choose one of the following items:

1. *There is a unique solution.*

2. *There is no solution.*

3. *There are infinitely many solutions.*

Quiz 40 Multiple Choice*: Suppose we have the reduced echelon form of the augmented matrix of a system of linear equations as follows:*

$$\begin{bmatrix} 1 & 2 & 0 & 0 & 0 & 5/4 \\ 0 & 0 & 1 & 5 & 0 & -1 \\ 0 & 0 & 0 & 0 & 1 & 2 \\ 0 & 0 & 0 & 0 & 0 & 1 \\ 0 & 0 & 0 & 0 & 0 & 0 \end{bmatrix}.$$

Then choose one of the following items:

1. *There is a unique solution.*

2. *There is no solution.*

3. *There are infinitely many solutions.*

Quiz 41 Multiple Choice*: Suppose we have the reduced echelon form of the augmented matrix of a system of linear equations as follows:*

$$\begin{bmatrix} 1 & 0 & 0 & 0 & 0 & -5 \\ 0 & 1 & 0 & 0 & 0 & 2 \\ 0 & 0 & 1 & 0 & 0 & 4 \\ 0 & 0 & 0 & 1 & 0 & 0 \\ 0 & 0 & 0 & 0 & 1 & 21 \end{bmatrix}.$$

Then choose one of the following items:

1. *There is a unique solution.*

2. *There is no solution.*

3. *There are infinitely many solutions.*

1.4.6   Regular Exercises

Exercise 1.38 *Suppose we have a system of linear equations such that*

$$\begin{aligned} x_1 & - & 6x_2 & = & 3 \\ -x_1 & + & 2x_2 & = & -1. \end{aligned}$$

Use Gaussian Elimination to solve the system of linear equations.

Exercise 1.39 *Suppose we have a system of linear equations such that*

$$\begin{aligned} -x_1 & - & 9x_2 & = & 1 \\ -10x_1 & + & 9x_2 & = & -10. \end{aligned}$$

Use Gaussian Elimination to solve the system of linear equations.

Exercise 1.40 *Suppose we have a system of linear equations such that*

$$\begin{aligned} -5x_1 & + & 5x_2 & - & x_3 & = & 57 \\ -7x_1 & - & 2x_2 & - & 4x_3 & = & 21 \\ x_1 & + & 3x_2 & + & 4x_3 & = & 3 \end{aligned}$$

Use Gaussian Elimination to solve the system of linear equations.

Exercise 1.41 *Suppose we have a system of linear equations such that*

$$\begin{aligned} 4x_1 & - & 10x_2 & + & 4x_3 & = & 66 \\ -4x_1 & + & 3x_2 & & & = & -27 \\ 9x_1 & + & 6x_2 & + & 2x_3 & = & -1 \end{aligned}$$

Use Gaussian Elimination to solve the system of linear equations.

Exercise 1.42 *Suppose we have a system of linear equations such that*

$$
\begin{aligned}
-5x_1 &+ 5x_2 &- x_3 &= 57 \\
-7x_1 &- 2x_2 &- 4x_3 &= 21 \\
x_1 &+ 3x_2 &+ 4x_3 &= 3
\end{aligned}
$$

Use Gaussian Elimination to solve the system of linear equations.

Exercise 1.43 *Suppose we have a system of linear equations such that*

$$
\begin{aligned}
5x_1 &+ 10x_2 &- 5x_3 &= -5 \\
&- 10x_2 &+ 3x_3 &= -5 \\
10x_1 &+ 6x_2 &- 9x_3 &= -1
\end{aligned}
$$

Use Gaussian Elimination to solve the system of linear equations.

Exercise 1.44 *Suppose we have a system of linear equations such that*

$$
\begin{aligned}
5x_1 &+ 10x_2 &- 5x_3 &= -5 \\
10x_1 &+ 6x_2 &- 9x_3 &= -1 \\
5x_1 &- 4x_2 &- 5x_3 &= 4
\end{aligned}
$$

Use Gaussian Elimination to solve the system of linear equations.

Exercise 1.45 *Suppose we have a system of linear equations such that*

$$
\begin{aligned}
5x_1 &+ 10x_2 &- 5x_3 &= -5 \\
10x_1 &+ 6x_2 &- 9x_3 &= -1 \\
5x_1 &- 4x_2 &- 5x_3 &= 3
\end{aligned}
$$

Use Gaussian Elimination to solve the system of linear equations.

Exercise 1.46 *Suppose we have a system of linear equations such that*

$$
\begin{aligned}
3x_1 &+ 6x_2 &+ 3x_3 &- 2x_4 &= 59 \\
7x_1 &- 10x_2 &- 2x_3 &- 8x_4 &= 80 \\
2x_1 &- 6x_2 &+ 2x_3 &- 4x_4 &= 28 \\
2x_1 &- 5x_2 &- 8x_3 &- x_4 &= 0
\end{aligned}
$$

Use Gaussian Elimination to solve the system of linear equations.

Exercise 1.47 *Suppose we have a system of linear equations such that*

$$
\begin{aligned}
5x_1 &+ x_2 &+ 10x_3 &+ 6x_4 &= 2 \\
-x_1 &- 7x_2 &- 6x_3 &+ 5x_4 &= 38 \\
-7x_1 &+ 4x_2 &- x_3 &+ x_4 &= 24 \\
-5x_1 &+ 3x_2 &- 4x_3 &- 8x_4 &= -18
\end{aligned}
$$

Use Gaussian Elimination to solve the system of linear equations.

Exercise 1.48 *Suppose we have a system of linear equations such that*

$$
\begin{array}{rcrcrcrcr}
 & & -14x_2 & - & 9x_3 & + & 11x_4 & = & 22 \\
-2x_1 & - & 4x_2 & - & 2x_3 & + & 6x_4 & = & 6 \\
2x_1 & - & 10x_2 & - & 7x_3 & + & 5x_4 & = & 16 \\
-6x_1 & - & 9x_2 & + & 8x_3 & + & 7x_4 & = & -73
\end{array}
$$

Use Gaussian Elimination to solve the system of linear equations.

Exercise 1.49 *Suppose we have a system of linear equations such that*

$$
\begin{array}{rcrcrcrcr}
8x_1 & + & 6x_2 & - & 4x_3 & - & 3x_4 & = & -75 \\
-5x_1 & + & 8x_2 & + & x_3 & + & 6x_4 & = & -7 \\
6x_1 & + & 5x_2 & + & 4x_3 & - & x_4 & = & -20 \\
-5x_1 & + & 10x_2 & + & 8x_3 & - & x_4 & = & -3
\end{array}
$$

Use Gaussian Elimination to solve the system of linear equations.

Exercise 1.50 *Suppose we have a system of linear equations such that*

$$
\begin{array}{rcrcrcrcr}
8x_1 & - & 9x_2 & + & x_3 & + & 10x_4 & = & 25 \\
-9x_1 & - & 7x_2 & - & 3x_3 & - & 7x_4 & = & 16 \\
7x_1 & - & 25x_2 & - & x_3 & + & 13x_4 & = & 60 \\
-x_1 & - & 16x_2 & - & 2x_3 & + & 3x_4 & = & 41
\end{array}
$$

Use Gaussian Elimination to solve the system of linear equations.

Exercise 1.51 *Find the following examples:*

1. *A system of three linear equations with three variables which has a unique solution.*

2. *A system of three linear equations with three variables which has infinitely many solutions.*

3. *A system of three linear equations with three variables which has no solution.*

Exercise 1.52 *Find the following examples:*

1. *A system of three linear equations with four variables which has a unique solution.*

2. *A system of three linear equations with four variables which has infinitely many solutions.*

3. *A system of three linear equations with four variables which has no solution.*

1.4.7   Lab Exercises

Lab Exercise 30 *Go back to Exercise 1.38. With* R, *do the following:*

1. *Solve the system of linear equations with the Solve() or solve() function.*

2. *Obtain the reduced echelon form of the augmented matrix of the system of linear equations using the echelon() function from the* matlib *package.*

Lab Exercise 31 *Go back to Exercise 1.39. With* R, *do the following:*

1. *Solve the system of linear equations with the Solve() or the solve() function.*

2. *Obtain the reduced echelon form of the augmented matrix of the system of linear equations using the echelon() function from the* matlib *package.*

Lab Exercise 32 *Go back to Exercise 1.40. With* R, *do the following:*

1. *Solve the system of linear equations with the Solve() or the solve() function.*

2. *Obtain the reduced echelon form of the augmented matrix of the system of linear equations using the echelon() function from the* matlib *package.*

Lab Exercise 33 *Go back to Exercise 1.41. With* R, *do the following:*

1. *Solve the system of linear equations with the Solve() or the solve() function.*

2. *Obtain the reduced echelon form of the augmented matrix of the system of linear equations using the echelon() function from the* matlib *package.*

Lab Exercise 34 *Go back to Exercise 1.42. With* R, *do the following:*

1. *Solve the system of linear equations with the Solve() or the solve() function.*

2. *Obtain the reduced echelon form of the augmented matrix of the system of linear equations using the echelon() function from the* matlib *package.*

Lab Exercise 35 *Go back to Exercise 1.43. With* R, *do the following:*

1. *Solve the system of linear equations with the Solve() or the solve() function.*

2. *Obtain the reduced echelon form of the augmented matrix of the system of linear equations using the echelon() function from the* matlib *package.*

Lab Exercise 36 *Go back to Exercise 1.44. With* R, *do the following:*

1. *Solve the system of linear equations with the Solve() or the solve() function.*

2. *Obtain the reduced echelon form of the augmented matrix of the system of linear equations using the echelon() function from the* matlib *package.*

Lab Exercise 37 *Go back to Exercise 1.45. With* R, *do the following:*

1. *Solve the system of linear equations with the Solve() or the solve() function.*

2. *Obtain the reduced echelon form of the augmented matrix of the system of linear equations using the echelon() function from the* matlib *package.*

Lab Exercise 38 *Go back to Exercise 1.46. With* R, *do the following:*

1. *Solve the system of linear equations with the Solve() or the solve() function.*

2. *Obtain the reduced echelon form of the augmented matrix of the system of linear equations using the echelon() function from the* matlib *package.*

Lab Exercise 39 *Go back to Exercise 1.47. With* R, *do the following:*

1. *Solve the system of linear equations with the Solve() or the solve() function.*

2. *Obtain the reduced echelon form of the augmented matrix of the system of linear equations using the echelon() function from the* matlib *package.*

Lab Exercise 40 *Go back to Exercise 1.48. With* R, *do the following:*

1. *Solve the system of linear equations with the Solve() or the solve() function.*

2. *Obtain the reduced echelon form of the augmented matrix of the system of linear equations using the echelon() function from the* matlib *package.*

Lab Exercise 41 *Go back to Exercise 1.49. With* R, *do the following:*

1. *Solve the system of linear equations with the Solve() or the solve() function.*

2. *Obtain the reduced echelon form of the augmented matrix of the system of linear equations using the echelon() function from the* matlib *package.*

Lab Exercise 42 *Go back to Exercise 1.50. With* R, *do the following:*

1. *Solve the system of linear equations with the Solve() or the solve() function.*

2. *Obtain the reduced echelon form of the augmented matrix of the system of linear equations using the echelon() function from the* matlib *package.*

1.4.8   Practical Applications

For this practical application, we go back to the example of a contingency table shown in Table 1.6 in the beginning of this section. The table was constructed by the data from the 1996 General Society Survey, National Opinion Research Center.

Recall that the measurements in Table 1.6 are income in terms of the US dollars and job satisfaction among black men in the US. Let us assign the numbers to the status of satisfactory. Let us assign the status "Very Dissatisfied" as 1, the status "Little Dissatisfied" as 2, the status "Moderately Satisfied" as 3 and the status "Very Satisfied" as 4. Similarly, for the income, let us assign the income $< 15,000$ as 1, the income $15,000 - 25,000$ as 2, the income $25,000 - 40,000$ as 3 and the income $> 40,000$ as 4. From the row and column sums computed from Table 1.6, we have a 4×4 table such that

					Total
	x_{11}	x_{12}	x_{13}	x_{14}	20
	x_{21}	x_{21}	x_{23}	x_{24}	22
	x_{31}	x_{32}	x_{33}	x_{34}	33
	x_{41}	x_{42}	x_{42}	$1x_{42}$	21
Total	4	13	43	36	96

where the variable x_{ij} is the number of people who answered the questions for the income as $i = 1, 2, 3, 4$ and for the satisfactory status as $j = 1, 2, 3, 4$. As we discussed before, in order to make sure that we can protect personal information, we perturb the column sums $(4, 13, 43, 63)$ as $(4.5, 14.2, 42.4, 62.9)$ and the row sums $(20, 22, 33, 21)$ as $(21.5, 20.9, 33.4, 22.1)$. Now the question is whether there exists a 4×4 table satisfying such row and column sums: This can be written as a system of linear equations such that

$$
\begin{aligned}
x_{11} + x_{12} + x_{13} + x_{14} &= 21.5 \\
x_{21} + x_{22} + x_{23} + x_{24} &= 20.9 \\
x_{31} + x_{32} + x_{33} + x_{34} &= 33.4 \\
x_{41} + x_{42} + x_{43} + x_{44} &= 22.1 \\
x_{11} + x_{21} + x_{31} + x_{41} &= 4.5 \\
x_{12} + x_{22} + x_{32} + x_{42} &= 14.2 \\
x_{13} + x_{23} + x_{33} + x_{43} &= 42.4 \\
x_{14} + x_{24} + x_{34} + x_{44} &= 62.9
\end{aligned}
$$

The coefficient matrix for this system is

$$
\begin{bmatrix}
1 & 1 & 1 & 1 & 0 & 0 & 0 & 0 & 0 & 0 & 0 & 0 & 0 & 0 & 0 & 0 \\
0 & 0 & 0 & 0 & 1 & 1 & 1 & 1 & 0 & 0 & 0 & 0 & 0 & 0 & 0 & 0 \\
0 & 0 & 0 & 0 & 0 & 0 & 0 & 0 & 1 & 1 & 1 & 1 & 0 & 0 & 0 & 0 \\
0 & 0 & 0 & 0 & 0 & 0 & 0 & 0 & 0 & 0 & 0 & 0 & 1 & 1 & 1 & 1 \\
1 & 0 & 0 & 0 & 1 & 0 & 0 & 0 & 1 & 0 & 0 & 0 & 1 & 0 & 0 & 0 \\
0 & 1 & 0 & 0 & 0 & 1 & 0 & 0 & 0 & 1 & 0 & 0 & 0 & 1 & 0 & 0 \\
0 & 0 & 1 & 0 & 0 & 0 & 1 & 0 & 0 & 0 & 1 & 0 & 0 & 0 & 1 & 0 \\
0 & 0 & 0 & 1 & 0 & 0 & 0 & 1 & 0 & 0 & 0 & 1 & 0 & 0 & 0 & 1
\end{bmatrix},
$$

which is a 8×16 matrix and the vector for the right hand side of the system of linear equations is

$$\begin{bmatrix} 21.5 \\ 20.9 \\ 33.4 \\ 22.1 \\ 4.5 \\ 14.2 \\ 42.4 \\ 62.9 \end{bmatrix},$$

which is a 1×8 matrix. Now we use the Solve() function from the R package `matlib`. First we define the coefficient matrix in R:

```
A <- matrix(c(1, 1, 1, 1, 0, 0, 0, 0, 0, 0, 0, 0, 0, 0,
0, 0, 0, 0, 0, 0, 1, 1, 1, 1, 0, 0, 0, 0, 0, 0, 0, 0,
0, 0, 0, 0, 0, 0, 0, 0, 1, 1, 1, 1, 0, 0, 0, 0, 0, 0, 0,
0, 0, 0, 0, 0, 0, 0, 0, 0, 1, 1, 1, 1, 1, 0, 0, 0, 1, 0,
0, 0, 1, 0, 0, 0, 1, 0, 0, 0, 0, 1, 0, 0, 0, 1, 0, 0,
0, 1, 0, 0, 0, 1, 0, 0, 0, 0, 1, 0, 0, 0, 1, 0, 0, 0,
1, 0, 0, 0, 1, 0, 0, 0, 0, 1, 0, 0, 0, 1, 0, 0, 0, 1,
0, 0, 0, 1 ), nrow = 8, ncol = 16, byrow = TRUE)
```

Then define the vector for the right hand side for the system

```
b <- c(21.5, 20.9, 33.4, 22.1, 4.5, 14.2, 42.4, 62.9)
```

Then we use the Solve() function. The following is the output from R:

```
> Solve(A, b)
x1        - 1*x6 - 1*x7 - 1*x8   - 1*x10 - 1*x11 - 1*x12   - 1*x14 - 1*x15 - 1*x16  =  -71.9
   x2       + x6              + x10                     + x14                  =   14.2
      x3         + x7              + x11              + x15                    =   42.4
         x4          + x8              + x12                  + x16            =   36.8
            x5 + x6 + x7 + x8                                                  =   20.9
                           x9 + x10 + x11 + x12                                =   33.4
                                              x13 + x14 + x15 + x16            =   22.1
                                                                   0  =  26.1
```

When we look at the last equation, it says $0 = 26.1$, which does not make sense. This is the form in Theorem 1.6. Therefore, there is no solution to the system of linear equations. Thus, we conclude that there does not exist such a table. This means that this perturbed information does not make sense for a statistical analysis for studying relationship. Thus perturbation does not work for releasing this information.

1.4.9 Supplements with `python` Code

In `python` we have a nice function rref() from the `sympy` package. In order to upload the package you type

```
from sympy import *
```

Let us find the reduced echelon form of the augmented matrix from Example 44. Recall we have the following system of linear equations:

$$
\begin{aligned}
x_2 &+ 3x_3 &- x_4 &= 1 \\
-x_1 + x_2 &- 4x_3 & &= 1 \\
x_1 &+ 2x_3 &+ 4x_4 &= 5 \\
x_2 & &- 4x_4 &= -2.
\end{aligned}
$$

Its augmented matrix is

$$
\begin{bmatrix}
0 & 1 & 3 & -1 & 1 \\
-1 & 1 & -4 & 0 & 1 \\
1 & 0 & 2 & 4 & 5 \\
0 & 1 & 0 & -4 & -2
\end{bmatrix}.
$$

First we define the augmented matrix in **python** as

```
M = Matrix([[0, 1, 3, -1, 1],
[-1, 1, -4, 0, 1],
[1, 0, 2, 4, 5],
[0, 1, 0, -4, -2]])
```

Then we call

```
M.rref()
```

Then **python** outputs

```
(Matrix([
[1, 0, 0, 0, 1],
[0, 1, 0, 0, 2],
[0, 0, 1, 0, 0],
[0, 0, 0, 1, 1]]), (0, 1, 2, 3))
```

The first output is the reduced echelon form of the augmented matrix and the second output is the location of the leading entries. In this example, $(0, 1, 2, 3)$ means the leading entries are located in the first column, the second column, the third column and the fourth column (recall that in **python** every index of the columns or rows of a matrix starts from 0). From this reduced echelon form of the augmented matrix we know that there is unique solution $(1, 2, 0, 1)$ to the system of linear equations.

1.5   Discussion

The death-month and birth-month problem for the "Introductory Example from Statistics" in this chapter is a 12×12 contingency table. To compute the sensitivity, we take the row sums and column sums of the table as we did for the working example in Section 1.4. For this system of linear equations, we have the 24×144 coefficient matrix. In reality, we also have constraints for non-negativity, i.e., $x_{ij} \geq 0$. Also note that all values for x_{ij} have to be an integer, so we also have integer constraints. If we add these constraints, then there are finitely many solutions and more than one solution for the system. Counting the number of solutions for such problems is very difficult and known to be #-P hard. However, understanding the system of linear equations is the first step toward solving this difficult problem.

2

Matrix Arithmetic

As you saw in the previous chapter, we can convert a system of linear equations into an augmented matrix and then we can solve the system of linear equations by operating on its augmented matrix. In addition to solving a system of linear equations, matrix operations can be useful in many areas. For example, as we will see in the Introductory Example section, these can be useful for image processing. In data science and computer science, image processing is one of the most popular and important areas. These operations relate to transforming images by **linear transformations**. In this chapter, we focus on matrix operations and properties of matrices. The walking example in this chapter is from statistics. Categorical data analysis is one of the classical methods of statistical analysis and the example is called a *two-way contingency table* from categorical data analysis.

2.1 Introductory Example from Statistics

Image processing is important in many areas. For example, if we want to protect someone's identity from a security camera, we might have to make some part of the image unclear. In machine learning, these processes can be important for learning. For example, if we want to train a machine to identify a suspicious person from a security camera, we need to train the machine to recognize the tendencies and characteristics of a suspicious person, and how these differ from an ordinary person. In this example, we use the `magick` package in R to demonstrate image processing without getting into details.

We need first to install the `magick` package (see the Appendix for how to install a package). Then we upload the `magick` package in R.

```
library(magick)
```

The picture shown in Figure 2.1 is the original colored image of a cat. In order to read the image in R from a local computer, we can do:

```
# Reading from a local computer.
inp_img <- image_read('DIRECTORY/FILE')
```

DOI: 10.1201/9781003042259-2

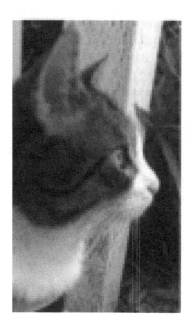

FIGURE 2.1
Original image of a cat for the example.

where "FILE" is the file name and "DIRECTORY" is the directory, where the file is located. If we want to read an image from a website, then you can call

```
inp_img <- image_read("url")
```

where "url" is the web address for the image. For example, if you would like to obtain the image in Figure 2.1, then type:

```
inp_img <- image_read("http://polytopes.net/Tora_color.png")
```

If we want to know information about the image, we can use the image_info() function. We can type as:

```
image_info(inp_img)
```

Then with this image, R outputs the following:

```
> image_info(inp_img)
  format width height colorspace matte filesize density
1   JPEG    600   1010       sRGB FALSE   117342    72x72
```

If we want to plot this image, you can simply call the plot() function. For this example we can type

```
plot(inp_img)
```

To modify the image, we can use the image_modulate() function. To modify the image as shown in the right side of Figure 2.1, we call the function

```
mod_img <- image_modulate(inp_img, brightness = 120, saturation = 20,
hue = 20)
```

then it will create a new image "mod_img". To plot them we type as:

```
plot(mod_img)
```

In the background of these processes, the magick package [31] uses **linear transformations** of a matrix. In image processing, we convert an image as a large matrix with numbers. Each pixel is an entry of a matrix. The value assigned to each entry is a scale of darkness and/or color number. As a result, any processing on an image is an operation on the matrix. For example, if we change colors, we assign different values in the entries of the matrix. If we rotate the image, this is equivalent to applying a linear map to the matrix. If we want to compress the image, we can truncate some values using a **principal component analysis**. These transformations are all based on matrix arithmetic.

2.2 Matrix Operations

Matrix operations are the foundation for linear algebra and important for solving a system of linear equations as well as linear transformations which we will discuss in later chapters. In the working example, we will show how matrix operations play role on a linear transformation in statistics.

2.2.1 Task Completion Checklist

- During the Lecture:

  1. Read the definition of matrix addition.
  2. Read the definition of matrix multiplication.
  3. Read the definition of scalar multiplication of a matrix.
  4. Read the definition of the transpose of a matrix.
  5. Learn how to do matrix addition.
  6. Learn how to do matrix multiplication.
  7. Learn how to do scalar multiplication of a matrix.

8. Learn how to compute the transpose of a matrix.

9. With R, learn how to do matrix addition.

10. With R, learn how to do matrix multiplication.

11. With R, learn how to do scalar multiplication of a matrix.

12. With R, learn how to compute the transpose of a matrix.

- After the Lecture:

    1. Take conceptual quizzes to make sure you understand the materials in this section.

    2. Do some regular exercises.

    3. Conduct lab exercises with R.

    4. Conduct practical applications with R.

    5. If you are interested in python, read the supplement in this section and conduct lab exercises and practical applications with python.

2.2.2   Working Examples

In this section first we look at the plot of a data set generated by R. Figure 2.2 shows the plot of the data set generated under a bivariate normal distribution. A bivariate normal distribution is a normal distribution (Gaussian distribution) with two random variables X, Y. The sample size of the data set is 1000.

If you look at the plot, note that the plot is not centered at the origin $(0, 0)$ and points are distributed like an ellipsoid. In many statistical analyses, we would like to move all points in a data set with a center at the origin and distributed like a circle instead of an ellipsoid as shown in Figure 2.3. In order to do so, we have to perform a **linear transformation** which we will discuss in more detail on linear transformations in Section 4.5. One of most important things we have to know for a linear transformation is a set of matrix operations. Without knowing matrix operations, we cannot perform a linear transformation of a data set. In the Practical Application for this section, we will show how we can transform the data points shown in Figure 2.2 to the data points shown in Figure 2.3 without going into detail on a linear transformation.

2.2.3   Matrix Operations

Matrix operations are the foundation of linear algebra, which we define here. First we define **matrix addition**. In order to do so, the number of row vectors of A and B have to be the same and the number of column vectors A and B have to be the same.

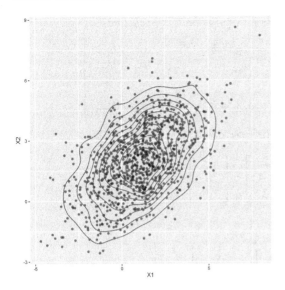

FIGURE 2.2
Plot of the data generated under a bivariate normal distribution.

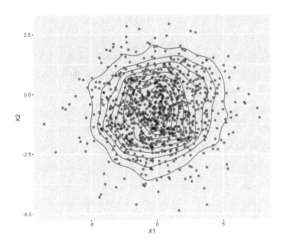

FIGURE 2.3
Plot of the data from Figure 2.3 after a linear transformation.

Definition 17 *Suppose we have $m \times n$ matrix A and B such that*

$$A = \begin{bmatrix} a_{11} & a_{12} & \cdots & a_{1n} \\ a_{21} & a_{22} & \cdots & a_{2n} \\ \vdots & \vdots & \vdots & \vdots \\ a_{m1} & a_{m2} & \cdots & a_{mn} \end{bmatrix}, B = \begin{bmatrix} b_{11} & b_{12} & \cdots & b_{1n} \\ b_{21} & b_{22} & \cdots & b_{2n} \\ \vdots & \vdots & \vdots & \vdots \\ b_{m1} & b_{m2} & \cdots & b_{mn} \end{bmatrix}$$

The **sum** of matrices A and B is an $m \times n$ matrix such that

$$A + B = \begin{bmatrix} a_{11} + b_{11} & a_{12} + b_{12} & \cdots & a_{1n} + b_{1n} \\ a_{21} + b_{21} & a_{22} + b_{22} & \cdots & a_{2n} + b_{2n} \\ \vdots & \vdots & \vdots & \vdots \\ a_{m1} + b_{m1} & a_{m2} + b_{m2} & \cdots & a_{mn} + b_{mn} \end{bmatrix}.$$

Example 47 *Suppose we have two vectors*

$$\begin{bmatrix} 2 \\ -1 \\ 3 \end{bmatrix}, \begin{bmatrix} -1 \\ 0 \\ 4 \end{bmatrix}.$$

Then the sum of these vectors is

$$\begin{bmatrix} 2 \\ -1 \\ 3 \end{bmatrix} + \begin{bmatrix} -1 \\ 0 \\ 4 \end{bmatrix} = \begin{bmatrix} 1 \\ -1 \\ 7 \end{bmatrix}.$$

Example 48 *Suppose we have two 2×3 matrices*

$$\begin{bmatrix} 3 & 0 & -5 \\ -1 & -3 & 4 \end{bmatrix}, \begin{bmatrix} -5 & 5 & 2 \\ 1 & -2 & 0 \end{bmatrix}.$$

Then the sum of these matrices is

$$\begin{bmatrix} 3 & 0 & -5 \\ -1 & -3 & 4 \end{bmatrix} + \begin{bmatrix} -5 & 5 & 2 \\ 1 & -2 & 0 \end{bmatrix} = \begin{bmatrix} -2 & 5 & -3 \\ 0 & -5 & 4 \end{bmatrix}.$$

Example 49 *From Example 47, suppose we have two vectors*

$$\begin{bmatrix} 2 \\ -1 \\ 3 \end{bmatrix}, \begin{bmatrix} -1 \\ 0 \\ 4 \end{bmatrix}.$$

In R, *we do the following: First we define two vectors:*

```
v1 <- c(2, -1, 3)
v2 <- c(-1, 0, 4)
```

Then the sum of these vectors is

```
v1 + v2
```

R *outputs the following*

```
> v1 + v2
[1]  1 -1  7
```

Example 50 *This is from Example 48. Suppose we have two* 2×3 *matrices*

$$\begin{bmatrix} 3 & 0 & -5 \\ -1 & -3 & 4 \end{bmatrix}, \begin{bmatrix} -5 & 5 & 2 \\ 1 & -2 & 0 \end{bmatrix}.$$

In R, *we do the following: First we define two matrices:*

```
A <- matrix(c(3, 0, -5, -1, -3, 4), nrow = 2, ncol = 3, byrow = TRUE)
B <- matrix(c(-5, 5, 2, 1, -2, 0), nrow = 2, ncol = 3, byrow = TRUE)
```

Then the sum of these matrices can be computed by

```
A + B
```

The output from R *is*

```
> A + B
     [,1] [,2] [,3]
[1,]   -2    5   -3
[2,]    0   -5    4
```

Definition 18 *Suppose we have a real number c and an* $m \times n$ *matrix*

$$A = \begin{bmatrix} a_{11} & a_{12} & \cdots & a_{1n} \\ a_{21} & a_{22} & \cdots & a_{2n} \\ \vdots & \vdots & \vdots & \vdots \\ a_{m1} & a_{m2} & \cdots & a_{mn} \end{bmatrix}.$$

The **scalar multiplication** *of c to A is*

$$c \cdot A = \begin{bmatrix} c \cdot a_{11} & c \cdot a_{12} & \cdots & c \cdot a_{1n} \\ c \cdot a_{21} & c \cdot a_{22} & \cdots & c \cdot a_{2n} \\ \vdots & \vdots & \vdots & \vdots \\ c \cdot a_{m1} & c \cdot a_{m2} & \cdots & c \cdot a_{mn} \end{bmatrix}.$$

Example 51 *Suppose we have a vector*

$$v = \begin{bmatrix} 2 \\ -1 \\ 3 \end{bmatrix}$$

and $c = -1$. *Then we have the scalar multiplication of* $c \cdot v$ *is*

$$c \cdot v = \begin{bmatrix} -2 \\ 1 \\ -3 \end{bmatrix}.$$

Example 52 *Suppose we have a 2 × 3 matrix*

$$A = \begin{bmatrix} 3 & 0 & -5 \\ -1 & -3 & 4 \end{bmatrix}$$

and $c = -3$. Then the scalar multiplication of $c \cdot A$ is

$$c \cdot A = \begin{bmatrix} -9 & 0 & 15 \\ 3 & 9 & -12 \end{bmatrix}.$$

Example 53 *In* R, *we can do the scalar multiplication from Example 52 as follows: First we define a matrix A:*

```
A <- matrix(c(3, 0, -5, -1, -3, 4), nrow = 2, ncol = 3, byrow = TRUE)
```

Then we can do the scalar multiplication in R *as*

```
-3 * A
```

Then R *outputs*

```
> -3 * A
     [,1] [,2] [,3]
[1,]   -9    0   15
[2,]    3    9  -12
```

We talked about the sum of two matrices. Now we are going to define the **multiplication** of two matrices. When we multiply two matrices, we have to be very careful of their dimension. The number of *column* vectors of the left matrix has to be the same as the number of *row* vectors of the right matrix. The multiplication of an $m \times n$ matrix and an $n \times k$ matrix is an $m \times k$ matrix.

Matrix multiplication is more complicated than matrix addition. Thus, we will show some examples first. First, we show a **dot product**, which is a special case of matrix multiplication.

Suppose we have two vectors, a 1×3 vector and a 3×1 vector, such that

$$v_1 = \begin{bmatrix} 1 & 2 & -3 \end{bmatrix}, v_2 = \begin{bmatrix} 0 \\ 1 \\ 2 \end{bmatrix}.$$

Note that the number of elements in both vectors are the same: more specifically, we have a $1 \times ③$ vector and a $③ \times 1$ vector, so that the number 3 is the same. In order to apply a dot product, they have to have the same number of entries in both vectors. In this example, the dot product $v_1 \cdot v_2$ is

$$v_1 \cdot v_2 = \begin{bmatrix} 1 & 2 & -3 \end{bmatrix} \cdot \begin{bmatrix} 0 \\ 1 \\ 2 \end{bmatrix} = 1 \cdot 0 + 2 \cdot 1 + (-3) \cdot 2 = -4.$$

In general, the **dot product** of two vectors

$$v_1 = \begin{bmatrix} x_1 & x_2 & \cdots & x_n \end{bmatrix}, v_2 = \begin{bmatrix} y_1 \\ y_2 \\ \vdots \\ y_n \end{bmatrix}$$

is

$$v_1 \cdot v_2 = x_1 \cdot y_1 + x_2 \cdot y_2 + \ldots + x_n \cdot y_n.$$

The multiplication of two matrices can be seen as a generalization of this dot product of two vectors. We will demonstrate matrix multiplication with an example. Suppose we have a 2×3 matrix and a 3×2 matrix such that

$$A = \begin{bmatrix} 1 & -2 & 3 \\ -3 & 2 & -1 \end{bmatrix}, B = \begin{bmatrix} 0 & 2 \\ 1 & 1 \\ 2 & 0 \end{bmatrix}.$$

Let $a_1 = [1, -2, 3]$ and $a_2 = [-3, 2, -1]$, i.e., a_1 is the first row vector of matrix A and a_2 is the second row vector of matrix A. Also let

$$b_1 = \begin{bmatrix} 0 \\ 1 \\ 2 \end{bmatrix}, b_2 = \begin{bmatrix} 2 \\ 1 \\ 0 \end{bmatrix}.$$

b_1 is the first column vector of matrix B and b_2 is the second column vector of matrix B.

Here, note the number of column vectors of A is the same as the number of row vectors of B. Also note that the dimension of the row vectors a_1, a_2 of A is the same as the dimension of column vectors b_1, b_2 of B. This is important for matrix multiplication. The matrix multiplication of A and B is a 2×2 matrix. Let this matrix be

$$C = \begin{bmatrix} c_{11} & c_{12} \\ c_{21} & c_{22} \end{bmatrix}.$$

For the entry in the first row and first column of C, we take the first row a_1 of A and the first column b_1 of B. Then we take the dot product of a_1 and b_1:

$$c_{11} = a_1 \cdot b_1 = 1 \cdot 0 + (-2) \cdot 1 + 3 \cdot 2 = 4.$$

So now we have

$$C = \begin{bmatrix} ④ & c_{12} \\ c_{21} & c_{22} \end{bmatrix}.$$

For the entry of the second row and first column of C, we take the dot product of the second row vector a_2 of A and the first column vector b_1 of B:

$$c_{21} = a_2 \cdot b_1 = (-3) \cdot 0 + 2 \cdot 1 + (-1) \cdot 2 = 0.$$

So now we have

$$C = \begin{bmatrix} 4 & c_{12} \\ ⓪ & c_{22} \end{bmatrix}.$$

For the entry of the first row and second column of C, we take the dot product of the first row vector a_1 of A and the second column vector b_2 of B:

$$c_{12} = a_1 \cdot b_2 = 1 \cdot 2 + (-2) \cdot 1 + 3 \cdot 0 = 0.$$

So now we have

$$C = \begin{bmatrix} 4 & ⓪ \\ 0 & c_{22} \end{bmatrix}.$$

Lastly, for the entry of the second row and second column of C, we take the dot product of the second row vector a_2 of A and the second column vector b_2 of B:

$$c_{22} = a_2 \cdot b_2 = (-3) \cdot 2 + 2 \cdot 1 + (-1) \cdot 0 = -4.$$

So now we have

$$C = \begin{bmatrix} 4 & 0 \\ 0 & ④ \end{bmatrix}.$$

In general, suppose we have an $m \times n$ matrix A and an $n \times k$ matrix B where the row vectors of A are a_1, a_2, a_m and the column vectors of B are b_1, b_2, ..., b_k. Then the matrix multiplication C of A and B is an $n \times k$ matrix such that

$$C = \begin{bmatrix} a_1 \cdot b_1 & a_1 \cdot b_2 & \cdots & a_1 \cdot b_k \\ a_2 \cdot b_1 & a_2 \cdot b_2 & \cdots & a_2 \cdot b_k \\ \vdots & \vdots & \vdots & \vdots \\ a_m \cdot b_1 & a_m \cdot b_2 & \cdots & a_m \cdot b_k \end{bmatrix}.$$

Definition 19 *Suppose we have $m \times n$ matrix A and $n \times k$ B such that*

$$A = \begin{bmatrix} a_{11} & a_{12} & \cdots & a_{1n} \\ a_{21} & a_{22} & \cdots & a_{2n} \\ \vdots & \vdots & \vdots & \vdots \\ a_{m1} & a_{m2} & \cdots & a_{mn} \end{bmatrix}, B = \begin{bmatrix} b_{11} & b_{12} & \cdots & b_{1k} \\ b_{21} & b_{22} & \cdots & b_{2k} \\ \vdots & \vdots & \vdots & \vdots \\ b_{n1} & b_{n2} & \cdots & b_{nk} \end{bmatrix}$$

*The **multiplication** of matrices A and B is an $m \times k$ matrix such that*

$$A \cdot B = \begin{bmatrix} \sum_{i=1}^{n} a_{1i} \cdot b_{i1} & \sum_{i=1}^{n} a_{1i} \cdot b_{i2} & \cdots & \sum_{i=1}^{n} a_{1i} \cdot b_{ik} \\ \sum_{i=1}^{n} a_{2i} \cdot b_{i1} & \sum_{i=1}^{n} a_{2i} \cdot b_{i2} & \cdots & \sum_{i=1}^{n} a_{2i} \cdot b_{ik} \\ \vdots & \vdots & \vdots & \vdots \\ \sum_{i=1}^{n} a_{mi} \cdot b_{i1} & \sum_{i=1}^{n} a_{mi} \cdot b_{i2} & \cdots & \sum_{i=1}^{n} a_{mi} \cdot b_{ik} \end{bmatrix}.$$

Remark 2.1 *We call the multiplication of a $1 \times n$ vector and an $n \times 1$ vector as a **dot product** of two vectors.*

Example 54 *From Example 47, suppose we have two vectors*

$$\begin{bmatrix} 2 & -1 & 3 \end{bmatrix}, \begin{bmatrix} -1 \\ 0 \\ 4 \end{bmatrix}.$$

Then the dot product of these vectors is

$$\begin{bmatrix} 2 & -1 & 3 \end{bmatrix} \cdot \begin{bmatrix} -1 \\ 0 \\ 4 \end{bmatrix} = 2 \cdot (-1) + (-1) \cdot 0 + 3 \cdot 4 = 10.$$

Example 55 *Suppose we have a 2 × 3 matrix and a 3 × 2 matrix*

$$\begin{bmatrix} 3 & 0 & -5 \\ -1 & -3 & 4 \end{bmatrix}, \begin{bmatrix} -5 & 5 \\ 2 & 1 \\ -2 & 0 \end{bmatrix}.$$

Then the multiplication of these matrices is

$$\begin{bmatrix} 3 & 0 & -5 \\ -1 & -3 & 4 \end{bmatrix} \cdot \begin{bmatrix} -5 & 5 \\ 2 & 1 \\ -2 & 0 \end{bmatrix}$$
$$= \begin{bmatrix} 3 \cdot (-5) + 0 \cdot 2 + (-5) \cdot (-2) & 3 \cdot 5 + 0 \cdot 1 + (-5) \cdot 0 \\ (-1) \cdot (-5) + (-3) \cdot 2 + 4 \cdot (-2) & (-1) \cdot 5 + (-3) \cdot 1 + 4 \cdot 0 \end{bmatrix}$$
$$= \begin{bmatrix} -5 & 15 \\ -9 & -8 \end{bmatrix}$$

Example 56 *From Example 54, suppose we have two vectors*

$$\begin{bmatrix} 2 & -1 & 3 \end{bmatrix}, \begin{bmatrix} -1 \\ 0 \\ 4 \end{bmatrix}.$$

First we define these vectors

```
v1 <- c(2, -1, 3)
v2 <- c(-1, 0, 4)
```

Then the dot product of these vectors in R *is*

```
v1 %*% v2
```

The output from R *is*

```
> v1 %*% v2
     [,1]
[1,]   10
```

Example 57 *This is from Example 55. Suppose we have a* 2×3 *matrix and a* 3×2 *matrix*

$$\begin{bmatrix} 3 & 0 & -5 \\ -1 & -3 & 4 \end{bmatrix}, \begin{bmatrix} -5 & 5 \\ 2 & 1 \\ -2 & 0 \end{bmatrix}.$$

In R, *we do the following: First we define two matrices:*

```
A <- matrix(c(3, 0, -5, -1, -3, 4), nrow = 2, ncol = 3, byrow = TRUE)
B <- matrix(c(-5, 5, 2, 1, -2, 0), nrow = 3, ncol = 2, byrow = TRUE)
```

Then the dot product of these matrices can be computed by

```
A %*% B
```

The output from R *is*

```
> A %*% B
     [,1] [,2]
[1,]   -5   15
[2,]   -9   -8
```

One of the important operations in matrices is called the **transpose** of a matrix. With this operation, we can be more flexible with other matrix operations.

Definition 20 *Suppose we have an* $m \times n$ *matrix*

$$A = \begin{bmatrix} a_{11} & a_{12} & \cdots & a_{1n} \\ a_{21} & a_{22} & \cdots & a_{2n} \\ \vdots & \vdots & \vdots & \vdots \\ a_{m1} & a_{m2} & \cdots & a_{mn} \end{bmatrix}.$$

The **transpose** *of the matrix* A *is an* $n \times m$ *matrix defined as*

$$A^T = \begin{bmatrix} a_{11} & a_{21} & \cdots & a_{m1} \\ a_{12} & a_{22} & \cdots & a_{m2} \\ \vdots & \vdots & \vdots & \vdots \\ a_{1n} & a_{2n} & \cdots & a_{mn} \end{bmatrix}.$$

Example 58 *Suppose we have a* 2×3 *matrix*

$$A = \begin{bmatrix} 4 & -1 & -5 \\ 0 & 1 & -2 \end{bmatrix}$$

The transpose of the matrix A *is a* 3×2 *matrix*

$$A^T = \begin{bmatrix} 4 & 0 \\ -1 & 1 \\ -5 & -2 \end{bmatrix}.$$

Example 59 *Suppose we have a 3×3 matrix*

$$A = \begin{bmatrix} 2 & 1 & 3 \\ 5 & -4 & 4 \\ -5 & -3 & -1 \end{bmatrix}$$

The transpose of the matrix A is a 3×3 matrix

$$A^T = \begin{bmatrix} 2 & 5 & -5 \\ 1 & -4 & -3 \\ 3 & 4 & -1 \end{bmatrix}.$$

Example 60 *We can use R to compute the transpose of a matrix using the t() function. With the matrix A from Example 58, we first create a matrix using the matrix() function:*

```
A <- matrix(c(4, -1, -5, 0, 1, -2), 2, 3, byrow = TRUE)
```

Then we type in R

```
t(A)
```

Then R outputs as follows:

```
> t(A)
     [,1] [,2]
[1,]    4    0
[2,]   -1    1
[3,]   -5   -2
```

We can see a system of linear equations as a sum of vectors. We will explain this with an example from Example 44. Suppose we have the following system of linear equations:

$$\begin{array}{rcrcrcrcr} & & x_2 & + & 3x_3 & - & x_4 & = & 1 \\ -x_1 & + & x_2 & - & 4x_3 & & & = & 1 \\ x_1 & & & + & 2x_3 & + & 4x_4 & = & 5 \\ & & x_2 & & & - & 4x_4 & = & -2. \end{array}$$

Its augmented matrix is

$$\begin{bmatrix} 0 & 1 & 3 & -1 & 1 \\ -1 & 1 & -4 & 0 & 1 \\ 1 & 0 & 2 & 4 & 5 \\ 0 & 1 & 0 & -4 & -2 \end{bmatrix}.$$

Let the first column vector be a_1, let the second column vector be a_2, the third column vector be a_3, the fourth column vector be a_4, and let the last column vector be b. Thus we have

$$a_1 = \begin{bmatrix} 0 \\ -1 \\ 1 \\ 0 \end{bmatrix}, a_2 = \begin{bmatrix} 1 \\ 1 \\ 0 \\ 1 \end{bmatrix}, a_3 = \begin{bmatrix} 3 \\ -4 \\ 2 \\ 0 \end{bmatrix}, a_4 = \begin{bmatrix} -1 \\ 0 \\ 4 \\ -4 \end{bmatrix}, b = \begin{bmatrix} 1 \\ 1 \\ 5 \\ -2 \end{bmatrix}.$$

Then the system of linear equations can be written as the sum of vectors such that

$$x_1 \cdot \begin{bmatrix} 0 \\ -1 \\ 1 \\ 0 \end{bmatrix} + x_2 \cdot \begin{bmatrix} 1 \\ 1 \\ 0 \\ 1 \end{bmatrix} + x_3 \cdot \begin{bmatrix} 3 \\ -4 \\ 2 \\ 0 \end{bmatrix} + x_4 \cdot \begin{bmatrix} -1 \\ 0 \\ 4 \\ -4 \end{bmatrix} = \begin{bmatrix} 1 \\ 1 \\ 5 \\ -2 \end{bmatrix},$$

which can be written as

$$x_1 \cdot a_1 + x_2 \cdot a_2 + x_3 \cdot a_3 + x_4 \cdot a_4 = b.$$

Let a matrix A be a 4×4 matrix whose first column vector is a_1, whose second vector is a_2, whose third column vector is a_3, and whose fourth column vector is a_4. Also let

$$x = \begin{bmatrix} x_1 \\ x_2 \\ x_3 \\ x_4 \end{bmatrix}.$$

We can now simplify as

$$x_1 \cdot \begin{bmatrix} 0 \\ -1 \\ 1 \\ 0 \end{bmatrix} + x_2 \cdot \begin{bmatrix} 1 \\ 1 \\ 0 \\ 1 \end{bmatrix} + x_3 \cdot \begin{bmatrix} 3 \\ -4 \\ 2 \\ 0 \end{bmatrix} + x_4 \cdot \begin{bmatrix} -1 \\ 0 \\ 4 \\ -4 \end{bmatrix} = A \cdot x.$$

The left hand side of the system of linear equations is the matrix multiplication of a 4×4 matrix A and a 4×1 matrix x. Thus, this system of linear equations can be written as

$$A \cdot x = b.$$

2.2.4 Checkmarks

- The definition of matrix addition.

- The definition of scalar multiplication.

- The definition of matrix multiplication.

- The definition of the transpose of a matrix.

- You can do matrix addition.

- You can do scalar multiplication.

- You can do matrix multiplication.

- You can compute the transpose of a matrix.

- You can do matrix addition with R.

- You can do scalar multiplication with R.

- You can do matrix multiplication with R.

- You can compute the transpose of a matrix R.

2.2.5 Conceptual Quizzes

Quiz 42 True or False: *If we have a 2×3 matrix A and a 3×2 matrix B, then we can compute the sum of A and B.*

Quiz 43 True or False: *If we have a 2×3 matrix A and a 2×3 matrix B, then we can compute the sum of A and B.*

Quiz 44 True or False: *If we have a 2×3 matrix A and a 3×3 matrix B, then we can compute the multiplication of A and B.*

Quiz 45 True or False: *If we have a 2×3 matrix A and a 2×3 matrix B, then we can compute the multiplication of A and B.*

Quiz 46 True or False: *If we have a 2×3 matrix A, then we can the transpose of the matrix A is a 2×3 matrix.*

Quiz 47 True or False: *If we have a 2×3 matrix A and a 3×2 matrix B, then we can compute the sum of A and B^T.*

Quiz 48 True or False: *If we have a 2×3 matrix A and a 2×3 matrix B, then we can compute the sum of A^T and B.*

Quiz 49 True or False: *If we have a 2×3 matrix A and a 3×3 matrix B, then we can compute the multiplication of A and B^T.*

Quiz 50 True or False: *If we have a 2×3 matrix A and a 2×3 matrix B, then we can compute the multiplication of A^T and B.*

Quiz 51 True or False: *Suppose we have a matrix*

$$\begin{bmatrix} -4 & 1 & -17 & 5 \\ 0 & 0 & -2 & 9 \\ 0 & -7 & 0 & 3 \end{bmatrix}$$

and a real number -1*. Then the scalar multiplication of them is*

$$\begin{bmatrix} 4 & -1 & 17 & -5 \\ 0 & 0 & 2 & -9 \\ 0 & 7 & 0 & -3 \end{bmatrix}.$$

Quiz 52 True or False: *Suppose we have a vector*

$$\begin{bmatrix} -4 & 1 & -17 & 5 \end{bmatrix}$$

and a real number -4*. Then the scalar multiplication of them is*

$$\begin{bmatrix} 8 & -3 & -21 & 1 \end{bmatrix}.$$

Quiz 53 True or False: *Suppose we have two vectors. The dot product of these two vectors is a real number.*

Quiz 54 True or False: *Suppose we have two matrices A and B. If we want to compute the matrix multiplication of A and B in* R *then we type:*

```
A * B
```

Quiz 55 True or False: *Suppose we have two matrices A and B. The matrix multiplication A · B is equal to B · A.*

Quiz 56 Multiple Choice: *Suppose we have the following matrices*

$$A = \begin{bmatrix} 5 & -4 & 0 \\ 1 & -1 & -3 \end{bmatrix}, B = \begin{bmatrix} 1 & -3 & -1 \\ 0 & 2 & -2 \end{bmatrix}.$$

Then $A + B$ *is*

1. 5

2.

$$\begin{bmatrix} 6 & -7 & -1 \\ 1 & 1 & -5 \end{bmatrix}$$

3.

$$\begin{bmatrix} 17 & -8 \\ 7 & 4 \end{bmatrix}$$

Quiz 57 Multiple Choice: *Suppose we have the following matrices*

$$A = \begin{bmatrix} 1 & 0 & 4 \\ -3 & -2 & 2 \end{bmatrix}, B = \begin{bmatrix} -2 & 0 \\ 5 & 3 \\ 1 & -3 \end{bmatrix}.$$

Then the matrix multiplication $A \cdot B$ is

1. 1

2.
$$\begin{bmatrix} -1 & 5 & 5 \\ -3 & 1 & -1 \end{bmatrix}$$

3.
$$\begin{bmatrix} 2 & -12 \\ -2 & -12 \end{bmatrix}$$

Quiz 58 Multiple Choice: *Suppose we have the following matrices*

$$A = \begin{bmatrix} 5 & -4 & 0 \\ 1 & -1 & -3 \end{bmatrix}, B = \begin{bmatrix} 1 & -3 & -1 \\ 0 & 2 & -2 \end{bmatrix}.$$

Then $A \cdot B^T$ is

1. 5

2.
$$\begin{bmatrix} 6 & -7 & -1 \\ 1 & 1 & -5 \end{bmatrix}$$

3.
$$\begin{bmatrix} 17 & -8 \\ 7 & 4 \end{bmatrix}$$

Quiz 59 Multiple Choice: *Suppose we have the following matrices*

$$A = \begin{bmatrix} 1 & 0 & 4 \\ -3 & -2 & 2 \end{bmatrix}, B = \begin{bmatrix} -2 & 0 \\ 5 & 3 \\ 1 & -3 \end{bmatrix}.$$

Then $A + B^T$ is

1. 1

2.
$$\begin{bmatrix} -1 & 5 & 5 \\ -3 & 1 & -1 \end{bmatrix}$$

3.
$$\begin{bmatrix} 2 & -12 \\ -2 & -12 \end{bmatrix}$$

Quiz 60 Multiple Choice: *Suppose we have the following vectors*

$$v_1 = \begin{bmatrix} 5 & -5 & -1 & 1 \end{bmatrix}, v_2 = \begin{bmatrix} 0 \\ 2 \\ 3 \\ 1 \end{bmatrix}.$$

Then the dot product $v_1 \cdot v_2$ is

1. -12

2.
$$\begin{bmatrix} 0 & -10 & -3 & 1 \end{bmatrix}$$

3.
$$\begin{bmatrix} 5 & -3 & 2 & 2 \end{bmatrix}$$

2.2.6   Regular Exercises

Exercise 2.1 *Suppose we have the following matrices with given numbers of row vectors and column vectors.*

$$
\begin{array}{cccccc}
A & B & C & D & E & F \\
(2 \times 3) & (3 \times 3) & (4 \times 3) & (2 \times 3) & (3 \times 2) & (2 \times 4)
\end{array}
$$

Determine the number of row vectors and the number of column vectors of the following:

1. C^T

2. $A + E^T$

3. $A \cdot B$

4. $A \cdot B + D$

5. $A \cdot B \cdot C^T$

6. $(A \cdot B \cdot C^T)^T$

7. $A \cdot B \cdot C^T + F$

Exercise 2.2 *Find real numbers a, b, c, d such that*

$$\begin{bmatrix} a+b & -a-2b \\ c & d+c \end{bmatrix} = \begin{bmatrix} 3 & -5 \\ 2 & 4 \end{bmatrix}.$$

Exercise 2.3 *Suppose we have matrices*

$$A = \begin{bmatrix} 2 & -5 & -4 \\ -1 & 4 & -3 \end{bmatrix}, B = \begin{bmatrix} 5 & 0 & 1 \\ -2 & -5 & -3 \end{bmatrix},$$

$$C = \begin{bmatrix} -2 & 0 & -3 \\ -1 & 3 & 5 \\ 2 & 4 & -5 \end{bmatrix}, D = \begin{bmatrix} 0 & 1 & 1 \\ 1 & 3 & 3 \\ 1 & 5 & 3 \end{bmatrix},$$

$$E = \begin{bmatrix} 0 & -2 & 4 & 2 \\ 3 & 5 & -3 & -4 \\ 3 & -4 & -1 & -1 \end{bmatrix}, F = \begin{bmatrix} 1 & -4 & -5 & -1 \\ -1 & 0 & -2 & 1 \\ 0 & 1 & 1 & 4 \end{bmatrix}.$$

Compute the following:

1. $(-2) \cdot A$

2. $A + B$

3. C^T

4. $A - 2 \cdot B$

5. $5 \cdot C + 3 \cdot D$

6. $A \cdot C$

7. $A \cdot C^T$

8. $A \cdot D - 2 \cdot B$

9. $-E + 3 \cdot F$

10. $-c \cdot F + 2 \cdot E$

11. $C \cdot D - E \cdot F^T$

Exercise 2.4 *We have the same matrices from Exercise 2.3. Then compute the following:*

1. $(A + B) \cdot C$

2. $A \cdot C + B \cdot C$

3. $(C + D) \cdot C$

4. $C \cdot (C + D)$

5. $(A + B) \cdot C^T$

6. $C \cdot (A + B)^T$

7. $C \cdot B^T + -A^T \cdot B$

8. $(3 \cdot C + E) + (2 \cdot E - F) \cdot F$

9. $-A^T \cdot B \cdot (C - 2 \cdot D) - (2 \cdot C + E) + ((-1) \cdot E + 2 \cdot F) \cdot F$

10. $4 \cdot (A + B) \cdot C - 4 \cdot (A \cdot C + B \cdot C)$

Exercise 2.5 *From Exercise 1.46, suppose we have a system of linear equations such that*

$$
\begin{array}{rrrrrrrrl}
3x_1 & + & 6x_2 & + & 3x_3 & - & 2x_4 & = & 59 \\
7x_1 & - & 10x_2 & - & 2x_3 & - & 8x_4 & = & 80 \\
2x_1 & - & 6x_2 & + & 2x_3 & - & 4x_4 & = & 28 \\
2x_1 & - & 5x_2 & - & 8x_3 & - & x_4 & = & 0.
\end{array}
$$

Find a matrix A and vectors x, b such that this system of linear equations is written as a matrix equation $A \cdot x = b$.

Exercise 2.6 *From Exercise 1.47, suppose we have a system of linear equations such that*

$$
\begin{array}{rrrrrrrrl}
5x_1 & + & x_2 & + & 10x_3 & + & 6x_4 & = & 2 \\
-x_1 & - & 7x_2 & - & 6x_3 & + & 5x_4 & = & 38 \\
-7x_1 & + & 4x_2 & - & x_3 & + & x_4 & = & 24 \\
-5x_1 & + & 3x_2 & - & 4x_3 & - & 8x_4 & = & -18.
\end{array}
$$

Find a matrix A and vectors x, b such that this system of linear equations is written as a matrix equation $A \cdot x = b$.

Exercise 2.7 *From Exercise 1.48, suppose we have a system of linear equations such that*

$$
\begin{array}{rrrrrrrrl}
& & -14x_2 & - & 9x_3 & + & 11x_4 & = & 22 \\
-2x_1 & - & 4x_2 & - & 2x_3 & + & 6x_4 & = & 6 \\
2x_1 & - & 10x_2 & - & 7x_3 & + & 5x_4 & = & 16 \\
-6x_1 & - & 9x_2 & + & 8x_3 & + & 7x_4 & = & -73.
\end{array}
$$

Find a matrix A and vectors x, b such that this system of linear equations is written as a matrix equation $A \cdot x = b$.

Exercise 2.8 *From Exercise 1.49, suppose we have a system of linear equations such that*

$$
\begin{array}{rrrrrrrrl}
8x_1 & + & 6x_2 & - & 4x_3 & - & 3x_4 & = & -75 \\
-5x_1 & + & 8x_2 & + & x_3 & + & 6x_4 & = & -7 \\
6x_1 & + & 5x_2 & + & 4x_3 & - & x_4 & = & -20 \\
-5x_1 & + & 10x_2 & + & 8x_3 & - & x_4 & = & -3.
\end{array}
$$

Find a matrix A and vectors x, b such that this system of linear equations is written as a matrix equation $A \cdot x = b$.

Exercise 2.9 *From Exercise 1.50, suppose we have a system of linear equations such that*

$$
\begin{array}{rcrcrcrcl}
8x_1 & - & 9x_2 & + & x_3 & + & 10x_4 & = & 25 \\
-9x_1 & - & 7x_2 & - & 3x_3 & - & 7x_4 & = & 16 \\
7x_1 & - & 25x_2 & - & x_3 & + & 13x_4 & = & 60 \\
-x_1 & - & 16x_2 & - & 2x_3 & + & 3x_4 & = & 41.
\end{array}
$$

Find a matrix A and vectors x, b such that this system of linear equations is written as a matrix equation $A \cdot x = b$.

2.2.7 Lab Exercises

Lab Exercise 43 *With R, do Exercise 2.1.*

Lab Exercise 44 *With R, do Exercise 2.2.*

Lab Exercise 45 *With R, do Exercise 2.3.*

Lab Exercise 46 *With R, do Exercise 2.4.*

2.2.8 Practical Applications

In order to create the data points for the plot shown in Figure 2.3, we need to compute the **eigen vectors** of the matrix sigma, the standard deviation for the bivariate normal distribution. The eigen vectors of a matrix will be discussed in a later chapter. This transformation using the eigen vectors is related to **principal component analysis**, one of the most popular unsupervised learning models in data science. For this practical application, without going into detail, we show how to transform to the data points shown in Figure 2.3. We will explain in more detail when we discuss eigen vectors and their eigen values in a later chapter.

For the working example, we generated the data points from a bivariate normal distribution. A bivariate normal distribution is defined by a mean and standard deviation, unlike a univariate normal distribution, however, its mean is defined by a two-dimensional vector and its standard deviation is defined by a 2×2 matrix. We use the R package `mvtnorm` [18] and we use the `matlib` package for matrix operations. First we upload all packages needed:

```
library(mvtnorm)
library(ggplot2)
library(matlib)
```

The `ggplot2` package [46] is for plotting.

Then, we define the mean and the standard deviation as:

```
## Standard deviation
sigma <- matrix(c(4,2,2,3), ncol = 2, nrow = 2)
## Mean
mu <- c(1, 2)
```

For this data we have the sample size $n = 10000$.

```
n <- 1000
```

The function set.seed() sets the seed, which is a value to start generating random numbers. The set.seed() function resets the values of the random numbers and random functions from the values previously obtained. We set a seed in order to reproduce the same outcome. If we set the same seed, then we can reproduce the same random numbers:

```
set.seed(123)
```

Finally, we generate the data points by

```
x <- rmvnorm(n = n, mean = mu, sigma = sigma)
```

To plot the data we set the points in a data frame:

```
d <- data.frame(x)
```

Using the **ggplot2** package we plot the data points as

```
p2 <- ggplot(d, aes(x = X1, y = X2)) +
  geom_point(alpha = .5) +
  geom_density_2d()

p2
```

This creates the plot shown in Figure 2.2.

First we translate all data points stored as a variable x by -mu $= [-1, -2]$ since the center of the distribution is at mu $= [1, 2]$.

```
y <- x - mu
```

We store these new data points as a variable y.

Then we compute the **eigen vectors** of sigma.

```
E <- eigen(sigma)
E$vectors
```

The eigen vectors of the matrix sigma is stored in E$vectors in the form of a matrix. Each column vector is an eigen vector.

Then we take matrix multiplication, the data set stored as a 10000×2 matrix multiplied by a 2×2 matrix, the transpose of the **inverse** of the matrix E. The inverse of a matrix will be discussed in the next section.

```
y <- y %*% t(inv(E$vectors))
```

As we did for the data set x, we set this data as a "data frame" using the data.frame() function:

```
dd <- data.frame(y)
```

Now we are going to plot by

```
p3 <- ggplot(dd, aes(x = X1, y = X2)) +
  geom_point(alpha = .5) +
  geom_density_2d()
p3
```

This creates the plot shown in Figure 2.3 in the beginning of this section.

2.2.9 Supplements with python Code

In python, we use the numpy package [19] for matrix operations. For matrix addition, we can use + operations. First, we need to call the package:

```
import numpy as np
```

We use matrices from Example 48. Suppose we have two 2×3 matrices

$$\begin{bmatrix} 3 & 0 & -5 \\ -1 & -3 & 4 \end{bmatrix}, \begin{bmatrix} -5 & 5 & 2 \\ 1 & -2 & 0 \end{bmatrix}.$$

We first define matrices in python:

```
A = np.matrix([[3, 0, -5], [-1, -3, 4]])
B = np.matrix([[-5, 5, 2], [1, -2, 0]])
```

For matrix addition, the sum of these matrices can be obtained by

```
A + B
```

For scalar multiplication, we will use Example 52. Suppose we have a 2×3 matrix

$$A = \begin{bmatrix} 3 & 0 & -5 \\ -1 & -3 & 4 \end{bmatrix}$$

and $c = -3$. Then in python we can define the matrix

```
A = np.matrix([[3, 0, -5], [-1, -3, 4]])
```

Then, scalar multiplication can be computed by

```
-3 * A
```

For matrix multiplication, we will use the following 2×3 matrix and 3×2 matrix such that

$$A = \begin{bmatrix} 1 & -2 & 3 \\ -3 & 2 & -1 \end{bmatrix}, B = \begin{bmatrix} 0 & 2 \\ 1 & 1 \\ 2 & 0 \end{bmatrix}.$$

First we define these matrices in python as

```
A = np.matrix([[1, -2, 3], [-3, 2, -1]])
B = np.matrix([[0, 2], [1, 1], [2, 0]])
```

Then, we can compute the matrix multiplication by

```
A * B
```

We can compute the transpose of a matrix with the transpose() function. For example, if we want to compute the transpose of the matrix A and B

$$A = \begin{bmatrix} 1 & -2 & 3 \\ -3 & 2 & -1 \end{bmatrix}, B = \begin{bmatrix} 0 & 2 \\ 1 & 1 \\ 2 & 0 \end{bmatrix}$$

then type

```
np.transpose(A)
np.transpose(B)
```

2.3   Properties of Matrix Operations and Matrix Inverse

In the previous section we saw some matrix operations and how to compute them with R. Matrix operations are computationally not stable because a computer can only store only a finite number of digits of a fraction. For example, $\frac{1}{3} = 0.333\cdots$, but in a computer they can only store a finite number of digits. In matrix operations, these fractional errors can be accumulated very quickly and they might give you an answer which might not be close to the true answer. In order to prevent these errors, it is nice to know theoretical properties of matrix operations. Therefore, in this section we show some properties of these matrix operations so that we can save some computational time. In general every computation in R is based on matrix operations, so R is robust against fractional errors. However, it is still useful to know these theoretical properties.

2.3.1 Task Completion Checklist

- During the Lecture:

  1. Read the definition of the inverse of a matrix.
  2. Read properties of matrix operations.
  3. Learn how to compute the inverse of a matrix.
  4. Learn how to apply properties of matrix operations to a matrix computation.
  5. With R, learn how to perform compute the inverse of a matrix.
  6. With R, learn how to apply properties of matrix operations to a matrix computation.

- After the Lecture:

  1. Take conceptual quizzes to make sure you understand the materials in this section.
  2. Do some regular exercises.
  3. Conduct lab exercises with R.
  4. Conduct practical applications with R.
  5. If you are interested in python, read the supplement in this section and conduct lab exercises and practical applications with python.

2.3.2 Working Examples

In order to solve a system of linear equations, we can use the inverse of a matrix A, the coefficient matrix, for the system of linear equations. As the working example for this section, we use the working example in Section 1.4, the Guinness data set [5]. Recall that this data set was collected from Guinness Ghana Ltd. and contains information on the supply of Malta Guinness from two production sites, Kaasi and Achimota, to nine key distributors geographically scattered in the regions of Ghana. Table 1.3 shows the demands from the distributors and supplies from the production sites.

In this example, we have a system of linear equations

$$
\begin{array}{rcll}
x_{11} + x_{12} + \ldots + x_{19} &=& 1298 \\
x_{21} + x_{22} + \ldots + x_{29} &=& 1948 \\
x_{11} + x_{21} &=& 465 \\
x_{12} + x_{22} &=& 605 \\
x_{13} + x_{23} &=& 451 \\
x_{14} + x_{24} &=& 338 \\
x_{15} + x_{25} &=& 260 \\
x_{16} + x_{26} &=& 183 \\
x_{17} + x_{27} &=& 282 \\
x_{18} + x_{28} &=& 127 \\
x_{19} + x_{29} &=& 535.
\end{array}
$$

The coefficient matrix for this system is

$$
\begin{bmatrix}
1 & 1 & 1 & 1 & 1 & 1 & 1 & 1 & 1 & 0 & 0 & 0 & 0 & 0 & 0 & 0 & 0 & 0 \\
0 & 0 & 0 & 0 & 0 & 0 & 0 & 0 & 0 & 1 & 1 & 1 & 1 & 1 & 1 & 1 & 1 & 1 \\
1 & 0 & 0 & 0 & 0 & 0 & 0 & 0 & 0 & 1 & 0 & 0 & 0 & 0 & 0 & 0 & 0 & 0 \\
0 & 1 & 0 & 0 & 0 & 0 & 0 & 0 & 0 & 0 & 1 & 0 & 0 & 0 & 0 & 0 & 0 & 0 \\
0 & 0 & 1 & 0 & 0 & 0 & 0 & 0 & 0 & 0 & 0 & 1 & 0 & 0 & 0 & 0 & 0 & 0 \\
0 & 0 & 0 & 1 & 0 & 0 & 0 & 0 & 0 & 0 & 0 & 0 & 1 & 0 & 0 & 0 & 0 & 0 \\
0 & 0 & 0 & 0 & 1 & 0 & 0 & 0 & 0 & 0 & 0 & 0 & 0 & 1 & 0 & 0 & 0 & 0 \\
0 & 0 & 0 & 0 & 0 & 1 & 0 & 0 & 0 & 0 & 0 & 0 & 0 & 0 & 1 & 0 & 0 & 0 \\
0 & 0 & 0 & 0 & 0 & 0 & 1 & 0 & 0 & 0 & 0 & 0 & 0 & 0 & 0 & 1 & 0 & 0 \\
0 & 0 & 0 & 0 & 0 & 0 & 0 & 1 & 0 & 0 & 0 & 0 & 0 & 0 & 0 & 0 & 1 & 0 \\
0 & 0 & 0 & 0 & 0 & 0 & 0 & 0 & 1 & 0 & 0 & 0 & 0 & 0 & 0 & 0 & 0 & 1
\end{bmatrix},
$$

which is an 11×18 matrix, and the vector for the right hand side of the system of linear equations is

$$
\begin{bmatrix}
1298 \\
1948 \\
465 \\
605 \\
451 \\
338 \\
260 \\
183 \\
282 \\
127 \\
535
\end{bmatrix},
$$

which is a 1×11 matrix. In Section 1.4, we used the Solve() function from the matlib package in R. Now we will solve this system of linear equations using the inverse of the coefficient matrix.

2.3.3   Properties of Matrix Operations and Matrix Inverse

In order to save some computational time, there are important theoretical properties of matrix operations. We begin with some properties of matrix operations in this section.

Theorem 2.2 *Suppose A, B, C are m × n matrices. Then we have the following rules:*

1. $A + B = B + A$ *(Commutative law for addition)*.

2. $A + (B + C) = (A + B) + C$ *(Associative law for addition)*.

Example 61 *Suppose we have*

$$
A = \begin{bmatrix} -4 & -3 & 3 \\ -1 & -3 & 2 \end{bmatrix}, B = \begin{bmatrix} 4 & -1 & 1 \\ 2 & -4 & 5 \end{bmatrix}, C = \begin{bmatrix} -5 & 2 & -4 \\ -5 & 5 & -1 \end{bmatrix}.
$$

Then we have

$$A + B = \begin{bmatrix} 0 & -4 & 4 \\ 1 & -7 & 7 \end{bmatrix}$$

and

$$B + A = \begin{bmatrix} 0 & -4 & 4 \\ 1 & -7 & 7 \end{bmatrix}.$$

Also

$$(A + B) + C = \begin{bmatrix} -5 & -2 & 0 \\ -4 & -2 & 6 \end{bmatrix}$$

and

$$A + (B + C) = \begin{bmatrix} -5 & -2 & 0 \\ -4 & -2 & 6 \end{bmatrix}.$$

Theorem 2.3 *Suppose A, B, C are $m \times n$ matrices and suppose a, b, c are constants. Then we have the following rules:*

1. $c \cdot (A + B) = c \cdot A + c \cdot B$

2. $c \cdot (A - B) = c \cdot A - c \cdot B$

3. $(a + b) \cdot C = a \cdot C + b \cdot C$

4. $(a - b) \cdot C = a \cdot C - b \cdot C$

Example 62 *Suppose we have*

$$A = \begin{bmatrix} -4 & -3 & 3 \\ -1 & -3 & 2 \end{bmatrix}, B = \begin{bmatrix} 4 & -1 & 1 \\ 2 & -4 & 5 \end{bmatrix}, C = \begin{bmatrix} -5 & 2 & -4 \\ -5 & 5 & -1 \end{bmatrix}$$

and $a = -1$, $b = 3$, $c = 5$. Then we have

$$c \cdot (A + B) = \begin{bmatrix} 0 & -20 & 20 \\ 5 & -35 & 35 \end{bmatrix}$$

and

$$c \cdot A + c \cdot B = \begin{bmatrix} 0 & -20 & 20 \\ 5 & -35 & 35 \end{bmatrix}.$$

$$(a + b) \cdot C = \begin{bmatrix} -10 & 4 & -8 \\ -10 & 10 & -2 \end{bmatrix}$$

and

$$a \cdot C + b \cdot C = \begin{bmatrix} -10 & 4 & -8 \\ -10 & 10 & -2 \end{bmatrix}.$$

Theorem 2.4 *Suppose A is an $m \times n$ matrix, B is an $n \times k$ matrix, and C is a $k \times s$ matrix. Then*

$$(A \cdot B) \cdot C = A \cdot (B \cdot C).$$

Example 63 *Suppose*

$$A = \begin{bmatrix} -4 & -3 & 3 \\ -1 & -3 & 2 \end{bmatrix}, B = \begin{bmatrix} -1 & -2 & 2 \\ -5 & -3 & -3 \\ 5 & 0 & 3 \end{bmatrix}, C = \begin{bmatrix} -4 & 5 \\ 4 & -5 \\ -3 & -4 \end{bmatrix}.$$

Then,

$$A \cdot B = \begin{bmatrix} 34 & 17 & 10 \\ 26 & 11 & 13 \end{bmatrix}$$

and

$$(A \cdot B) \cdot C = \begin{bmatrix} -98 & 45 \\ -99 & 23 \end{bmatrix}.$$

Also

$$B \cdot C = \begin{bmatrix} -10 & -3 \\ 17 & 2 \\ -29 & 13 \end{bmatrix}.$$

and

$$A \cdot (B \cdot C) = \begin{bmatrix} -98 & 45 \\ -99 & 23 \end{bmatrix}.$$

Theorem 2.5 *Suppose A is an $m \times n$ matrix, and B and C are $n \times k$ matrices. Then*

1. $A \cdot (B + C) = A \cdot B + A \cdot C$

2. $A \cdot (B - C) = A \cdot B - A \cdot C.$

Example 64 *Suppose*

$$A = \begin{bmatrix} -4 & -3 & 3 \\ -1 & -3 & 2 \end{bmatrix}, B = \begin{bmatrix} 2 & -2 \\ 0 & 2 \\ -5 & 1 \end{bmatrix}, C = \begin{bmatrix} -4 & 5 \\ 4 & -5 \\ -3 & -4 \end{bmatrix}.$$

$$B + C = \begin{bmatrix} -2 & 3 \\ 4 & -3 \\ -8 & -3 \end{bmatrix}$$

and

$$A \cdot (B + C) = \begin{bmatrix} -28 & -12 \\ -26 & 0 \end{bmatrix}.$$

Also

$$A \cdot B = \begin{bmatrix} -23 & 5 \\ -12 & -2 \end{bmatrix}, A \cdot C = \begin{bmatrix} -5 & -17 \\ -14 & 2 \end{bmatrix}$$

and

$$A \cdot B + A \cdot C = \begin{bmatrix} -28 & -12 \\ -26 & 0 \end{bmatrix}.$$

Theorem 2.6 *Suppose A is an $m \times n$ matrix, and B and C are $k \times m$ matrices. Then*

1. $(B + C) \cdot A = B \cdot A + C \cdot A$

2. $(B - C) \cdot A = B \cdot A - C \cdot A$.

Example 65 *Suppose*

$$A = \begin{bmatrix} -4 & -3 & 3 \\ -1 & -3 & 2 \end{bmatrix}, B = \begin{bmatrix} 2 & -2 \\ 0 & 2 \\ -5 & 1 \end{bmatrix}, C = \begin{bmatrix} -4 & 5 \\ 4 & -5 \\ -3 & -4 \end{bmatrix}.$$

Then,

$$B + C = \begin{bmatrix} -2 & 3 \\ 4 & -3 \\ -8 & -3 \end{bmatrix}$$

and

$$(B + C) \cdot A = \begin{bmatrix} 5 & -3 & 0 \\ -13 & -3 & 6 \\ 35 & 33 & -30 \end{bmatrix}.$$

Also

$$B \cdot A = \begin{bmatrix} -6 & 0 & 2 \\ -2 & -6 & 4 \\ 19 & 12 & -13 \end{bmatrix}, C \cdot A = \begin{bmatrix} 11 & -3 & -2 \\ -11 & 3 & 2 \\ 16 & 21 & -17 \end{bmatrix}$$

and

$$B \cdot A + C \cdot A = \begin{bmatrix} 5 & -3 & 0 \\ -13 & -3 & 6 \\ 35 & 33 & -30 \end{bmatrix}.$$

Theorem 2.7 *Suppose A is an $m \times n$ matrix, B is an $n \times k$ matrix, and b, c are constants. Then*

1. $c \cdot (A \cdot B) = (c \cdot A) \cdot B = A \cdot (c \cdot B)$

2. $(b \cdot c) \cdot A = b \cdot (c \cdot A) = C \cdot (b \cdot A)$.

Example 66 *Suppose*

$$A = \begin{bmatrix} -4 & -3 & 3 \\ -1 & -3 & 2 \end{bmatrix}, B = \begin{bmatrix} 2 & -2 \\ 0 & 2 \\ -5 & 1 \end{bmatrix}.$$

Also $c = 3$. Then,

$$c \cdot (A \cdot B) = \begin{bmatrix} -69 & 15 \\ -36 & -6 \end{bmatrix}$$

and

$$A \cdot (c \cdot B) = \begin{bmatrix} -69 & 15 \\ -36 & -6 \end{bmatrix}.$$

Now, one of the most important concepts is the **inverse** of a matrix. This concept is very important for solving a system of linear equations.

Definition 21 *Suppose A is a square matrix, i.e., an $m \times m$ matrix. Then the **inverse** of a matrix A is an $m \times m$ matrix A^{-1} such that*

$$A \cdot A^{-1} = A^{-1} \cdot A = I_m,$$

*where I_m is the identity matrix of size m. If a matrix A has the inverse, then we say a matrix A is **invertible**.*

Note that not all square matrices have their inverse.

Example 67 *Suppose*

$$A = \begin{bmatrix} -1 & 4 \\ 1 & -3 \end{bmatrix}, A^{-1} = \begin{bmatrix} 3 & 4 \\ 1 & 1 \end{bmatrix}.$$

Then we have

$$A \cdot A^{-1} = A^{-1} \cdot A = \begin{bmatrix} 1 & 0 \\ 0 & 1 \end{bmatrix}.$$

Theorem 2.8 *Suppose A is an invertible matrix. Then*

$$(A^{-1})^{-1} = A.$$

Theorem 2.9 *Suppose we have a 2×2 matrix A such that*

$$A = \begin{bmatrix} a & b \\ c & d \end{bmatrix},$$

where a, b, c, d are real numbers. Then A is invertible if and only if $a \cdot d - b \cdot c \neq 0$. If A is invertible, then the inverse of the matrix A is

$$A^{-1} = \frac{1}{a \cdot d - b \cdot c} \cdot \begin{bmatrix} d & -b \\ -c & a \end{bmatrix}.$$

Example 68 *Suppose*

$$A = \begin{bmatrix} -1 & 4 \\ 1 & -3 \end{bmatrix}.$$

Then, we have

$$(-1) \cdot (-3) - 4 \cdot 1 = -1.$$

Therefore,

$$A^{-1} = \frac{1}{-1} \cdot \begin{bmatrix} -3 & -4 \\ -1 & -1 \end{bmatrix} = \begin{bmatrix} 3 & 4 \\ 1 & 1 \end{bmatrix}.$$

Example 69 *In* R, *if we want to compute the inverse of a matrix we can use the inv() function from the* matlib *package. For example, if we have a matrix*

$$A = \begin{bmatrix} 1 & 2 & 3 \\ -2 & 3 & -2 \\ -1 & 2 & 1 \end{bmatrix},$$

then we can type:

```
library(matlib)
A <- matrix(c(1,-2,-1,2,3,2,3,-2,1), nrow = 3, ncol = 3)
inv(A)
```

Then, R *returns*

```
> inv(A)
            [,1]        [,2]        [,3]
[1,]  0.58333333  0.3333333 -1.0833333
[2,]  0.33333333  0.3333333 -0.3333333
[3,] -0.08333333 -0.3333333  0.5833333
```

If we want to see the output in terms of rational numbers, then we can use the fractions() function from the MASS *package [42].*

```
library(MASS)
fractions(inv(A))
```

Then R *outputs*

```
> fractions(inv(A))
       [,1]   [,2]   [,3]
[1,]   7/12   1/3  -13/12
[2,]    1/3   1/3    -1/3
[3,]  -1/12  -1/3    7/12
```

Remark 2.10 *In* R *we can use the solve() function to find the inverse of a matrix instead of the inv() function. For example, with the matrix*

$$A = \begin{bmatrix} 1 & 2 & 3 \\ -2 & 3 & -2 \\ -1 & 2 & 1 \end{bmatrix},$$

we can do:

```
library(matlib)
A <- matrix(c(1,-2,-1,2,3,2,3,-2,1), nrow = 3, ncol = 3)
solve(A)
```

Then we obtained the same result as the inv() function:

```
> solve(A)
         [,1]         [,2]        [,3]
[1,]  0.58333333   0.3333333  -1.0833333
[2,]  0.33333333   0.3333333  -0.3333333
[3,] -0.08333333  -0.3333333   0.5833333
```

Remark 2.11 *There are some square matrices which do not have inverse matrices, i.e., they are not invertible. This relates to the existence of a unique solution to the system of linear equations.*

Theorem 2.12 *Suppose we have a system of n many linear equations in n variables such that*

$$A \cdot x = b$$

where A is an $n \times n$ matrix, x is an n-dimensional vector, and b is an n-dimensional vector. Then, if A is invertible, this system of linear equations has a unique solution.

Example 70 *This is from Example 39. The system of linear equations is:*

$$
\begin{array}{rcrcrcr}
x_1 & + & 2x_2 & + & 3x_3 & = & 6 \\
-2x_1 & + & 3x_2 & - & 2x_3 & = & -1 \\
-x_1 & + & 2x_2 & + & x_3 & = & 2.
\end{array}
$$

Its coefficient matrix is

$$
A = \begin{bmatrix} 1 & 2 & 3 \\ -2 & 3 & -2 \\ -1 & 2 & 1 \end{bmatrix}
$$

and its inverse is

$$
A^{-1} = \begin{bmatrix} 7/12 & 1/3 & -13/12 \\ 1/3 & 1/3 & -1/3 \\ -1/12 & -1/3 & 7/12 \end{bmatrix}.
$$

Let

$$
b = \begin{bmatrix} 6 \\ -1 \\ 2 \end{bmatrix}.
$$

Then the solution of this system of linear equations is

$$
A^{-1} \cdot b = \begin{bmatrix} 1 \\ 1 \\ 1 \end{bmatrix}.
$$

Now we discuss properties of matrix operations with the transpose of a matrix.

Theorem 2.13 *Suppose we have a matrix A. Then*

$$(A^T)^T = A.$$

Example 71 *Suppose we have a matrix*

$$A = \begin{bmatrix} -4 & -3 & 3 \\ -1 & -3 & 2 \end{bmatrix}.$$

Then the transpose of A is

$$A^T = \begin{bmatrix} -4 & -1 \\ -3 & -3 \\ 3 & 2 \end{bmatrix}.$$

Then the transpose of A^T is

$$(A^T)^T = \begin{bmatrix} -4 & -3 & 3 \\ -1 & -3 & 2 \end{bmatrix}.$$

Theorem 2.14 *Suppose A and B are $m \times n$ matrices, Then*

1. $(A + B)^T = A^T + B^T$

2. $(A - B)^T = A^T - B^T.$

Example 72 *Suppose we have*

$$A = \begin{bmatrix} -4 & -3 & 3 \\ -1 & -3 & 2 \end{bmatrix}, B = \begin{bmatrix} 4 & -1 & 1 \\ 2 & -4 & 5 \end{bmatrix}.$$

Then we have

$$A + B = \begin{bmatrix} 0 & -4 & 4 \\ 1 & -7 & 7 \end{bmatrix}$$

$$(A + B)^T = \begin{bmatrix} 0 & 1 \\ -4 & -7 \\ 4 & 7 \end{bmatrix}.$$

Also

$$A^T = \begin{bmatrix} -4 & -1 \\ -3 & -3 \\ 3 & 2 \end{bmatrix}$$

and

$$B^T = \begin{bmatrix} 4 & 2 \\ -1 & -4 \\ 1 & 5 \end{bmatrix}.$$

Then we have

$$A^T + B^T = \begin{bmatrix} 0 & 1 \\ -4 & -7 \\ 4 & 7 \end{bmatrix}.$$

Theorem 2.15 *Suppose A is an $m \times n$ matrix and B is an $n \times k$ matrix, and c is a constant. Then*

1. $(c \cdot A)^T = c \cdot A^T$

2. $(A \cdot B)^T = B^T \cdot A^T$.

Example 73 *Suppose*

$$A = \begin{bmatrix} -4 & -3 & 3 \\ -1 & -3 & 2 \end{bmatrix}, B = \begin{bmatrix} 2 & -2 \\ 0 & 2 \\ -5 & 1 \end{bmatrix}.$$

Also let $c = -1$. Then

$$(-1) \cdot A = \begin{bmatrix} 4 & 3 & -3 \\ 1 & 3 & -2 \end{bmatrix}$$

and then

$$((-1) \cdot A)^T = \begin{bmatrix} 4 & 1 \\ 3 & 3 \\ -3 & -2 \end{bmatrix}.$$

Also we have

$$A^T = \begin{bmatrix} -4 & -1 \\ -3 & -3 \\ 3 & 2 \end{bmatrix}$$

and

$$(-1) \cdot A^T = \begin{bmatrix} 4 & 1 \\ 3 & 3 \\ -3 & -2 \end{bmatrix}.$$

Also we have

$$(A \cdot B) = \begin{bmatrix} -23 & 5 \\ -12 & -2 \end{bmatrix}$$

and

$$(A \cdot B)^T = \begin{bmatrix} -23 & -12 \\ 5 & -2 \end{bmatrix}.$$

$$A^T = \begin{bmatrix} -4 & -1 \\ -3 & -3 \\ 3 & 2 \end{bmatrix}$$

and

$$B^T = \begin{bmatrix} 2 & 0 & 5 \\ -2 & 2 & 1 \end{bmatrix}.$$

Thus

$$B^T \cdot A^T = \begin{bmatrix} -23 & -12 \\ 5 & -2 \end{bmatrix}.$$

Theorem 2.16 *Suppose A is a square invertible matrix. Then*

$$(A^T)^{-1} = (A^{-1})^T.$$

Example 74 *Suppose we have*

$$A = \begin{bmatrix} -1 & 4 \\ 1 & -3 \end{bmatrix}.$$

Also we have

$$A^{-1} = \begin{bmatrix} 3 & 4 \\ 1 & 1 \end{bmatrix}$$

and

$$(A^{-1})^T = \begin{bmatrix} 3 & 1 \\ 4 & 1 \end{bmatrix}.$$

Also we have

$$A^T = \begin{bmatrix} -1 & 1 \\ 4 & -3 \end{bmatrix}$$

and

$$(A^T)^{-1} = \begin{bmatrix} 3 & 1 \\ 4 & 1 \end{bmatrix}.$$

Theorem 2.17 *If A is a square and symmetric matrix, then*

$$A^T = A.$$

Example 75 *Suppose we have*

$$A = \begin{bmatrix} 3 & 1 & 4 \\ 1 & 5 & 0 \\ 4 & 0 & -2 \end{bmatrix}.$$

Then

$$A^T = \begin{bmatrix} 3 & 1 & 4 \\ 1 & 5 & 0 \\ 4 & 0 & -2 \end{bmatrix}.$$

Thus $A^T = A$.

2.3.4   Checkmarks

- The definition of the inverse of a matrix.

- The properties of matrix operations.

- You can compute the inverse of a matrix.

- You can apply properties of matrix operations to a matrix computation.

- With R, you can perform compute the inverse of a matrix.

- With R, you can apply properties of matrix operations to a matrix computation.

2.3.5   Conceptual Quizzes

Quiz 61 True or False: *Suppose A is an invertible matrix. Then $(A^{-1})^{-1} = A$.*

Quiz 62 True or False: *Suppose A is an $m \times n$ matrix. Then $(A^T)^T = A$.*

Quiz 63 True or False: *Suppose we have $m \times m$ matrices A, B. Then $A \cdot B = B \cdot A$.*

Quiz 64 True or False: *Suppose we have $m \times m$ matrices A, B. Then $A + B = B + A$.*

Quiz 65 True or False: *Suppose we have $m \times m$ matrices A, B, C. Then $A \cdot (B \cdot C) = C \cdot (B \cdot A)$.*

Quiz 66 True or False: *Suppose we have $m \times m$ matrices A, B, C. Then $(A + B) + C = A + (B + C)$.*

Quiz 67 True or False: *Suppose we have $m \times m$ matrices A, B, C. Then $A + (B + C) = C + (B + A)$.*

Quiz 68 True or False: *Suppose we have $m \times m$ matrices A, B, C. Then $A \cdot (B + C) = (B + C) \cdot A$.*

Quiz 69 True or False: *Suppose we have $m \times m$ matrices A, B, C. Then $A \cdot (B + C) = A \cdot C + A \cdot B$.*

Quiz 70 True or False: *Suppose we have $m \times m$ matrices A, B, C and a is a constant. Then $a \cdot A \cdot (B + C) = A \cdot (a \cdot B + a \cdot C)$.*

Quiz 71 True or False: *Suppose we have $m \times m$ matrices A, B, C and a is a constant. Then $a \cdot A \cdot (B \cdot C) = A \cdot (a \cdot B \cdot a \cdot C)$.*

Quiz 72 True or False: *Suppose we have $m \times m$ invertible matrices A, B, C. Then $A^{-1} \cdot A \cdot (B + C) = (B + C) \cdot A \cdot A^{-1}$.*

Quiz 73 True or False: *Suppose we have $m \times m$ invertible matrices A, B, C. Then $(A^{-1} + A) \cdot (B + C) = (B + C) \cdot (A + A^{-1})$.*

Quiz 74 True or False: *Suppose we have $m \times m$ matrices A, B, C. Then $A^T \cdot A \cdot (B + C) = (B + C) \cdot A \cdot A^T$.*

Quiz 75 True or False: *Suppose we have $m \times m$ matrices A, B, C. Then $(A^T + A) \cdot (B + C) = (B + C) \cdot (A + A^T)$.*

Quiz 76 True or False: *Suppose we have $m \times m$ invertible matrices A, B, C. Then $A^{-1} \cdot A \cdot (B + C) = (B + C) \cdot A \cdot A^{-1}$.*

Quiz 77 True or False: *Suppose we have $m \times m$ invertible matrices A, B, C. Then $(A \cdot B \cdot C)^{-1} = (C^{-1} \cdot B^{-1} \cdot A^{-1})$.*

Quiz 78 True or False: *Suppose we have $m \times m$ matrices A, B, C. Then $(A \cdot B \cdot C)^T = (C^T \cdot B^T \cdot A^T)$.*

2.3.6 Regular Exercises

Exercise 2.10 *Suppose we have matrices*

$$A = \begin{bmatrix} -5 & -1 & -3 \\ 4 & -5 & 5 \\ -3 & 5 & -2 \end{bmatrix}, B = \begin{bmatrix} -5 & -2 & 5 \\ -2 & -4 & -2 \\ -2 & -3 & 4 \end{bmatrix}.$$

Using these matrices, verify Theorem 2.2.

Exercise 2.11 *Suppose we have matrices*

$$A = \begin{bmatrix} -4 & 5 & 2 \\ -5 & 0 & -1 \\ 5 & -5 & 2 \end{bmatrix}, B = \begin{bmatrix} -2 & -1 & 4 \\ 2 & 5 & -5 \\ 0 & -3 & 5 \end{bmatrix}, C = \begin{bmatrix} 1 & -2 & -2 \\ -5 & -4 & 4 \\ 4 & 2 & 3 \end{bmatrix}$$

and $a = 2$, $b = -1$, $c = 3$. Using these matrices, verify Theorem 2.3.

Exercise 2.12 *Suppose we have matrices*

$$A = \begin{bmatrix} -3 & -5 & -3 \\ 4 & 0 & -5 \\ -2 & -5 & 5 \end{bmatrix}, B = \begin{bmatrix} -1 & 1 & -5 \\ 0 & -5 & 1 \\ -1 & -2 & -2 \end{bmatrix}, C = \begin{bmatrix} -5 & 3 & 0 \\ 2 & 1 & 5 \\ -3 & -5 & 0 \end{bmatrix}.$$

Using these matrices, verify Theorem 2.4.

Exercise 2.13 *Suppose we have matrices*

$$A = \begin{bmatrix} -2 & -3 & 3 \\ 1 & 0 & 2 \\ -2 & 3 & -3 \end{bmatrix}, B = \begin{bmatrix} -2 & 4 & 0 \\ 4 & -4 & 1 \\ 1 & -5 & 1 \end{bmatrix}, C = \begin{bmatrix} 4 & 0 & 1 \\ -5 & 0 & 3 \\ -1 & 4 & 5 \end{bmatrix}.$$

Using these matrices, verify Theorem 2.5.

Exercise 2.14 *Suppose we have matrices*

$$A = \begin{bmatrix} 3 & -2 & -1 \\ 2 & -5 & 2 \\ -3 & 2 & 5 \end{bmatrix}, B = \begin{bmatrix} -5 & 0 & 5 \\ -5 & -5 & -4 \\ -3 & 3 & -1 \end{bmatrix}, C = \begin{bmatrix} 2 & -4 & -2 \\ -1 & -5 & -1 \\ 1 & 5 & -2 \end{bmatrix}$$

Using these matrices, verify Theorem 2.6.

Exercise 2.15 *Suppose we have matrices*

$$A = \begin{bmatrix} 2 & -3 & 2 \\ 4 & 1 & -4 \\ -1 & 4 & -4 \end{bmatrix}, B = \begin{bmatrix} 0 & 5 & -5 \\ 5 & 2 & -4 \\ -5 & 0 & 3 \end{bmatrix}$$

and $b = 4$, $c = 2$. Using these matrices, verify Theorem 2.7.

Exercise 2.16 *Suppose we have matrices*

$$A = \begin{bmatrix} 2 & -3 & -2 \\ -3 & 5 & -2 \\ 0 & -2 & 4 \end{bmatrix}, B = \begin{bmatrix} -4 & -1 & 4 \\ 0 & -1 & 5 \\ 5 & 2 & 3 \end{bmatrix}.$$

Using these matrices, verify Theorem 2.14.

Exercise 2.17 *Suppose we have matrices*

$$A = \begin{bmatrix} 4 & -5 & -1 \\ -4 & 5 & 3 \\ 0 & -5 & -5 \end{bmatrix}, B = \begin{bmatrix} -1 & 1 & 1 \\ 1 & -1 & 0 \\ 3 & 2 & 3 \end{bmatrix}$$

and c = 3. Using these matrices, verify Theorem 2.15.

Exercise 2.18 *Suppose we have*

$$A = \begin{bmatrix} 3 & 0 \\ -3 & 3 \end{bmatrix}.$$

Compute the inverse of A if one exists.

Exercise 2.19 *Suppose we have*

$$A = \begin{bmatrix} -5 & -4 \\ 1 & -5 \end{bmatrix}.$$

Compute the inverse of A if one exists.

Exercise 2.20 *Suppose we have*

$$A = \begin{bmatrix} 3 & 4 \\ -2 & 5 \end{bmatrix}.$$

Compute the inverse of A if one exists.

Exercise 2.21 *Suppose we have a matrix*

$$A = \begin{bmatrix} 3 & 2 \\ -2 & 1 \end{bmatrix}.$$

Using this matrix, verify Theorem 2.16.

Exercise 2.22 *Suppose we have a matrix*

$$A = \begin{bmatrix} 3 & 2 & 0 \\ 2 & -1 & 1 \\ 0 & 1 & 5 \end{bmatrix}.$$

Using this matrix, verify Theorem 2.17.

Exercise 2.23 *Suppose we have a $n \times n$ matrix A such that*

$$A = \begin{bmatrix} a_{11} & 0 & \cdots & 0 & 0 \\ 0 & a_{22} & \cdots & 0 & 0 \\ \vdots & \vdots & \ddots & \vdots & \vdots \\ 0 & 0 & \cdots & a_{n-1n-1} & 0 \\ 0 & 0 & \cdots & 0 & a_{nn} \end{bmatrix},$$

where a_{ij} are not equal to 0 for all $i = 1, \ldots n$ and $j = 1, \ldots n$. For example, if $n = 3$, we can have

$$A = \begin{bmatrix} 3 & 0 & 0 \\ 0 & -1 & 0 \\ 0 & 0 & 5 \end{bmatrix}.$$

Is A an invertible matrix? If so compute A^{-1}.

Exercise 2.24 *Construct examples for the following cases:*

1. *A 4×4 matrix A such that*
$$A^T = A.$$

2. *A 4×4 matrix A such that*

$$A^{-1} = A.$$

3. *A 4×4 matrix A with all non-zero entries which is not invertible.*

4. *4×4 matrices B and C such that*

$$B \cdot C = C \cdot B.$$

5. *4×4 matrices B, C, D such that*

$$(B \cdot C) \cdot D = D \cdot (C \cdot B).$$

2.3.7 Lab Exercises

Lab Exercise 47 *Verify Theorem 2.2 with* R.

Lab Exercise 48 *Verify Theorem 2.3 with* R.

Lab Exercise 49 *Verify Theorem 2.4 with* R.

Lab Exercise 50 *Verify Theorem 2.5 with* R.

Lab Exercise 51 *Verify Theorem 2.6 with* R.

Lab Exercise 52 *Verify Theorem 2.7 with* R.

Lab Exercise 53 *Verify Theorem 2.14 with* R.

Lab Exercise 54 *Verify Theorem 2.15 with* R.

Lab Exercise 55 *With the function inv() from the* matlib *package, compute the inverse of the matrix*

$$A = \begin{bmatrix} -4 & 2 & 4 \\ 2 & 4 & 1 \\ 3 & -4 & -5 \end{bmatrix}.$$

Lab Exercise 56 *With the function inv() from the* matlib *package, compute the inverse of the matrix*

$$A = \begin{bmatrix} 2 & -5 & 1 & -2 \\ 2 & 5 & 5 & -3 \\ 3 & -4 & -2 & 5 \\ 3 & -2 & -1 & 3 \end{bmatrix}.$$

Lab Exercise 57 *With the function inv() from the* matlib *package, compute the inverse of the matrix*

$$A = \begin{bmatrix} -5 & 5 & 0 & -1 & 4 \\ -2 & 5 & -3 & -5 & 5 \\ 2 & 3 & -1 & -3 & -2 \\ 1 & 3 & 2 & -2 & -3 \\ -1 & -4 & -1 & -3 & -4 \end{bmatrix}.$$

2.3.8   Practical Applications

In order to apply the inverse of the coefficient matrix for a system of linear equations, the coefficient matrix has to be square and invertible. Note that the coefficient matrix for this system of linear equations is not square. So first let's make it a square matrix. Recall that we have a system of linear equations such that

$$
\begin{array}{rclcrcl}
x_{11} & + & x_{12} & +\ldots+ & x_{19} & = & 1298 \\
x_{21} & + & x_{22} & +\ldots+ & x_{29} & = & 1948 \\
& & x_{11} & + & x_{21} & = & 465 \\
& & x_{12} & + & x_{22} & = & 605 \\
& & x_{13} & + & x_{23} & = & 451 \\
& & x_{14} & + & x_{24} & = & 338 \\
& & x_{15} & + & x_{25} & = & 260 \\
& & x_{16} & + & x_{26} & = & 183 \\
& & x_{17} & + & x_{27} & = & 282 \\
& & x_{18} & + & x_{28} & = & 127 \\
& & x_{19} & + & x_{29} & = & 535.
\end{array}
$$

The coefficient matrix for this system is

$$
\begin{bmatrix}
1 & 1 & 1 & 1 & 1 & 1 & 1 & 1 & 1 & 0 & 0 & 0 & 0 & 0 & 0 & 0 & 0 & 0 \\
0 & 0 & 0 & 0 & 0 & 0 & 0 & 0 & 0 & 1 & 1 & 1 & 1 & 1 & 1 & 1 & 1 & 1 \\
1 & 0 & 0 & 0 & 0 & 0 & 0 & 0 & 0 & 1 & 0 & 0 & 0 & 0 & 0 & 0 & 0 & 0 \\
0 & 1 & 0 & 0 & 0 & 0 & 0 & 0 & 0 & 0 & 1 & 0 & 0 & 0 & 0 & 0 & 0 & 0 \\
0 & 0 & 1 & 0 & 0 & 0 & 0 & 0 & 0 & 0 & 0 & 1 & 0 & 0 & 0 & 0 & 0 & 0 \\
0 & 0 & 0 & 1 & 0 & 0 & 0 & 0 & 0 & 0 & 0 & 0 & 1 & 0 & 0 & 0 & 0 & 0 \\
0 & 0 & 0 & 0 & 1 & 0 & 0 & 0 & 0 & 0 & 0 & 0 & 0 & 1 & 0 & 0 & 0 & 0 \\
0 & 0 & 0 & 0 & 0 & 1 & 0 & 0 & 0 & 0 & 0 & 0 & 0 & 0 & 1 & 0 & 0 & 0 \\
0 & 0 & 0 & 0 & 0 & 0 & 1 & 0 & 0 & 0 & 0 & 0 & 0 & 0 & 0 & 1 & 0 & 0 \\
0 & 0 & 0 & 0 & 0 & 0 & 0 & 1 & 0 & 0 & 0 & 0 & 0 & 0 & 0 & 0 & 1 & 0 \\
0 & 0 & 0 & 0 & 0 & 0 & 0 & 0 & 1 & 0 & 0 & 0 & 0 & 0 & 0 & 0 & 0 & 1
\end{bmatrix},
$$

which is an 11×18 matrix and the vector for the right hand side of the system of linear equations is

$$
\begin{bmatrix}
1298 \\
1948 \\
465 \\
605 \\
451 \\
338 \\
260 \\
183 \\
282 \\
127 \\
535
\end{bmatrix},
$$

which is a 1×11 matrix.

Since the coefficient matrix is an 11×18 matrix, we need 7 more equations to make the coefficient matrix square. However, this is not enough in this case: if we compute the reduced echelon form of the coefficient matrix using the rref() function from the **pracma** package [7], we get

```
> library(pracma)
> rref(A)
       [,1] [,2] [,3] [,4] [,5] [,6] [,7] [,8] [,9] [,10] [,11] [,12] [,13]
 [1,]    1    0    0    0    0    0    0    0    0     0    -1    -1    -1
 [2,]    0    1    0    0    0    0    0    0    0     0     1     0     0
 [3,]    0    0    1    0    0    0    0    0    0     0     0     1     0
 [4,]    0    0    0    1    0    0    0    0    0     0     0     0     1
 [5,]    0    0    0    0    1    0    0    0    0     0     0     0     0
 [6,]    0    0    0    0    0    1    0    0    0     0     0     0     0
 [7,]    0    0    0    0    0    0    1    0    0     0     0     0     0
 [8,]    0    0    0    0    0    0    0    1    0     0     0     0     0
 [9,]    0    0    0    0    0    0    0    0    1     0     0     0     0
[10,]    0    0    0    0    0    0    0    0    0     1     1     1     1
[11,]    0    0    0    0    0    0    0    0    0     0     0     0     0
```

	[,14]	[,15]	[,16]	[,17]	[,18]
[1,]	-1	-1	-1	-1	-1
[2,]	0	0	0	0	0
[3,]	0	0	0	0	0
[4,]	0	0	0	0	0
[5,]	1	0	0	0	0
[6,]	0	1	0	0	0
[7,]	0	0	1	0	0
[8,]	0	0	0	1	0
[9,]	0	0	0	0	1
[10,]	1	1	1	1	1
[11,]	0	0	0	0	0

The last row of the reduced echelon form of the coefficient matrix is all zeros. This means that we need 8 additional new equations to have a unique solution for the system of linear equations.

In the practical applications in Sec 1.4, we computed the reduced echelon form using the Solve() function. However, this is not easy to use for this practical application, so instead we use the rref() function of the augmented matrix of the system of linear equations.

Using the cbind() function which combines the matrix A and a vector b by margin in terms of columns (see Appendix), we have

```
C <- cbind(A, b)
```

R outputs

```
> C
                                          b
[1,]  1 1 1 1 1 1 1 1 1 0 0 0 0 0 0 0 0 0 1298
[2,]  0 0 0 0 0 0 0 0 0 1 1 1 1 1 1 1 1 1 1948
[3,]  1 0 0 0 0 0 0 0 0 1 0 0 0 0 0 0 0 0  465
[4,]  0 1 0 0 0 0 0 0 0 0 1 0 0 0 0 0 0 0  605
[5,]  0 0 1 0 0 0 0 0 0 0 0 1 0 0 0 0 0 0  451
[6,]  0 0 0 1 0 0 0 0 0 0 0 0 1 0 0 0 0 0  338
[7,]  0 0 0 0 1 0 0 0 0 0 0 0 0 1 0 0 0 0  260
[8,]  0 0 0 0 0 1 0 0 0 0 0 0 0 0 1 0 0 0  183
[9,]  0 0 0 0 0 0 1 0 0 0 0 0 0 0 0 1 0 0  282
[10,] 0 0 0 0 0 0 0 1 0 0 0 0 0 0 0 0 1 0  127
[11,] 0 0 0 0 0 0 0 0 1 0 0 0 0 0 0 0 0 1  535
```

You can ignore the top "b" in the output. Then we create a new augmented matrix for this system:

```
E <- rref(C)
```

R outputs

```
> E
```

																			b
[1,]	1	0	0	0	0	0	0	0	0	0	-1	-1	-1	-1	-1	-1	-1	-1	-1483
[2,]	0	1	0	0	0	0	0	0	0	0	1	0	0	0	0	0	0	0	605
[3,]	0	0	1	0	0	0	0	0	0	0	0	1	0	0	0	0	0	0	451
[4,]	0	0	0	1	0	0	0	0	0	0	0	0	1	0	0	0	0	0	338
[5,]	0	0	0	0	1	0	0	0	0	0	0	0	0	1	0	0	0	0	260
[6,]	0	0	0	0	0	1	0	0	0	0	0	0	0	0	1	0	0	0	183
[7,]	0	0	0	0	0	0	1	0	0	0	0	0	0	0	0	1	0	0	282
[8,]	0	0	0	0	0	0	0	1	0	0	0	0	0	0	0	0	1	0	127
[9,]	0	0	0	0	0	0	0	0	1	0	0	0	0	0	0	0	0	1	535
[10,]	0	0	0	0	0	0	0	0	0	1	1	1	1	1	1	1	1	1	1948
[11,]	0	0	0	0	0	0	0	0	0	0	0	0	0	0	0	0	0	0	0

Since the last row of this new augmented matrix has all zeros, we delete the last row by "-" in the index as follows:

```
E <- E[-11,]
```

Now the new matrix "E" is the augmented matrix for this system of linear equations:

```
> E
```

																			b
[1,]	1	0	0	0	0	0	0	0	0	0	-1	-1	-1	-1	-1	-1	-1	-1	-1483
[2,]	0	1	0	0	0	0	0	0	0	0	1	0	0	0	0	0	0	0	605
[3,]	0	0	1	0	0	0	0	0	0	0	0	1	0	0	0	0	0	0	451
[4,]	0	0	0	1	0	0	0	0	0	0	0	0	1	0	0	0	0	0	338
[5,]	0	0	0	0	1	0	0	0	0	0	0	0	0	1	0	0	0	0	260
[6,]	0	0	0	0	0	1	0	0	0	0	0	0	0	0	1	0	0	0	183
[7,]	0	0	0	0	0	0	1	0	0	0	0	0	0	0	0	1	0	0	282
[8,]	0	0	0	0	0	0	0	1	0	0	0	0	0	0	0	0	1	0	127
[9,]	0	0	0	0	0	0	0	0	1	0	0	0	0	0	0	0	0	1	535
[10,]	0	0	0	0	0	0	0	0	0	1	1	1	1	1	1	1	1	1	1948

Now, how can we add 8 more equations? Recall that this system of linear equations comes from Table 1.3. This problem can be formulated as a contingency table, where x_{ij} is the number of beers transferred from a production site i to a distributor j. For distributors, we assign numbers for simplicity: let $1 = $ FTA, $2 = $ RICKY, $3 = $ OBIBAJK, $4 = $ KADOM, $5 = $ NAATO, $6 = $ LESK, $7 = $ DCEE, $8 = $ JOEMAN, $9 = $ KBOA. Also for the production sites, let $1 = $ Achimota, $2 = $ Kaasi.

Then we can formulate a contingency table as

Product/Dist	1	2	3	4	5	6	7	8	9	Total
1	x_{11}	x_{12}	x_{13}	x_{14}	x_{15}	x_{16}	x_{17}	x_{18}	x_{19}	1298
2	x_{11}	x_{12}	x_{13}	x_{14}	x_{15}	x_{16}	x_{17}	x_{18}	x_{19}	1948
Total	465	605	451	338	260	183	282	127	535	3246

Then if we fill $x_{11}, x_{12}, \ldots, x_{18}$ for some numbers, the rest of the variables can be obtained from the difference with the total number of beers.

So suppose we have

$$x_{11} = 266, x_{12} = 223, x_{13} = 140, x_{14} = 264, x_{15} = 137, x_{16} = 67,$$
$$x_{17} = 130, x_{18} = 24.$$

These values are assigned randomly using a method called *sequential importance sampling*. Then we add these equations to the new system of linear equations computed from the augmented matrix. Note that the coefficient matrix for only these 8 equations is the identity matrix of size 8. The eye() function created the identity matrix. All we have to do is to add the row vectors to the augmented matrix for the original system of linear equations:

```
G1 <- eye(8)
G2 <- matrix(rep(0, 80), 8, 10)
b2 <- c(266, 223, 140, 264, 137, 67, 130, 24)
G <- cbind(G1, G2, b2)
M <- rbind(E, G)
```

Then you will see the augmented matrix of the new system of linear equations is

```
> M
                                                         b
 [1,] 1 0 0 0 0 0 0 0 0 0 -1 -1 -1 -1 -1 -1 -1 -1 -1483
 [2,] 0 1 0 0 0 0 0 0 0 0  1  0  0  0  0  0  0  0   605
 [3,] 0 0 1 0 0 0 0 0 0 0  0  1  0  0  0  0  0  0   451
 [4,] 0 0 0 1 0 0 0 0 0 0  0  0  1  0  0  0  0  0   338
 [5,] 0 0 0 0 1 0 0 0 0 0  0  0  0  1  0  0  0  0   260
 [6,] 0 0 0 0 0 1 0 0 0 0  0  0  0  0  1  0  0  0   183
 [7,] 0 0 0 0 0 0 1 0 0 0  0  0  0  0  0  1  0  0   282
 [8,] 0 0 0 0 0 0 0 1 0 0  0  0  0  0  0  0  1  0   127
 [9,] 0 0 0 0 0 0 0 0 1 0  0  0  0  0  0  0  0  1   535
[10,] 0 0 0 0 0 0 0 0 0 1  1  1  1  1  1  1  1  1  1948
[11,] 1 0 0 0 0 0 0 0 0 0  0  0  0  0  0  0  0  0   266
[12,] 0 1 0 0 0 0 0 0 0 0  0  0  0  0  0  0  0  0   223
[13,] 0 0 1 0 0 0 0 0 0 0  0  0  0  0  0  0  0  0   140
[14,] 0 0 0 1 0 0 0 0 0 0  0  0  0  0  0  0  0  0   264
[15,] 0 0 0 0 1 0 0 0 0 0  0  0  0  0  0  0  0  0   137
[16,] 0 0 0 0 0 1 0 0 0 0  0  0  0  0  0  0  0  0    67
[17,] 0 0 0 0 0 0 1 0 0 0  0  0  0  0  0  0  0  0   130
[18,] 0 0 0 0 0 0 0 1 0 0  0  0  0  0  0  0  0  0    24
```

Now, we check if the coefficient matrix of this new system is invertible and if so, we compute its inverse. We can remove the last column of the augmented matrix by the command M[, -19]. Then we use the inv() function to see if it is invertible.

```
> inv(M[,-19])
     [,1] [,2] [,3] [,4] [,5] [,6] [,7] [,8] [,9] [,10] [,11] [,12] [,13]
       0    0    0    0    0    0    0    0    0    0     1     0     0
       0    0    0    0    0    0    0    0    0    0     0     1     0
       0    0    0    0    0    0    0    0    0    0     0     0     1
       0    0    0    0    0    0    0    0    0    0     0     0     0
       0    0    0    0    0    0    0    0    0    0     0     0     0
       0    0    0    0    0    0    0    0    0    0     0     0     0
       0    0    0    0    0    0    0    0    0    0     0     0     0
       0    0    0    0    0    0    0    0    0    0     0     0     0
       1    1    1    1    1    1    1    1    1    0    -1    -1    -1
       1    0    0    0    0    0    0    0    0    1    -1     0     0
       0    1    0    0    0    0    0    0    0    0     0    -1     0
       0    0    1    0    0    0    0    0    0    0     0     0    -1
       0    0    0    1    0    0    0    0    0    0     0     0     0
       0    0    0    0    1    0    0    0    0    0     0     0     0
       0    0    0    0    0    1    0    0    0    0     0     0     0
       0    0    0    0    0    0    1    0    0    0     0     0     0
       0    0    0    0    0    0    0    1    0    0     0     0     0
      -1   -1   -1   -1   -1   -1   -1   -1    0    0     1     1     1
     [,14]  [,15]  [,16]  [,17]  [,18]
       0      0      0      0      0
       0      0      0      0      0
       0      0      0      0      0
       1      0      0      0      0
       0      1      0      0      0
       0      0      1      0      0
       0      0      0      1      0
       0      0      0      0      1
      -1     -1     -1     -1     -1
       0      0      0      0      0
       0      0      0      0      0
       0      0      0      0      0
      -1      0      0      0      0
       0     -1      0      0      0
       0      0     -1      0      0
       0      0      0     -1      0
       0      0      0      0     -1
       1      1      1      1      1
```

It seems that it is. Recall that the unique solution can be obtained by multiplying the inverse of the coefficient matrix by the vector for the right hand side. In this case, the right hand side can be obtained by the command M[, 19] since it is the 19th column of the matrix M. So now we use the inverse of the coefficient matrix to solve the system of linear equations as

```
inv(M[, -19]) %*% M[, 19]
```

Then, R outputs the following:

```
> inv(M[, -19]) %*% M[, 19]
   [,1]
  266
  223
  140
  264
  137
   67
  130
   24
   47
  199
  382
  311
   74
  123
  116
  152
  103
  488
```

2.3.9   Supplements with python Code

In python, we use the numpy package for matrix operations. For computing the inverse of a matrix, we can use the inv() function from numpy. First, we need to call the package and import the inv() function:

```
import numpy as np
from numpy.linalg import inv
```

Suppose we have a 3×3 matrix

$$A = \begin{bmatrix} 3 & 0 & -5 \\ -1 & -3 & 4 \\ -5 & 5 & 2 \end{bmatrix}.$$

We first define the matrix in python:

```
A = np.matrix([[3, 0, -5], [-1, -3, 4], [-5, 5, 2]])
```

Then call the inv() function as:

```
inv(A)
```

Then python outputs:

```
>>> inv(A)
matrix([[-1.18181818, -1.13636364, -0.68181818],
        [-0.81818182, -0.86363636, -0.31818182],
        [-0.90909091, -0.68181818, -0.40909091]])
```

2.4   Elementary Matrices

Elementary row operations are important for computing the reduced echelon form of a matrix and relate to matrix operations on **elementary matrices**. In this section we discuss elementary matrices and how to obtain the inverse of a matrix using elementary row operations.

2.4.1   Task Completion Checklist

• During the Lecture:

  1. Read definition of an elementary matrix.

  2. Learn how to compute the inverse of a matrix using a sequence of elementary row operations.

  3. Try some examples without the help of a computer.

  4. With R, learn how to perform computational examples in this section.

• After the Lecture:

  1. Take conceptual quizzes to make sure you understand the materials in this section.

  2. Do some regular exercises.

  3. Conduct lab exercises with R.

  4. Conduct practical applications with R.

  5. If you are interested in **python**, read the supplement in this section and conduct lab exercises and practical applications with **python**.

2.4.2   Working Examples

We will use the same example from Section 1.4, the Guinness data set [5]. Recall that this data set was collected from Guinness Ghana Ltd. and contains information on the supply of Malta Guinness from two production sites, Kaasi and Achimota, to nine key distributors geographically scattered in the regions of Ghana. Table 1.3 shows the demands from the distributors and supplies from the production sites. In the practical application in Section 2.3 we assigned numbers: $1 = \text{FTA}, 2 = \text{RICKY}, 3 = \text{OBIBAJK}, 4 = \text{KADOM}, 5 = \text{NAATO}, 6 = \text{LESK}, 7 = \text{DCEE}, 8 = \text{JOEMAN}, 9 = \text{KBOA}$. Also, for the production sites, let $1 = \text{Achimota}$ and $2 = \text{Kaasi}$. We formulated as a contingency table:

Product/Dist	1	2	3	4	5	6	7	8	9	Total
1	x_{11}	x_{12}	x_{13}	x_{14}	x_{15}	x_{16}	x_{17}	x_{18}	x_{19}	1298
2	x_{11}	x_{12}	x_{13}	x_{14}	x_{15}	x_{16}	x_{17}	x_{18}	x_{19}	1948
Total	465	605	451	338	260	183	282	127	535	3246

Then we filled the numbers $x_{11}, x_{12}, \ldots, x_{18}$

$$x_{11} = 266, x_{12} = 223, x_{13} = 140, x_{14} = 264, x_{15} = 137,$$
$$x_{16} = 67, x_{17} = 130, x_{18} = 24.$$

Now we will learn how to obtain the inverse of the coefficient matrix for this problem via elementary row operations.

2.4.3   Elementary Matrices

In this section we define an elementary matrix which relates to one elementary row operation. Then we discuss how we can obtain the inverse of an inversible matrix using a sequence of elementary row operations. First we define an elementary matrix.

Definition 22 *An **elementary matrix** is an $n \times n$ square matrix obtained from the identity matrix of size n operated by one elementary row operation.*

Example 76 *Here are some examples:*

$$\begin{bmatrix} 2 & 0 & 0 & 0 \\ 0 & 1 & 0 & 0 \\ 0 & 0 & 1 & 0 \\ 0 & 0 & 0 & 1 \end{bmatrix}$$ *Multiply the first row by 2.*

$$\begin{bmatrix} 1 & 0 & 0 & 0 \\ 0 & 0 & 1 & 0 \\ 0 & 1 & 0 & 0 \\ 0 & 0 & 0 & 1 \end{bmatrix}$$ *Exchange the second row and third row.*

$$\begin{bmatrix} 1 & 0 & 0 & 0 \\ 0 & 1 & 0 & -2 \\ 0 & 0 & 1 & 0 \\ 0 & 0 & 0 & 1 \end{bmatrix}$$ *Add (-2) times the fourth row to the second row.*

An elementary row operation on a matrix is in fact equivalent to the matrix multiplication of an elementary matrix on the matrix from the right side. We use the example from Section 1.4 to demonstrate how it works.

Suppose we have the following system of 3 linear equations with the variables x_1, x_2, x_3:

$$
\begin{array}{rcrcrcr}
x_1 & - & x_2 & + & 4x_3 & = & 1 \\
2x_1 & & & - & x_3 & = & -1.5 \\
-x_1 & + & x_2 & & & = & 2.
\end{array}
$$

Recall the augmented matrix of this system is

$$
\begin{bmatrix}
1 & -1 & 4 & 1 \\
2 & 0 & -1 & -1.5 \\
-1 & 1 & 0 & 2
\end{bmatrix}.
$$

Here we will go through step-by-step how an elementary row operation is a matrix multiplication of an elementary matrix on the matrix.

1. We add the first equation and the third equation. This new equation becomes a new third equation:

$$
\begin{array}{rcrcrcr}
x_1 & - & x_2 & + & 4x_3 & = & 1 \\
-x_1 & + & x_2 & & & = & 2 \\
\hline
& & & & 4x_3 & = & 3
\end{array}
$$

This is equivalent to

$$
\begin{bmatrix}
1 & 0 & 0 \\
0 & 1 & 0 \\
1 & 0 & 1
\end{bmatrix}
\cdot
\begin{bmatrix}
1 & -1 & 4 & 1 \\
2 & 0 & -1 & -1.5 \\
-1 & 1 & 0 & 2
\end{bmatrix}
=
\begin{bmatrix}
1 & -1 & 4 & 1 \\
2 & 0 & -1 & -1.5 \\
0 & 0 & 4 & 3
\end{bmatrix}.
$$

2. Then we divide the third equation by 4. The new augmented matrix is:

$$
\begin{bmatrix}
1 & 0 & 0 \\
0 & 1 & 0 \\
0 & 0 & 1/4
\end{bmatrix}
\cdot
\begin{bmatrix}
1 & -1 & 4 & 1 \\
2 & 0 & -1 & -1.5 \\
0 & 0 & 4 & 3
\end{bmatrix}
=
\begin{bmatrix}
1 & -1 & 4 & 1 \\
2 & 0 & -1 & -1.5 \\
0 & 0 & 1 & 0.75
\end{bmatrix}.
$$

3. Now we add the second equation and the third equation. This becomes the new second equation:

$$
\begin{bmatrix}
1 & 0 & 0 \\
0 & 1 & 1 \\
0 & 0 & 1
\end{bmatrix}
\cdot
\begin{bmatrix}
1 & -1 & 4 & 1 \\
2 & 0 & -1 & -1.5 \\
0 & 0 & 1 & 0.75
\end{bmatrix}
=
\begin{bmatrix}
1 & -1 & 4 & 1 \\
2 & 0 & 0 & -0.75 \\
0 & 0 & 1 & 0.75
\end{bmatrix}.
$$

4. Now we divide the second equation by 2. The new augmented matrix is:

$$
\begin{bmatrix}
1 & 0 & 0 \\
0 & 1/2 & 0 \\
0 & 0 & 1
\end{bmatrix}
\cdot
\begin{bmatrix}
1 & -1 & 4 & 1 \\
2 & 0 & 0 & -0.75 \\
0 & 0 & 1 & 0.75
\end{bmatrix}
=
\begin{bmatrix}
1 & -1 & 4 & 1 \\
1 & 0 & 0 & -0.375 \\
0 & 0 & 1 & 0.75
\end{bmatrix}.
$$

134   *Linear Algebra and Its Applications with R*

5. Exchange the first equation and the second equation:

$$
\begin{bmatrix} 0 & 1 & 0 \\ 1 & 0 & 0 \\ 0 & 0 & 1 \end{bmatrix} \cdot \begin{bmatrix} 1 & -1 & 4 & 1 \\ 1 & 0 & 0 & -0.375 \\ 0 & 0 & 1 & 0.75 \end{bmatrix} = \begin{bmatrix} 1 & 0 & 0 & -0.375 \\ 1 & -1 & 4 & 1 \\ 0 & 0 & 1 & 0.75 \end{bmatrix}.
$$

6. Now we take the second equation and subtract the first equation. This equation becomes the new second equation.

$$
\begin{bmatrix} 1 & 0 & 0 \\ -1 & 1 & 0 \\ 0 & 0 & 1 \end{bmatrix} \cdot \begin{bmatrix} 1 & 0 & 0 & -0.375 \\ 1 & -1 & 4 & 1 \\ 0 & 0 & 1 & 0.75 \end{bmatrix} = \begin{bmatrix} 1 & 0 & 0 & -0.375 \\ 0 & -1 & 4 & 1.375 \\ 0 & 0 & 1 & 0.75 \end{bmatrix}.
$$

7. Similarly we take the second equation and subtract 4 times the third equation. This becomes the new second equation:

$$
\begin{bmatrix} 1 & 0 & 0 \\ 0 & 1 & -4 \\ 0 & 0 & 1 \end{bmatrix} \cdot \begin{bmatrix} 1 & 0 & 0 & -0.375 \\ 0 & -1 & 4 & 1.375 \\ 0 & 0 & 1 & 0.75 \end{bmatrix} = \begin{bmatrix} 1 & 0 & 0 & -0.375 \\ 0 & -1 & 0 & -1.625 \\ 0 & 0 & 1 & 0.75 \end{bmatrix}.
$$

8. The last step is to multiply the second equation by -1. Then we have the solution:

$$
\begin{bmatrix} 1 & 0 & 0 \\ 0 & -1 & 0 \\ 0 & 0 & 1 \end{bmatrix} \cdot \begin{bmatrix} 1 & 0 & 0 & -0.375 \\ 0 & -1 & 0 & -1.625 \\ 0 & 0 & 1 & 0.75 \end{bmatrix} = \begin{bmatrix} 1 & 0 & 0 & -0.375 \\ 0 & 1 & 0 & 1.625 \\ 0 & 0 & 1 & 0.75 \end{bmatrix}.
$$

This leads to the following theorem:

Theorem 2.18 *Suppose we have a matrix A and let R be the reduced echelon form of A. Then we can write R as a multiplication of elementary matrices and A such that*

$$
R = E_k \cdot E_{k-1} \cdot \ldots \cdot E_1 \cdot A.
$$

Also, we have the following theorem on the inverse of an elementary matrix.

Theorem 2.19 *The inverse of an elementary matrix is also an elementary matrix.*

Now we discuss how we can obtain the inverse of an invertible matrix using a sequence of elementary row operations.

Definition 23 *Suppose we have two $m \times n$ matrices A and B. Then*

$$
[A|B]
$$

is an $m \times (2n)$ matrix combining the matrices A and B in column-wise.

Example 77 *Suppose we have*

$$A = \begin{bmatrix} 1 & 0 & 2 \\ 0 & 1 & 1 \end{bmatrix} \text{ and } B = \begin{bmatrix} 1 & 2 & 3 \\ 4 & 5 & 6 \end{bmatrix}.$$

Then

$$[A|B] = \begin{bmatrix} 1 & 0 & 2 & | & 1 & 2 & 3 \\ 0 & 1 & 1 & | & 4 & 5 & 6 \end{bmatrix}.$$

Then, we have the following theorem to find the inverse of an $n \times n$ invertible matrix:

Theorem 2.20 *Suppose A is an $n \times n$ invertible matrix. Also, let*

$$M = [A|I_n]$$

where I_n is the identity matrix of size n. Then, the reduced row echelon form R of M forms as

$$R = \left[I_n|A^{-1}\right].$$

Example 78 *Suppose*

$$A = \begin{bmatrix} -1 & 4 \\ 1 & -3 \end{bmatrix}.$$

Then we have

$$M = \begin{bmatrix} -1 & 4 & | & 1 & 0 \\ 1 & -3 & | & 0 & 1 \end{bmatrix}.$$

Then we have the reduced echelon form of M is:

$$\begin{bmatrix} 1 & 0 & | & 3 & 4 \\ 0 & 1 & | & 1 & 1 \end{bmatrix}.$$

Note that the inverse of A is

$$A^{-1} = \begin{bmatrix} 3 & 4 \\ 1 & 1 \end{bmatrix}.$$

Example 79 *In R, we can use the rref() function from the **pracma** package to compute the inverse of an invertible matrix. For example, suppose we have a matrix*

$$A = \begin{bmatrix} 1 & 3 & 2 & 1 \\ 2 & 5 & -2 & -3 \\ 2 & 2 & -3 & -3 \\ -4 & -2 & -1 & 4 \end{bmatrix}.$$

Then we have

$$M = \begin{bmatrix} 1 & 3 & 2 & 1 & | & 1 & 0 & 0 & 0 \\ 2 & 5 & -2 & -3 & | & 0 & 1 & 0 & 0 \\ 2 & 2 & -3 & -3 & | & 0 & 0 & 1 & 0 \\ -4 & -2 & -1 & 4 & | & 0 & 0 & 0 & 1 \end{bmatrix}.$$

First we define a matrix A in R:

```
library(matlib)
A <- matrix(c(1, 3, 2, 1, 2, 5, -2, -3, 2, 2, -3, -3,
-4, -2, -1, 4), nrow = 4, ncol = 4, byrow=TRUE)
```

Then we create a matrix M with the function diag():

```
M <- cbind(A, diag(4))
```

Now we upload the **pracma** *package.*

```
library(pracma)
```

Then call the rref() function.

```
R <- rref(M)
```

If you want to obtain the inverse of the matrix A, then we take the fifth, sixth, seventh and eighth columns of R.

```
Ainv <- R[, 5:8]
```

If we type "Ainv" in R, *it shows*

```
> Ainv
            [,1]        [,2]       [,3]         [,4]
[1,]  0.58024691 -0.7037037  0.8641975 -0.02469136
[2,]  0.04938272  0.2592593 -0.1604938  0.06172840
[3,] -0.14814815  0.2222222 -0.5185185 -0.18518519
[4,]  0.56790123 -0.5185185  0.6543210  0.20987654
```

To check if it is the inverse of the matrix A, we can use the inv() function to obtain the inverse of A:

```
> inv(A)
            [,1]        [,2]       [,3]         [,4]
[1,]  0.58024691 -0.7037037  0.8641975 -0.02469136
[2,]  0.04938272  0.2592593 -0.1604938  0.06172840
[3,] -0.14814815  0.2222222 -0.5185185 -0.18518519
[4,]  0.56790123 -0.5185185  0.6543210  0.20987654
```

2.4.4   Checkmarks

• The definition of an elementary matrix.

• You can write an elementary row operation on a matrix as a matrix multiplication of an elementary matrix and the matrix.

• You can compute the inverse of a matrix using the multiplication of elementary matrices.

• With R, perform compute the inverse of a matrix using the rref() function.

2.4.5 Conceptual Quizzes

Quiz 79 True or False: *An elementary matrix can be obtained by an elementary row operation on the identity matrix of size n.*

Quiz 80 True or False: *An elementary row operation on a matrix is equivalent to matrix multiplication of an elementary matrix on the matrix from the left side.*

Quiz 81 True or False: *An identity matrix of size n is an elementary matrix.*

Quiz 82 True or False: *A matrix*

$$\begin{bmatrix} 1 & 0 & 2 & 0 \\ 0 & 1 & 0 & 0 \\ 0 & 0 & 1 & 0 \\ 0 & 0 & 0 & 1 \end{bmatrix}$$

is an elementary matrix.

Quiz 83 True or False: *A matrix*

$$\begin{bmatrix} 0 & 0 & 1 & 0 \\ 0 & 1 & 0 & 0 \\ 1 & 0 & 0 & 0 \\ 0 & 0 & 0 & 1 \end{bmatrix}$$

is an elementary matrix.

Quiz 84 True or False: *A matrix*

$$\begin{bmatrix} 1 & 0 & 0 & 0 \\ 0 & 1 & 0 & 0 \\ 0 & 0 & 3 & 1 \\ 0 & 0 & 0 & 1 \end{bmatrix}$$

is an elementary matrix.

Quiz 85 True or False: *The inverse of an elementary matrix is an elementary matrix.*

2.4.6 Regular Exercises

Exercise 2.25 *Go back to Exercise 1.38. Compute the reduced echelon form of the augmented matrix as a product of elementary matrices and the augmented matrix.*

Exercise 2.26 *Go back to Exercise 1.39. Compute the reduced echelon form of the augmented matrix as a product of elementary matrices and the augmented matrix.*

Exercise 2.27 *Go back to Exercise 1.40. Compute the reduced echelon form of the augmented matrix as a product of elementary matrices and the augmented matrix.*

Exercise 2.28 *Go back to Exercise 1.41. Compute the reduced echelon form of the augmented matrix as a product of elementary matrices and the augmented matrix.*

Exercise 2.29 *Go back to Exercise 1.42. Compute the reduced echelon form of the augmented matrix as a product of elementary matrices and the augmented matrix.*

Exercise 2.30 *Go back to Exercise 1.43. Compute the reduced echelon form of the augmented matrix as a product of elementary matrices and the augmented matrix.*

Exercise 2.31 *Go back to Exercise 1.44. Compute the reduced echelon form of the augmented matrix as a product of elementary matrices and the augmented matrix.*

Exercise 2.32 *Go back to Exercise 1.45. Compute the reduced echelon form of the augmented matrix as a product of elementary matrices and the augmented matrix.*

Exercise 2.33 *Go back to Exercise 1.46. Compute the reduced echelon form of the augmented matrix as a product of elementary matrices and the augmented matrix.*

Exercise 2.34 *Go back to Exercise 1.47. Compute the reduced echelon form of the augmented matrix as a product of elementary matrices and the augmented matrix.*

Exercise 2.35 *Go back to Exercise 1.48. Compute the reduced echelon form of the augmented matrix as a product of elementary matrices and the augmented matrix.*

Exercise 2.36 *Go back to Exercise 1.49. Compute the reduced echelon form of the augmented matrix as a product of elementary matrices and the augmented matrix.*

Exercise 2.37 *Go back to Exercise 1.50. Compute the reduced echelon form of the augmented matrix as a product of elementary matrices and the augmented matrix.*

Exercise 2.38 *Suppose A is a 4×4 matrix. Construct a matrix by adding -3 the second row and the fourth row of the identity matrix I_4. Then compute $E \cdot A$. What is the effect on the matrix A? What does it look like?*

Exercise 2.39 *Suppose A is a 4×4 matrix. Construct a matrix by exchanging the second row and fourth row of the identity matrix I_4. Then compute $E \cdot A$. What is the effect on the matrix A? What does it look like?*

Exercise 2.40 *Construct 4×4 elementary matrices such that:*

1. *The first row of the 4×4 matrix A is multiplied by -1.*

2. *The second row and fourth row of the 4×4 matrix A are exchanged.*

Exercise 2.41 *Suppose we have*

$$A = \begin{bmatrix} 2 & 5 \\ -1 & 3 \\ 2 & 4 \end{bmatrix}$$

and

$$C = \begin{bmatrix} 1 & 3 & 1 \\ 1 & 0 & -1 \end{bmatrix}.$$

Verify

$$C \cdot A = I_2.$$

Is A invertible?

Exercise 2.42 *Suppose we have*

$$A = \begin{bmatrix} 1/ & -2/3 & 2/3 \\ 2/3 & 2/3 & 1/3 \end{bmatrix},$$

and $D = A^T$. Verify

$$A \cdot D = I_2.$$

Is A invertible?

2.4.7 Lab Exercises

Lab Exercise 58 *1. Compute the inverse of the coefficient matrix of the system of linear equations from Exercise 1.38 with the rref() function.*

2. *Using the inverse matrix of the coefficient matrix, solve the system of linear equations.*

Lab Exercise 59 *1. Compute the inverse of the coefficient matrix of the system of linear equations from Exercise 1.39 with the rref() function.*

2. Using the inverse matrix of the coefficient matrix, solve the system of linear equations.

Lab Exercise 60 *1. Compute the inverse of the coefficient matrix of the system of linear equations from Exercise 1.40 with the rref() function.*

2. Using the inverse matrix of the coefficient matrix, solve the system of linear equations.

Lab Exercise 61 *1. Compute the inverse of the coefficient matrix of the system of linear equations from Exercise 1.41 with the rref() function.*

2. Using the inverse matrix of the coefficient matrix, solve the system of linear equations.

Lab Exercise 62 *1. Compute the inverse of the coefficient matrix of the system of linear equations from Exercise 1.42 with the rref() function.*

2. Using the inverse matrix of the coefficient matrix, solve the system of linear equations.

Lab Exercise 63 *1. Compute the inverse of the coefficient matrix of the system of linear equations from Exercise 1.43 with the rref() function.*

2. Using the inverse matrix of the coefficient matrix, solve the system of linear equations.

Lab Exercise 64 *1. Compute the inverse of the coefficient matrix of the system of linear equations from Exercise 1.44 with the rref() function.*

2. Using the inverse matrix of the coefficient matrix, solve the system of linear equations.

Lab Exercise 65 *1. Compute the inverse of the coefficient matrix of the system of linear equations from Exercise 1.45 with the rref() function.*

2. Using the inverse matrix of the coefficient matrix, solve the system of linear equations.

Lab Exercise 66 *1. Compute the inverse of the coefficient matrix of the system of linear equations from Exercise 1.46 with the rref() function.*

2. Using the inverse matrix of the coefficient matrix, solve the system of linear equations.

Lab Exercise 67 *1. Compute the inverse of the coefficient matrix of the system of linear equations from Exercise 1.47 with the rref() function.*

2. Using the inverse matrix of the coefficient matrix, solve the system of linear equations.

Lab Exercise 68 *1. Compute the inverse of the coefficient matrix of the system of linear equations from Exercise 1.48 with the rref() function.*

2. Using the inverse matrix of the coefficient matrix, solve the system of linear equations.

Lab Exercise 69 *1. Compute the inverse of the coefficient matrix of the system of linear equations from Exercise 1.49 with the rref() function.*

2. Using the inverse matrix of the coefficient matrix, solve the system of linear equations.

Lab Exercise 70 *1. Compute the inverse of the coefficient matrix of the system of linear equations from Exercise 1.50 with the rref() function.*

2. Using the inverse matrix of the coefficient matrix, solve the system of linear equations.

Lab Exercise 71 *In R, the runif() function generates random number(s) uniformly. Using the runif() function, create a 5×5 matrix with random numbers sampled from the interval between 0 and 1. The R code to create such a matrix is*

```
A <- matrix(runif(25), 5, 5)
```

Then, find the inverse of the matrix A if it is invertible using the inv() function from the **pracma** *package. If elements in the matrix A are random enough, most likely the matrix A is invertible. If not, then try to type one more time.*
Then compute

$$B \cdot A$$

where B is the output from the inv() function. What do you see? Is it the identity matrix I_5? If not why not?
Now compute the inverse of A using the rref() function and compute

$$C \cdot A$$

where C is the output from the rref() function. What do you see? Is it the identity matrix I_5? If not why not?

Lab Exercise 72 *In R, the sample() function generates random integers(s) from a finite set. Using the sample() function create a 5×5 matrix with random numbers sampled from the interval between -100 and 100. The R code to create such a matrix is*

```
A <- matrix(sample(-100:100, 25, replace=TRUE), 5, 5)
```

*Then, find the inverse of the matrix A if it is invertible using the inv()
function from the* **pracma** *package. If elements in the matrix A are random
enough, most likely the matrix A is invertible. If not, then try to type one
more time.*

Then compute

$$B \cdot A$$

*where B is the output from the inv() function. What do you see? Is it the
identity matrix I_5? If not, why not?*

Now compute the inverse of A using the rref() function and compute

$$C \cdot A$$

*where C is the output from the rref() function. What do you see? Is it the
identity matrix I_5? If not, why not?*

2.4.8   Practical Applications

Now we solve the system of linear equations for the working example by using
the rref function to compute the inverse of the coefficient matrix. Recall from
the practical application in Section 1.4, we define the coefficient matrix for
the original problem

```
A <- matrix(c(1 , 1 , 1 , 1 , 1 , 1 , 1 , 1 , 1 , 0 ,
0 , 0 , 0 , 0 , 0 , 0 , 0 , 0 , 0 , 0 , 0 , 0 , 0 , 0 ,
0 , 0 , 0 , 1 , 1 , 1 , 1 , 1 , 1 , 1 , 1 , 1 , 1 , 0 ,
0 , 0 , 0 , 0 , 0 , 0 , 0 , 1 , 0 , 0 , 0 , 0 , 0 , 0 ,
0 , 0 , 0 , 1 , 0 , 0 , 0 , 0 , 0 , 0 , 0 , 0 , 1 , 0 ,
0 , 0 , 0 , 0 , 0 , 0 , 0 , 0 , 1 , 0 , 0 , 0 , 0 , 0 ,
0 , 0 , 0 , 1 , 0 , 0 , 0 , 0 , 0 , 0 , 0 , 0 , 0 , 1 ,
0 , 0 , 0 , 0 , 0 , 0 , 0 , 0 , 1 , 0 , 0 , 0 , 0 , 0 ,
0 , 0 , 0 , 0 , 1 , 0 , 0 , 0 , 0 , 0 , 0 , 0 , 0 , 1 ,
0 , 0 , 0 , 0 , 0 , 0 , 0 , 0 , 0 , 1 , 0 , 0 , 0 , 0 ,
0 , 0 , 0 , 0 , 1 , 0 , 0 , 0 , 0 , 0 , 0 , 0 , 0 , 0 ,
1 , 0 , 0 , 0 , 0 , 0 , 0 , 0 , 0 , 1 , 0 , 0 , 0 , 0 ,
0 , 0 , 0 , 0 , 0 , 1 , 0 , 0 , 0 , 0 , 0 , 0 , 0 , 0 ,
1 , 0 , 0 , 0 , 0 , 0 , 0 , 0 , 0 , 0 , 1 , 0 , 0 , 0 ,
0 , 0 , 0 , 0 , 0 , 1), nrow = 11, ncol = 18, byrow = TRUE)
```

and we write the right hand side in **R**:

```
b <- c(1298, 1948, 465, 605, 451,  338,  260,  183,  282, 127, 535)
```

First we create the augmented matrix

```
C <- cbind(A, b)
```

To eliminate one row from this system, as we did in the previous section,
we apply the rref() function from the **pracma** package. Then we have:

```
library(pracma)
E <- rref(C)
```

E is the reduced echelon form of the augmented matrix.

```
> E
                                                                    b
 [1,] 1 0 0 0 0 0 0 0 0 0 -1 -1 -1 -1 -1 -1 -1 -1 -1483
 [2,] 0 1 0 0 0 0 0 0 0 0  1  0  0  0  0  0  0  0   605
 [3,] 0 0 1 0 0 0 0 0 0 0  0  1  0  0  0  0  0  0   451
 [4,] 0 0 0 1 0 0 0 0 0 0  0  0  1  0  0  0  0  0   338
 [5,] 0 0 0 0 1 0 0 0 0 0  0  0  0  1  0  0  0  0   260
 [6,] 0 0 0 0 0 1 0 0 0 0  0  0  0  0  1  0  0  0   183
 [7,] 0 0 0 0 0 0 1 0 0 0  0  0  0  0  0  1  0  0   282
 [8,] 0 0 0 0 0 0 0 1 0 0  0  0  0  0  0  0  1  0   127
 [9,] 0 0 0 0 0 0 0 0 1 0  0  0  0  0  0  0  0  1   535
[10,] 0 0 0 0 0 0 0 0 0 1  1  1  1  1  1  1  1  1  1948
[11,] 0 0 0 0 0 0 0 0 0 0  0  0  0  0  0  0  0  0     0
```

Since the last row of this matrix has all zeros, we remove the last row vector.

```
E <- E[-11,]
```

As we did in the previous section, we obtain the augmented matrix for the new system of linear equations with additional equations:

$$x_{11} = 266, x_{12} = 223, x_{13} = 140, x_{14} = 264, x_{15} = 137, x_{16} = 67,$$
$$x_{17} = 130, x_{18} = 24.$$

```
G1 <- eye(8)
G2 <- matrix(rep(0, 80), 8, 10)
b2 <- c(266, 223, 140, 264, 137, 67, 130, 24)
G <- cbind(G1, G2, b2)
M <- rbind(E, G)
```

Note that the eye() function outputs the identity matrix. The command "M[, -19]" creates the coefficient matrix for the system of linear equations.

```
M2 <- M[,-19]
```

Now we want to obtain the inverse of this matrix using the rref() function. Since this matrix is an 18×18 matrix, we add the identity matrix of size 18.

```
M3 <- cbind(M2, diag(18))
M4 <- rref(M3)
```

Then we have

```
> M4
```

```
[1,]  1 0 0 0 0 0 0 0 0 0 0 0 0 0 0 0 0 0   0  0  0  0  0  0  0  0  0  0 0
[2,]  0 1 0 0 0 0 0 0 0 0 0 0 0 0 0 0 0 0   0  0  0  0  0  0  0  0  0  0 0
[3,]  0 0 1 0 0 0 0 0 0 0 0 0 0 0 0 0 0 0   0  0  0  0  0  0  0  0  0  0 0
[4,]  0 0 0 1 0 0 0 0 0 0 0 0 0 0 0 0 0 0   0  0  0  0  0  0  0  0  0  0 0
[5,]  0 0 0 0 1 0 0 0 0 0 0 0 0 0 0 0 0 0   0  0  0  0  0  0  0  0  0  0 0
[6,]  0 0 0 0 0 1 0 0 0 0 0 0 0 0 0 0 0 0   0  0  0  0  0  0  0  0  0  0 0
[7,]  0 0 0 0 0 0 1 0 0 0 0 0 0 0 0 0 0 0   0  0  0  0  0  0  0  0  0  0 0
[8,]  0 0 0 0 0 0 0 1 0 0 0 0 0 0 0 0 0 0   0  0  0  0  0  0  0  0  0  0 0
[9,]  0 0 0 0 0 0 0 0 1 0 0 0 0 0 0 0 0 0   1  1  1  1  1  1  1  1  1  0
[10,] 0 0 0 0 0 0 0 0 0 1 0 0 0 0 0 0 0 0   1  0  0  0  0  0  0  0  0  1
[11,] 0 0 0 0 0 0 0 0 0 0 1 0 0 0 0 0 0 0   0  1  0  0  0  0  0  0  0  0
[12,] 0 0 0 0 0 0 0 0 0 0 0 1 0 0 0 0 0 0   0  0  1  0  0  0  0  0  0  0
[13,] 0 0 0 0 0 0 0 0 0 0 0 0 1 0 0 0 0 0   0  0  0  1  0  0  0  0  0  0
[14,] 0 0 0 0 0 0 0 0 0 0 0 0 0 1 0 0 0 0   0  0  0  0  1  0  0  0  0  0
[15,] 0 0 0 0 0 0 0 0 0 0 0 0 0 0 1 0 0 0   0  0  0  0  0  1  0  0  0  0
[16,] 0 0 0 0 0 0 0 0 0 0 0 0 0 0 0 1 0 0   0  0  0  0  0  0  1  0  0  0
[17,] 0 0 0 0 0 0 0 0 0 0 0 0 0 0 0 0 1 0   0  0  0  0  0  0  0  1  0  0
[18,] 0 0 0 0 0 0 0 0 0 0 0 0 0 0 0 0 0 1  -1 -1 -1 -1 -1 -1 -1 -1 -1  0  0
```

```
[1,]   1  0  0  0  0  0  0  0
[2,]   0  1  0  0  0  0  0  0
[3,]   0  0  1  0  0  0  0  0
[4,]   0  0  0  1  0  0  0  0
[5,]   0  0  0  0  1  0  0  0
[6,]   0  0  0  0  0  1  0  0
[7,]   0  0  0  0  0  0  1  0
[8,]   0  0  0  0  0  0  0  1
[9,]  -1 -1 -1 -1 -1 -1 -1 -1
[10,] -1  0  0  0  0  0  0  0
[11,]  0 -1  0  0  0  0  0  0
[12,]  0  0 -1  0  0  0  0  0
[13,]  0  0  0 -1  0  0  0  0
[14,]  0  0  0  0 -1  0  0  0
[15,]  0  0  0  0  0 -1  0  0
[16,]  0  0  0  0  0  0 -1  0
[17,]  0  0  0  0  0  0  0 -1
[18,]  0  0  0  1  1  1  1  1
```

Note that the 19th column vector to the 36th column vector form the inverse of the matrix M2. Thus we have

```
Minv <- M4[, 19:36]
```

Then the solution of the system of linear equations for this example is

```
Minv %*% M[,19]
```

where "M[,19]" is the vector for the right hand side of the system of linear equations. R outputs the solution as

```
> Minv %*% M[,19]
       [,1]
 [1,]  266
 [2,]  223
 [3,]  140
 [4,]  264
 [5,]  137
 [6,]   67
 [7,]  130
 [8,]   24
 [9,]   47
[10,]  199
[11,]  382
[12,]  311
[13,]   74
[14,]  123
[15,]  116
[16,]  152
[17,]  103
[18,]  488
```

2.4.9 Supplements with python Code

Here we show how to obtain the inverse of a matrix using a sequence of elementary operations.

In python, we use the numpy package for matrix operations. For combining matrices, we can use the concatenate() function from numpy. First, we need to call the package and import the inv() function:

```
import numpy as np
```

Now we use the rref() function to get the reduced echelon form of this matrix:

```
M.rref()
```

Suppose we have a 3×3 matrix

$$A = \begin{bmatrix} 3 & 0 & -5 \\ -1 & -3 & 4 \\ -5 & 5 & 2 \end{bmatrix}.$$

We first define a matrix in python:

```
A = np.matrix([[3, 0, -5], [-1, -3, 4], [-5, 5, 2]])
```

Then the np.identity() function creates the identity matrix.

```
I3 = np.identity(3)
```

Then we combine them column-wise with the concatenate() function:

```
M = np.concatenate((A, I3), axis=1)
```

The argument for "axis" defines if you want to combine the matrices by row or column; "axis=0" combines matrices by row and "axis=1" combines them by column.

```
>>> M
matrix([[ 3.,   0.,  -5.,   1.,   0.,   0.],
        [-1.,  -3.,   4.,   0.,   1.,   0.],
        [-5.,   5.,   2.,   0.,   0.,   1.]])
```

Now we have to convert M to the matrix type for the **sympy** package [28]. Then we call the rref() function as:

```
M2 = Matrix(M)
M2.rref()
```

Then **python** outputs

```
>>> M2.rref()
(Matrix([
[1, 0, 0,  -1.18181818181818,  -1.13636363636364, -0.681818181818182],
[0, 1, 0, -0.818181818181818, -0.863636363636364, -0.318181818181818],
[0, 0, 1, -0.909090909090909, -0.681818181818182, -0.409090909090909]]),
(0, 1, 2))
```

2.5 Discussion

Go back to the image processing example. As we discussed in Section 2.1, the picture on the left side of Figure 2.1 is translated to a matrix. In order to see the matrix, let us use the **png** package [40]. To upload the package we type:

```
library(png)
```

Then, we read the png image:

```
x <- readPNG("Tora_color.png")
```

The png file can be found in the author's webpage. Then in order to see how many row vectors and column vectors it has, we can type:

```
dim(x)
```

Then, it outputs the following:

```
> dim(x)
[1] 133   77    4
```

This means that this image has 133×77 pixels and it has four colors in each pixel: red, green, blue, and alpha for semi-transparency. If you type "x[,,1]" then you can see the 133×77 matrix with some numbers varying from zero to one. The head() function shows the first 6 rows. Here we show the first 6 rows and 7 columns of the matrix:

```
> head(x[,,1])
        [,1]      [,2]      [,3]      [,4]      [,5]      [,6]      [,7]
[1,] 1.0000000 1.0000000 1.0000000 1.0000000 1.0000000 1.00000000 1.0000000
[2,] 1.0000000 1.0000000 1.0000000 1.0000000 1.0000000 1.00000000 1.0000000
[3,] 0.9803922 0.9803922 0.9803922 0.9764706 0.9490196 0.93333333 0.9333333
[4,] 0.7019608 0.7019608 0.7137255 0.6549020 0.3058824 0.08627451 0.1019608
[5,] 0.7098039 0.7019608 0.7137255 0.6549020 0.3019608 0.09803922 0.1294118
[6,] 0.7019608 0.7019608 0.7137255 0.6588235 0.3450980 0.16078431 0.1215686
```

This 133×77 matrix for "x[,,1]" presents scale of the strength of the red color. Similarly, "x[,,2]" does the same for the green color, "x[,,3]" for the blue color, and "x[,,4]" for the alpha channel.

3

Determinants

A **determinant** of a square matrix is a numerical summary of the matrix. This can be used for computing the inverse of a matrix if it is invertible and can also be used for solving a system of linear equations. Computing the determinant of a square matrix has applications in many areas; one is from astronomy. As the Introductory Example for this chapter, we will consider Kepler's first laws of planetary motion.

3.1 Introductory Example from Astronomy

Johannes Kepler discovered the mathematical formula for a planet to orbit around its sun and these results were published in 1609 and 1619. Basically Kepler formulated the orbit of a planet as an ellipsoid focused at the sun. In an ellipsoid there are multiple foci and in the case of an orbit of a planet, the sun is one of the foci.

For this problem we consider a general formula for finding a **conic** section. A conic section is the intersection of a cone with a linear plane. Some examples of conic sections are shown in Figure 3.1.

A formula of a conic section with two variables, i.e., in the two-dimension, is given as follows:

$$a \cdot x^2 + b \cdot x \cdot y + c \cdot y^2 + d \cdot x + e \cdot y + f = 0,$$

where a, b, c, d, e, f are constants.

In order to draw a conic section, we can use the `conics` package in R [15]. In order to install this package, you have to use the install.packages() function. Then we use the library() function to upload the package:

```
library(conics)
```

Here, with the `conics` package, we will plot the orbit of the ellipsoid with the following polynomial:

$$2 \cdot x^2 + 2 \cdot x \cdot y + 2 \cdot y^2 - 20 \cdot x - 28 \cdot y + 10 = 0. \qquad (3.1)$$

In R we first define a vector representing all coefficients of the polynomial:

DOI: 10.1201/9781003042259-3   149

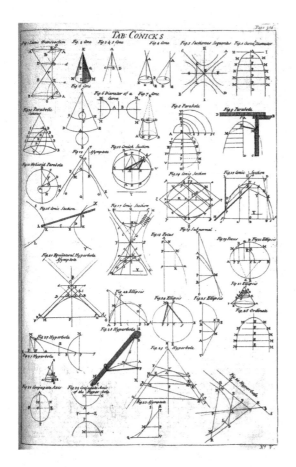

FIGURE 3.1
Table of conics [9].

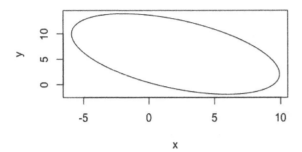

FIGURE 3.2
The plot computed by the conicPlot() function from the `conics` package for the ellipsoid defined by the polynomial in (3.1).

```
v <- c(2, 2, 2, -20, -28, 10)
```

Then we use the conicPlot() function to plot the ellipsoid:

```
conicPlot(v)
```

Then R outputs the plot shown in Figure 3.2.

In reality, when a researcher only observes coordinates of the planet in orbit, they record these coordinates. From these observations, they want to compute the polynomial to describe the orbit in the form of (3.1). For this example, this problem can be written as follows: Given five observed points (x_1, y_1), (x_2, y_2), (x_3, y_3), (x_4, y_4), (x_5, y_5), which are the roots of the polynomial, we want to find the coefficients of the polynomial a, b, c, d, e, f by computing the *determinant* of a 6×6 matrix. Thus, in Section 3.6, we will show how to construct this 6×6 matrix to find the polynomial in (3.1) from a set of given points (recorded coordinates in the orbit) using the determinant of a square matrix.

3.2 Determinants

The determinant of a square matrix contains information of the matrix. For example, from the determinant of a square matrix, we know whether it is invertible or not. It is also used to compute the inverse of a square matrix if it is invertible, and also solve a system of linear equations via **Cramer's rule**.

There are several ways to define the determinant of a square matrix. In this section we will define the determinant of a square matrix with a recursive method.

3.2.1 Task Completion Checklist

- During the Lecture:

  1. Learn the definition of the determinant of a square matrix.
  2. Learn how to compute the determinant of a matrix.
  3. Try some examples without the help of a computer.
  4. With R, learn how to perform computational examples in this section.

- After the Lecture:

  1. Take conceptual quizzes to make sure you understand the materials in this section.
  2. Do some regular exercises.
  3. Conduct lab exercises with R.
  4. Conduct practical applications with R.
  5. If you are interested in python, read the supplement in this section and conduct lab exercises and practical applications with python.

3.2.2 Working Examples

In the Practical Applications in Section 2.3, we worked on the Guinness data set with additional information collected from Guinness Ghana Ltd. on the supply of Malta Guinness from two production sites, Kaasi and Achimota, to nine key distributors geographically scattered in the regions of Ghana. Table 1.3 shows the demands from the distributors and supplies from the production sites.

Recall that here we have a system of linear equations

$$
\begin{aligned}
x_{11} + x_{12} + \ldots + x_{19} &= 1298 \\
x_{21} + x_{22} + \ldots + x_{29} &= 1948 \\
x_{11} + x_{21} &= 465 \\
x_{12} + x_{22} &= 605 \\
x_{13} + x_{23} &= 451 \\
x_{14} + x_{24} &= 338 \\
x_{15} + x_{25} &= 260 \\
x_{16} + x_{26} &= 183 \\
x_{17} + x_{27} &= 282 \\
x_{18} + x_{28} &= 127 \\
x_{19} + x_{29} &= 535,
\end{aligned}
$$

where $1 = $ FTA, $2 = $ RICKY, $3 = $ OBIBAJK, $4 = $ KADOM, $5 = $ NAATO, $6 = $ LESK, $7 = $ DCEE, $8 = $ JOEMAN, $9 = $ KBOA for the distributors and $1 = $ Achimota, $2 = $ Kaasi for the production sites. The coefficient matrix for this system is

$$
\begin{bmatrix}
1 & 1 & 1 & 1 & 1 & 1 & 1 & 1 & 1 & 0 & 0 & 0 & 0 & 0 & 0 & 0 & 0 & 0 \\
0 & 0 & 0 & 0 & 0 & 0 & 0 & 0 & 0 & 1 & 1 & 1 & 1 & 1 & 1 & 1 & 1 & 1 \\
1 & 0 & 0 & 0 & 0 & 0 & 0 & 0 & 0 & 1 & 0 & 0 & 0 & 0 & 0 & 0 & 0 & 0 \\
0 & 1 & 0 & 0 & 0 & 0 & 0 & 0 & 0 & 0 & 1 & 0 & 0 & 0 & 0 & 0 & 0 & 0 \\
0 & 0 & 1 & 0 & 0 & 0 & 0 & 0 & 0 & 0 & 0 & 1 & 0 & 0 & 0 & 0 & 0 & 0 \\
0 & 0 & 0 & 1 & 0 & 0 & 0 & 0 & 0 & 0 & 0 & 0 & 1 & 0 & 0 & 0 & 0 & 0 \\
0 & 0 & 0 & 0 & 1 & 0 & 0 & 0 & 0 & 0 & 0 & 0 & 0 & 1 & 0 & 0 & 0 & 0 \\
0 & 0 & 0 & 0 & 0 & 1 & 0 & 0 & 0 & 0 & 0 & 0 & 0 & 0 & 1 & 0 & 0 & 0 \\
0 & 0 & 0 & 0 & 0 & 0 & 1 & 0 & 0 & 0 & 0 & 0 & 0 & 0 & 0 & 1 & 0 & 0 \\
0 & 0 & 0 & 0 & 0 & 0 & 0 & 1 & 0 & 0 & 0 & 0 & 0 & 0 & 0 & 0 & 1 & 0 \\
0 & 0 & 0 & 0 & 0 & 0 & 0 & 0 & 1 & 0 & 0 & 0 & 0 & 0 & 0 & 0 & 0 & 1
\end{bmatrix},
$$

which is an 11×18 matrix and the vector for the right hand side of the system of linear equations is

$$
\begin{bmatrix}
1298 \\
1948 \\
465 \\
605 \\
451 \\
338 \\
260 \\
183 \\
282 \\
127 \\
535
\end{bmatrix},
$$

which is a 1×11 matrix. Then we generated additional information via sequential importance sampling as

$x_{11} = 266, x_{12} = 223, x_{13} = 140, x_{14} = 264, x_{15} = 137, x_{16} = 67,$
$x_{17} = 130, x_{18} = 24.$

Now, using the determinant of the coefficient matrix for the system of linear equations, we check whether the system of linear equations has a unique solution or not.

3.3 Introduction of Determinants

In Section 2.3, we discussed that there are square matrices which are invertible and there are square matrices which are not invertible. In order to check whether a square matrix is invertible or not, we can use the **determinant** of a square matrix. If the determinant of a square matrix is zero, then it is not invertible. Otherwise, the matrix is invertible. In this section we will define the determinant of a square matrix in a recursive way.

Recall that in Section 2.3 we defined the inverse of a square matrix. If we have a 2×2 matrix such that

$$A = \left[\begin{array}{cc} a_{11} & a_{12} \\ a_{21} & a_{22} \end{array} \right],$$

where $a_{11}, a_{12}, a_{21}, a_{22}$ are real numbers, then A is invertible if and only if $a_{11} \cdot a_{22} - a_{12} \cdot a_{21} \neq 0$. For a 2×2 matrix A, the **determinant** of A is

$$a_{11} \cdot a_{22} - a_{12} \cdot a_{21}.$$

In this book, we notate the determinant of a square matrix A as $\det(A)$.

Suppose we have a 3×3 matrix A such that

$$A = \left[\begin{array}{ccc} a_{11} & a_{12} & a_{13} \\ a_{21} & a_{22} & a_{23} \\ a_{31} & a_{32} & a_{33} \end{array} \right].$$

Now we will compute $\det(A)$. Now let A_{ij} be the square matrix computed from the matrix A by deleting the ith row and the jth column. For example, A_{12} is a 2×2 matrix computed from A by deleting the 1st row and the 2nd column. Specifically,

$$A_{12} = \left[\begin{array}{cc} a_{21} & a_{23} \\ a_{31} & a_{33} \end{array} \right].$$

Then we have

$$\det(A) = a_{11} \cdot \det(A_{11}) - a_{12} \cdot \det(A_{12}) + a_{13} \det(A_{13}).$$

Example 80 *Suppose we have a 3×3 matrix A such that*

$$A = \left[\begin{array}{ccc} -2 & 4 & -5 \\ -1 & -1 & 1 \\ -5 & 0 & -3 \end{array} \right].$$

Then we have

$$A_{11} = \left[\begin{array}{cc} -1 & 1 \\ 0 & -3 \end{array} \right], A_{12} = \left[\begin{array}{cc} -1 & 1 \\ -5 & -3 \end{array} \right], A_{13} = \left[\begin{array}{cc} -1 & -1 \\ -5 & 0 \end{array} \right].$$

Also we have

$$\det(A_{11}) = (-1) \cdot (-3) - 1 \cdot 0 = 3$$
$$\det(A_{12}) = (-1) \cdot (-3) - 1 \cdot (-5) = 8$$
$$\det(A_{13}) = (-1) \cdot 0 - (-1) \cdot (-5) = -5.$$

Then we have

$$\begin{aligned} \det(A) &= a_{11} \cdot \det(A_{11}) - a_{12} \cdot \det(A_{12}) + a_{13} \det(A_{13}) \\ &= (-2) \cdot 3 - 4 \cdot 8 + (-5) \cdot (-5) = -13. \end{aligned}$$

Using the notion of A_{ij}, we have the definition of determinants of square matrices.

Definition 24 *Suppose we have an $n \times n$ square matrix A such that*

$$A = \begin{bmatrix} a_{11} & a_{12} & \cdots & a_{1n} \\ a_{21} & a_{22} & \cdots & a_{2n} \\ \vdots & \vdots & \vdots & \vdots \\ a_{n1} & a_{n2} & \cdots & a_{nn} \end{bmatrix}$$

and let A_{ij} be the $(n-1) \times (n-1)$ matrix computed from the matrix A by deleting the ith row and the ith column. Then the **determinant** *of A is*

$$\det(A) = a_{11} \cdot A_{11} - a_{12} \cdot A_{12} + \ldots + (-1)^{n+1} \cdot a_{1n} \cdot A_{1n}.$$

In R we can use the det() function to compute the determinant of a square matrix. We go back to Example 80. Suppose we have a 3×3 matrix such that

$$A = \begin{bmatrix} -2 & 4 & -5 \\ -1 & -1 & 1 \\ -5 & 0 & -3 \end{bmatrix}.$$

First we define the matrix in R:

```
A <- matrix(c(-2, 4, -5, -1, -1, 1, -5, 0, -3), nrow = 3, ncol = 3)
```

Then we use the det() function:

```
det(A)
```

Then R returns:

```
> det(A)
[1] -13
```

Theorem 3.1 *Suppose we have an $n \times n$ square matrix A such that*

$$A = \begin{bmatrix} a_{11} & a_{12} & \cdots & a_{1n} \\ a_{21} & a_{22} & \cdots & a_{2n} \\ \vdots & \vdots & \vdots & \vdots \\ a_{n1} & a_{n2} & \cdots & a_{nn} \end{bmatrix}$$

and let A_{ij} be the $(n-1) \times (n-1)$ matrix computed from the matrix A by deleting the ith row and the jth column. Then the determinant of A is

$$\det(A) = a_{11} \cdot A_{11} - a_{21} \cdot A_{21} + \ldots + (-1)^{n+1} \cdot a_{n1} \cdot A_{n1}.$$

Example 81 *This is from Example 80. Suppose we have a 3×3 matrix A such that*

$$A = \begin{bmatrix} -2 & 4 & -5 \\ -1 & -1 & 1 \\ -5 & 0 & -3 \end{bmatrix}.$$

Then we have

$$A_{11} = \begin{bmatrix} -1 & 1 \\ 0 & -3 \end{bmatrix}, A_{21} = \begin{bmatrix} 4 & -5 \\ 0 & -3 \end{bmatrix}, A_{31} = \begin{bmatrix} 4 & -5 \\ -1 & 1 \end{bmatrix}.$$

Also we have

$$\det(A_{11}) = (-1)\cdot(-3) - 1\cdot 0 = 3$$
$$\det(A_{21}) = 4\cdot(-3) - 0\cdot(-5) = -12$$
$$\det(A_{31}) = 4\cdot 1 - (-1)\cdot(-5) = -1.$$

Then we have

$$\begin{aligned} \det(A) &= a_{11}\cdot\det(A_{11}) - a_{21}\cdot\det(A_{21}) + a_{31}\det(A_{31}) \\ &= (-2)\cdot 3 - (-1)\cdot(-12) + (-5)\cdot(-1) = -13. \end{aligned}$$

Definition 25 *The (i,j)th **cofactor** of a square matrix A is*

$$C_{ij} = (-1)^{i+j}\cdot\det(A_{ij}).$$

Using cofactors of a square matrix A we can compute the determinant of the matrix A.

Theorem 3.2 *Suppose we have a $n \times n$ matrix A such that*

$$A = \begin{bmatrix} a_{11} & a_{12} & \cdots & a_{1n} \\ a_{21} & a_{22} & \cdots & a_{2n} \\ \vdots & \vdots & \vdots & \vdots \\ a_{n1} & a_{n2} & \cdots & a_{nn} \end{bmatrix}.$$

Then we have

$$\det(A) = a_{11}\cdot C_{11} + a_{12}\cdot C_{12} + \ldots + a_{1n}\cdot C_{1n}$$
$$\det(A) = a_{11}\cdot C_{11} + a_{21}\cdot C_{21} + \ldots + a_{n1}\cdot C_{n1}.$$

Example 82 *Suppose we have a 3×3 matrix A such that*

$$A = \begin{bmatrix} 0 & -3 & 0 \\ 2 & -1 & 0 \\ 5 & 4 & -1 \end{bmatrix}.$$

Then we have

$$C_{11} = (-1)^{1+1}\cdot\det\left(\begin{bmatrix} -1 & 0 \\ 4 & -1 \end{bmatrix}\right),$$
$$C_{12} = (-1)^{1+2}\cdot\det\left(\begin{bmatrix} 2 & 0 \\ 5 & -1 \end{bmatrix}\right),$$
$$C_{13} = (-1)^{1+3}\cdot\det\left(\begin{bmatrix} 2 & -1 \\ 5 & 4 \end{bmatrix}\right).$$

Also we have

$$C_{11} = (-1)^2 \cdot ((-1)\cdot(-1) - 0\cdot 4) = 1$$
$$C_{12} = (-1)^3 \cdot (2\cdot(-1) - 0\cdot 5) = 2$$
$$C_{13} = (-1)^4 \cdot (2\cdot 4 - (-1)\cdot 5) = 13.$$

Then we have

$$\det(A) = a_{11}\cdot\det(A_{11}) - a_{12}\cdot\det(A_{12}) + a_{13}\det(A_{13})$$
$$= 0\cdot 1 + (-3)\cdot 2 + 0\cdot 13 = -6.$$

Theorem 3.3 *If a square matrix A is triangular, then $\det(A)$ is the product of entries in its diagonal.*

Example 83 *Suppose we have a 4×4 matrix such that*

$$A = \begin{bmatrix} -3 & -3 & 1 & 0 \\ 0 & 3 & 3 & 1 \\ 0 & 0 & 1 & -5 \\ 0 & 0 & 0 & 2 \end{bmatrix}.$$

Then the determinant of A is

$$\det(A) = (-3)\cdot 3\cdot 1\cdot 2 = -18.$$

3.3.1 Checkmarks

- The definition of the determinant of a square matrix.
- You can compute the determinant of a square matrix.
- You can compute the determinant of a square matrix using cofactors of the matrix.
- With R, you can compute the determinant of a square matrix using the det() function.

3.3.2 Conceptual Quizzes

Quiz 86 True or False: *The determinant of a square matrix A is not zero if A is invertible.*

Quiz 87 True or False: *The determinant of a square matrix A cannot be bigger than the biggest entry of A.*

Quiz 88 True or False: *The determinant of a square matrix A is the product of the entries in its diagonal.*

Quiz 89 True or False: *The (i,j)th cofactor of a square matrix A is*

$$C_{ij} = (-1)^{j+i}\cdot\det(A_{ij}).$$

Quiz 90 True or False: *The (i,j)th cofactor of a square matrix A is*

$$C_{ij} = (-1)^{i+j} \cdot \det(A_{ji}).$$

Quiz 91 True or False: *We can compute the determinant of a matrix with any size.*

Quiz 92 True or False: *Suppose we have $n \times n$ matrices A and B. Then*

$$\det(A + B) = \det(A) + \det(B).$$

Quiz 93 True or False: *Suppose we have $n \times n$ matrices A and B. Then*

$$\det(A \cdot B) = \det(A) \cdot \det(B).$$

3.3.3 Regular Exercises

Exercise 3.1 *Suppose we have a 3×3 matrix such that*

$$A = \begin{bmatrix} 1 & 2 & -1 \\ 0 & -4 & -3 \\ 1 & -3 & 5 \end{bmatrix}.$$

Compute the determinant of A.

Exercise 3.2 *Suppose we have a 3×3 matrix such that*

$$A = \begin{bmatrix} -3 & 0 & -1 \\ 1 & 2 & -1 \\ 3 & -3 & -2 \end{bmatrix}.$$

Compute the determinant of A.

Exercise 3.3 *Suppose we have a 3×3 matrix such that*

$$A = \begin{bmatrix} -3 & 0 & 3 \\ -1 & 2 & 2 \\ 1 & 1 & -1 \end{bmatrix}.$$

Compute the determinant of A.

Exercise 3.4 *Suppose we have a 3×3 matrix such that*

$$A = \begin{bmatrix} 1 & -2 & 2 \\ 0 & 1 & 0 \\ 3 & 0 & -1 \end{bmatrix}.$$

Compute the determinant of A.

Exercise 3.5 *Suppose we have a 4 × 4 matrix such that*

$$A = \begin{bmatrix} 2 & 4 & 3 & 0 \\ 1 & -4 & -1 & -2 \\ 0 & -2 & 1 & 2 \\ -1 & -3 & 2 & -3 \end{bmatrix}.$$

Compute the determinant of A.

Exercise 3.6 *Suppose we have a 4 × 4 matrix such that*

$$A = \begin{bmatrix} 1 & 4 & 0 & -4 \\ -2 & 2 & -1 & -2 \\ -2 & 1 & 3 & -3 \\ -4 & -4 & -1 & -2 \end{bmatrix}.$$

Compute the determinant of A.

Exercise 3.7 *Suppose we have a 4 × 4 matrix such that*

$$A = \begin{bmatrix} 3 & 3 & 3 & -1 \\ 2 & -1 & 1 & 0 \\ 1 & -4 & -2 & 1 \\ 3 & -1 & -3 & -2 \end{bmatrix}.$$

Compute the determinant of A.

Exercise 3.8 *Suppose we have a 4 × 4 matrix such that*

$$A = \begin{bmatrix} -3 & -2 & 4 & 1 \\ -3 & -3 & -3 & -1 \\ -1 & 2 & 0 & 4 \\ 1 & -2 & -2 & -3 \end{bmatrix}.$$

Compute the determinant of A.

Exercise 3.9 *Consider the matrix in Exercise 3.1. Then, compute the determinant of the matrix after exchanging the first and the second rows of the matrix A.*

Exercise 3.10 *Consider the matrix in Exercise 3.2. Compute the determinant of the matrix after exchanging the first and the second columns of the matrix A.*

Exercise 3.11 *Consider the matrix in Exercise 3.3. Compute the determinant of the matrix after exchanging the first and the third rows of the matrix A.*

Exercise 3.12 *Consider the matrix in Exercise 3.4. Compute the determinant of the matrix after exchanging the first and the third rows of the matrix A.*

Exercise 3.13 *Consider the matrix in Exercise 3.5. Compute the determinant of the matrix after exchanging the first and the second rows of the matrix A.*

Exercise 3.14 *Consider the matrix in Exercise 3.6. Compute the determinant of the matrix after exchanging the first and the second columns of the matrix A.*

Exercise 3.15 *Consider the matrix in Exercise 3.7. Compute the determinant of the matrix after exchanging the first and the third rows of the matrix A.*

Exercise 3.16 *Consider the matrix in Exercise 3.8. Compute the determinant of the matrix after exchanging the first and the fourth rows of the matrix A.*

Exercise 3.17 *Go back to Exercise 1.38. Compute the determinant of the coefficient matrix.*

Exercise 3.18 *Go back to Exercise 1.39. Compute the determinant of the coefficient matrix.*

Exercise 3.19 *Go back to Exercise 1.40. Compute the determinant of the coefficient matrix.*

Exercise 3.20 *Go back to Exercise 1.41. Compute the determinant of the coefficient matrix.*

Exercise 3.21 *Go back to Exercise 1.42. Compute the determinant of the coefficient matrix.*

Exercise 3.22 *Go back to Exercise 1.43. Compute the determinant of the coefficient matrix.*

Exercise 3.23 *Go back to Exercise 1.44. Compute the determinant of the coefficient matrix.*

Exercise 3.24 *Go back to Exercise 1.45. Compute the determinant of the coefficient matrix.*

Exercise 3.25 *Go back to Exercise 1.46. Compute the determinant of the coefficient matrix.*

Exercise 3.26 *Go back to Exercise 1.47. Compute the determinant of the coefficient matrix.*

Exercise 3.27 *Go back to Exercise 1.48. Compute the determinant of the coefficient matrix.*

Exercise 3.28 *Go back to Exercise 1.49. Compute the determinant of the coefficient matrix.*

Exercise 3.29 *Go back to Exercise 1.50. Compute the determinant of the coefficient matrix.*

Exercise 3.30 *Suppose*

$$A = \begin{bmatrix} a & b \\ c & d \end{bmatrix}.$$

Then, consider the following elementary matrices and verify that $\det(E \cdot A) = \det(E) \cdot \det(A)$.

1.

$$E = \begin{bmatrix} 0 & 1 \\ 1 & 0 \end{bmatrix}$$

2.

$$E = \begin{bmatrix} 1 & 0 \\ 0 & k \end{bmatrix}$$

3.

$$E = \begin{bmatrix} 1 & k \\ 0 & 1 \end{bmatrix}$$

4.

$$E = \begin{bmatrix} 1 & 0 \\ k & 1 \end{bmatrix}$$

Exercise 3.31 *Suppose*

$$A = \begin{bmatrix} a & b \\ c & d \end{bmatrix}$$

and suppose k is a scalar. Compute $\det(k \cdot A)$.

3.3.4   Lab Exercises

Lab Exercise 73 *Using the det() function in* R, *compute the determinant of the matrix in Exercise 3.1.*

Lab Exercise 74 *Using the det() function in* R, *compute the determinant of the matrix in Exercise 3.2.*

Lab Exercise 75 *Using the det() function in* R, *compute the determinant of the matrix in Exercise 3.3.*

Lab Exercise 76 *Using the det() function in* R, *compute the determinant of the matrix in Exercise 3.4.*

Lab Exercise 77 *Using the det() function in* R, *compute the determinant of the matrix in Exercise 3.5.*

Lab Exercise 78 *Using the det() function in* R, *compute the determinant of the matrix in Exercise 3.6.*

Lab Exercise 79 *Using the det() function in* R, *compute the determinant of the matrix in Exercise 3.7.*

Lab Exercise 80 *Using the det() function in* R, *compute the determinant of the matrix in Exercise 3.8.*

Lab Exercise 81 *Using the det() function in* R, *compute the determinant of the matrix in Exercise 3.9.*

Lab Exercise 82 *Using the det() function in* R, *compute the determinant of the matrix in Exercise 3.10.*

Lab Exercise 83 *Using the det() function in* R, *compute the determinant of the matrix in Exercise 3.11.*

Lab Exercise 84 *Using the det() function in* R, *compute the determinant of the matrix in Exercise 3.12.*

Lab Exercise 85 *Using the det() function in* R, *compute the determinant of the matrix in Exercise 3.13.*

Lab Exercise 86 *Using the det() function in* R, *compute the determinant of the matrix in Exercise 3.14.*

Lab Exercise 87 *Using the det() function in* R, *compute the determinant of the matrix in Exercise 3.15.*

Lab Exercise 88 *Using the det() function in* R, *compute the determinant of the matrix in Exercise 3.16.*

Lab Exercise 89 *Using the det() function in* R, *compute the determinant of the matrix in Exercise 3.17.*

Lab Exercise 90 *Using the det() function in* R, *compute the determinant of the matrix in Exercise 3.19.*

Lab Exercise 91 *Using the det() function in* R, *compute the determinant of the matrix in Exercise 3.20.*

Lab Exercise 92 *Using the det() function in* R, *compute the determinant of the matrix in Exercise 3.21.*

Lab Exercise 93 *Using the det() function in* R, *compute the determinant of the matrix in Exercise 3.22.*

Lab Exercise 94 *Using the det() function in* R, *compute the determinant of the matrix in Exercise 3.23.*

Lab Exercise 95 *Using the det() function in* R, *compute the determinant of the matrix in Exercise 3.24.*

Lab Exercise 96 *Using the det() function in* R, *compute the determinant of the matrix in Exercise 3.25.*

Lab Exercise 97 *Using the det() function in* R, *compute the determinant of the matrix in Exercise 3.26.*

Lab Exercise 98 *Using the det() function in* R, *compute the determinant of the matrix in Exercise 3.27.*

Lab Exercise 99 *Using the det() function in* R, *compute the determinant of the matrix in Exercise 3.28.*

Lab Exercise 100 *Using the det() function in* R, *compute the determinant of the matrix in Exercise 3.29.*

Lab Exercise 101 *In* R, *the sample() function generates random integer(s) from a finite set. Using the sample() function, create a 5×5 matrix A with random numbers sampled from the interval between -10 to 10. The* R *code to create such a matrix is*

```
A <- matrix(sample(-10:10, 25, replace=TRUE), 5, 5)
```

Using the det() function in R, *compute the determinant of the following matrices:*

1. $-A$,

2. $2 \cdot A$,

3. $5 \cdot A$, *and*

4. $-10 \cdot A$.

Construct a conjecture how these determinants and the determinant of the matrix A are related.

Lab Exercise 102 *Again using the sample() function in* R, *create a* 5 × 5 *matrix A with random numbers sampled from the interval between* −10 *to* 10.

Using the det() function in R, *compute the determinant of the following matrices:*

1. A^2,

2. $(2 \cdot A)^2$,

3. $(5 \cdot A)^2$, *and*

4. $(-10 \cdot A)^2$.

Construct a conjecture how these determinants and the determinant of the matrix A are related.

Lab Exercise 103 *Again using the sample() function in* R, *create a* 5 × 5 *invertible matrix A with random numbers sampled from the interval between* −10 *to* 10. *Using the det() function in* R, *compute the determinant of* A^{-1}. *Construct a conjecture how these determinants are related.*

3.3.5 Practical Applications

In the Practical Applications in Section 2.3, we computed the coefficient matrix for the system of linear equations, which is encoded M2 in our R code. M2 looks like this in R:

```
> M2

 [1,]  1 0 0 0 0 0 0 0 0 0  -1 -1 -1 -1 -1 -1 -1 -1
 [2,]  0 1 0 0 0 0 0 0 0 0   1  0  0  0  0  0  0  0
 [3,]  0 0 1 0 0 0 0 0 0 0   0  1  0  0  0  0  0  0
 [4,]  0 0 0 1 0 0 0 0 0 0   0  0  1  0  0  0  0  0
 [5,]  0 0 0 0 1 0 0 0 0 0   0  0  0  1  0  0  0  0
 [6,]  0 0 0 0 0 1 0 0 0 0   0  0  0  0  1  0  0  0
 [7,]  0 0 0 0 0 0 1 0 0 0   0  0  0  0  0  1  0  0
 [8,]  0 0 0 0 0 0 0 1 0 0   0  0  0  0  0  0  1  0
 [9,]  0 0 0 0 0 0 0 0 1 0   0  0  0  0  0  0  0  1
[10,]  0 0 0 0 0 0 0 0 0 1   1  1  1  1  1  1  1  1
[11,]  1 0 0 0 0 0 0 0 0 0   0  0  0  0  0  0  0  0
[12,]  0 1 0 0 0 0 0 0 0 0   0  0  0  0  0  0  0  0
[13,]  0 0 1 0 0 0 0 0 0 0   0  0  0  0  0  0  0  0
[14,]  0 0 0 1 0 0 0 0 0 0   0  0  0  0  0  0  0  0
[15,]  0 0 0 0 1 0 0 0 0 0   0  0  0  0  0  0  0  0
[16,]  0 0 0 0 0 1 0 0 0 0   0  0  0  0  0  0  0  0
[17,]  0 0 0 0 0 0 1 0 0 0   0  0  0  0  0  0  0  0
[18,]  0 0 0 0 0 0 0 1 0 0   0  0  0  0  0  0  0  0
```

Recall that we have a unique solution to the system of linear equations and now we will compute the determinant of this coefficient matrix using the det() function in R:

```
det(M2)
```

Then we have

```
> det(M2)
[1] 1
```

Therefore we have the determinant of the coefficient matrix equal to 1. Since this is not zero, we have a unique solution for the system of linear equations.

3.3.6   Supplements with python Code

In python, we can use the function linalg.det() from the numpy package to compute the determinant of a square matrix. First we need to upload the numpy package:

```
import numpy as np
```

Then we define a matrix as the np.array type. We are going to use a 3×3 matrix from Example 80. Suppose we have a 3×3 matrix such that

$$A = \begin{bmatrix} -2 & 4 & -5 \\ -1 & -1 & 1 \\ -5 & 0 & -3 \end{bmatrix}.$$

Then we define this matrix in python:

```
A = np.array([[-2, 4, -5], [-1, -1, 1], [-5, 0, -3]])
```

If we type "A" in python it returns:

```
>>> A
array([[-2,  4, -5],
       [-1, -1,  1],
       [-5,  0, -3]])
```

Then we use the np.linalg.det() function to compute the determinant of A:

```
np.linalg.det(A)
```

and python returns:

```
>>> np.linalg.det(A)
-13.0
```

3.4   Properties of Determinants

In order to compute the inverse of an invertible matrix, you need to compute the determinant of the matrix and divide each entry by the determinant. As we discussed earlier, when we operate arithmetic operations in rational numbers in a computer, a computer has only finite precision. For example, if we type "1/3" in R, R records this number as 0.3333333333333333 (16 digits). Even though we need infinite precision for numerical computations we have to truncate fractions in a computer. This accumulates errors during a computation and in the end we might have a value which is far from the right answer. This type of computational error is significant in division. In order to avoid such errors, we use properties of determinants to avoid arithmetic operations as much as we can, which we will discuss in this section.

3.4.1   Task Completion Checklist

- During the Lecture:

  1. Read about properties of determinants.
  2. Learn how to apply properties of the determinant of a matrix.
  3. Try some examples without the help of a computer.
  4. With R, learn how to perform computational examples in this section.

- After the Lecture:

  1. Take conceptual quizzes to make sure you understand the materials in this section.
  2. Do some regular exercises.
  3. Conduct lab exercises with R.
  4. Conduct practical applications with R.
  5. If you are interested in python, read the supplement in this section and conduct lab exercises and practical applications with python.

3.4.2   Working Examples

In this section we work on applications of determinants to computations of areas and volumes. Computing areas and volumes of parallelepipeds (high-dimensional parallelograms) and tetrahedrons (high-dimensional triangles) has many applications in engineering. In order to build a blueprint for a building or a ship, you need to partition a complicated object into parallelepipeds or/and tetrahedrons to measure the volume or area of the object. Therefore fast computation is important. Therefore, in this working example we will discuss how to efficiently compute volumes of a parallelepiped and a tetrahedron.

In this example we consider a tetrahedron with its vertices:

$$(4, 0, 0) \quad (1, 2, 5) \quad (3, 5, 1) \quad (0, 3, 0).$$

In R, we can visualize this tetrahedron using the `plotly` package [37]. First we upload the library using the library() function:

```
library(plotly)
```

Then we create vectors of these four points in the three-dimension as:

```
x <- c(4, 1, 3, 0)
y <- c(0, 2, 5, 3)
z <- c(0, 5, 1, 0)
```

Then we define color intensity:

```
intensity <- c(0, 0.33, 0.66, 1)
```

Then we use the plot_ly() function to visualize the tetrahedron:

```
p<- plot_ly(x = x, y = y, z = z,
            type = "mesh3d",
            intensity = intensity,
            showscale = TRUE
      )
p
```

Then R outputs the plot shown in Figure 3.3.

If you want to see this figure from different angle, you can drag the pointer and rotate the figure. This can be very useful to better visualize the object. In the end of this section we will discuss how to compute the volume of this tetrahedron using the determinant of a square matrix.

3.4.3   Properties of Determinants

As we discussed in the previous section, the determinant of a square matrix has information of the matrix such as whether it is invertible or not. The following theorem is the information on its inverse.

Theorem 3.4 *A square matrix A is invertible if and only if* $\det(A) \neq 0$.

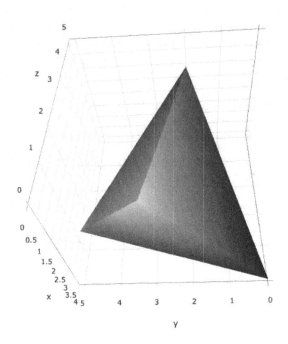

FIGURE 3.3
A tetrahedron plotted by the plot_ly() function in the `plotly` package in R.

This theorem implies the following corollary on the system of linear equations. This corollary was used in the practical applications in the previous section.

Corollary 3.5 *Suppose we have a system of n linear equations on n variables. Then the $n \times n$ coefficient matrix of the system of linear equations has a unique solution if and only if the determinant of the coefficient matrix of the system of linear equations is not equal to zero.*

Example 84 *This is from Example 44. Suppose we have the following system of linear equations:*

$$
\begin{array}{rcrcrcrcr}
 & & x_2 & + & 3x_3 & - & x_4 & = & 1 \\
-x_1 & + & x_2 & - & 4x_3 & & & = & 1 \\
x_1 & & & + & 2x_3 & + & 4x_4 & = & 5 \\
 & & x_2 & & & - & 4x_4 & = & -2.
\end{array}
$$

Its coefficient matrix is

$$
\begin{bmatrix}
0 & 1 & 3 & -1 \\
-1 & 1 & -4 & 0 \\
1 & 0 & 2 & 4 \\
0 & 1 & 0 & -4
\end{bmatrix}.
$$

We are going to use the det() function in R. *As we saw in the previous section, first we define the matrix in* R:

```
A <- matrix(c(0, 1, 3, -1, -1, 1, -4, 0, 1, 0, 2, 4, 0, 1, 0, -4),
nrow = 4, ncol = 4, byrow = TRUE)
```

Then we use the det() function:

```
det(A)
```

Then R *returns:*

```
> det(A)
[1] 30
```

Therefore, there exists a unique solution for the system of linear equations.

Theorem 3.6 *This is again from Example 44. Suppose A and B are $n \times n$ matrices. Then*

$$\det(A \cdot B) = \det(A) \cdot \det(B).$$

Example 85 *This is from Example 44. Suppose we have the following system of linear equations:*

$$
\begin{aligned}
x_2 + 3x_3 - x_4 &= 1 \\
-x_1 + x_2 - 4x_3 &= 1 \\
x_1 + 2x_3 + 4x_4 &= 5 \\
x_2 - 4x_4 &= -2.
\end{aligned}
$$

Its coefficient matrix is

$$
A = \begin{bmatrix}
0 & 1 & 3 & -1 \\
-1 & 1 & -4 & 0 \\
1 & 0 & 2 & 4 \\
0 & 1 & 0 & -4
\end{bmatrix}.
$$

Then we exchange the first row and the third row of the system. This is equivalent to multiplying an elementary matrix

$$
E = \begin{bmatrix}
0 & 0 & 1 & 0 \\
0 & 1 & 0 & 0 \\
1 & 0 & 0 & 0 \\
0 & 0 & 0 & 1
\end{bmatrix}
$$

from the left side. The determinant of E is −1. Therefore

$$\det(E \cdot A) = \det(E) \cdot \det(A) = (-1) \cdot (30) = -30.$$

Theorem 3.7 *Suppose A is a square matrix. Then*

$$\det(A^T) = \det(A).$$

Example 86 *This is from Example 44. Suppose we have the following system of linear equations:*

$$
\begin{array}{rcrcrcrcr}
 & & x_2 & + & 3x_3 & - & x_4 & = & 1 \\
-x_1 & + & x_2 & - & 4x_3 & & & = & 1 \\
x_1 & & & + & 2x_3 & + & 4x_4 & = & 5 \\
 & & x_2 & & & - & 4x_4 & = & -2.
\end{array}
$$

Its coefficient matrix is

$$
A = \begin{bmatrix}
0 & 1 & 3 & -1 \\
-1 & 1 & -4 & 0 \\
1 & 0 & 2 & 4 \\
0 & 1 & 0 & -4
\end{bmatrix}.
$$

Then we consider the transpose of the matrix A. We are going to use the t() function to compute the transpose of A and the det() function to compute the determinant of A in R.
   First we define the matrix in R:

```
A <- matrix(c(0, 1, 3, -1, -1, 1, -4, 0, 1, 0, 2, 4, 0, 1, 0, -4),
nrow = 4, ncol = 4, byrow = TRUE)
```

Then we use the t() function to compute the transpose of A and the det() function to compute the determinant of A:

```
det(t(A))
```

Then R *returns:*

```
> det(t(A))
[1] 30
```

Theorem 3.8 *Suppose we have a square matrix A and suppose A is invertible. Then*

$$\det(A^{-1}) = \frac{1}{\det(A)}.$$

Example 87 *This is again from Example 44. Suppose we have the following system of linear equations:*

$$
\begin{array}{rcrcrcrcr}
 & & x_2 & + & 3x_3 & - & x_4 & = & 1 \\
-x_1 & + & x_2 & - & 4x_3 & & & = & 1 \\
x_1 & & & + & 2x_3 & + & 4x_4 & = & 5 \\
 & & x_2 & & & - & 4x_4 & = & -2.
\end{array}
$$

Its coefficient matrix is

$$
A = \begin{bmatrix}
0 & 1 & 3 & -1 \\
-1 & 1 & -4 & 0 \\
1 & 0 & 2 & 4 \\
0 & 1 & 0 & -4
\end{bmatrix}.
$$

Then we consider the transpose of the matrix A. We are going to use the inv() function from the **pracma** *package to compute the inverse of A and the det() function to compute the determinant of A in* R.

First we upload the **pracma** *package using the library() function and we define the matrix in* R:

```
library(pracma)
A <- matrix(c(0, 1, 3, -1, -1, 1, -4, 0, 1, 0, 2, 4, 0, 1, 0, -4),
nrow = 4, ncol = 4, byrow = TRUE)
```

Then we use the inv() function to compute the inverse of A and the det() function to compute the determinant of A:

```
det(inv(A))
```

Then R *returns:*

```
> det(inv(A))
[1] 0.03333333
```

Note that if we type "1/30" in R *it returns:*

```
> 1/30
[1] 0.03333333
```

From Theorem 3.4, the determinant of the product of square matrices is the product of the determinants of these matrices. Recall that computing the reduced echelon form of the augmented matrix is a product of elementary matrices on the augmented matrix by Theorem 2.18. Thus computing the determinant of the coefficient matrix after the Gaussian Elimination can be easily computed if we know the determinant of an elementary matrix. The following theorem is about the determinant of an elementary row matrix:

Theorem 3.9 *If an elementary matrix E represents an Exchange operation, then*

$$\det(E) = -1.$$

If an elementary matrix E represents a Scaling operation, i.e., multiplying a row by a constant c, then

$$\det(E) = c.$$

If an elementary matrix E represents a Replacement operation, i.e., multiplying a row by a constant c and adding the row to another row, then

$$\det(E) = 1.$$

Example 88 *Let us consider an example in Section 2.4. Suppose we have the following system of 3 linear equations with the variables x_1, x_2, x_3:*

$$
\begin{array}{rcrcrcr}
x_1 & - & x_2 & + & 4x_3 & = & 1 \\
2x_1 & & & - & x_3 & = & -1.5 \\
-x_1 & + & x_2 & & & = & 2.
\end{array}
$$

Recall that the augmented matrix of this system is:

$$
B = \begin{bmatrix}
1 & -1 & 4 & 1 \\
2 & 0 & -1 & -1.5 \\
-1 & 1 & 0 & 2
\end{bmatrix}.
$$

The reduced echelon form of the augmented matrix can be computed by the product of elementary matrices to the augmented matrix of the system of linear equations:

$$E_8 \cdot E_7 \cdot E_6 \cdot E_5 \cdot E_4 \cdot E_3 \cdot E_2 \cdot E_1 \cdot B$$

where

$$E_1 = \begin{bmatrix} 1 & 0 & 0 \\ 0 & -1 & 0 \\ 0 & 0 & 1 \end{bmatrix},$$

$$E_2 = \begin{bmatrix} 1 & 0 & 0 \\ 0 & 1 & -4 \\ 0 & 0 & 1 \end{bmatrix},$$

$$E_3 = \begin{bmatrix} 1 & 0 & 0 \\ -1 & 1 & 0 \\ 0 & 0 & 1 \end{bmatrix},$$

$$E_4 = \begin{bmatrix} 0 & 1 & 0 \\ 1 & 0 & 0 \\ 0 & 0 & 1 \end{bmatrix},$$

$$E_5 = \begin{bmatrix} 1 & 0 & 0 \\ 0 & 1/2 & 0 \\ 0 & 0 & 1 \end{bmatrix},$$

$$E_6 = \begin{bmatrix} 1 & 0 & 0 \\ 0 & 1 & 1 \\ 0 & 0 & 1 \end{bmatrix},$$

$$E_7 = \begin{bmatrix} 1 & 0 & 0 \\ 0 & 1 & 0 \\ 0 & 0 & 1/4 \end{bmatrix},$$

$$E_8 = \begin{bmatrix} 1 & 0 & 0 \\ 0 & 1 & 0 \\ 1 & 0 & 1 \end{bmatrix}.$$

Let A be the coefficient matrix such that

$$A = \begin{bmatrix} 1 & -1 & 4 \\ 2 & 0 & -1 \\ -1 & 1 & 0 \end{bmatrix}.$$

Then, we check the determinant of the matrix such that

$$E_8 \cdot E_7 \cdot E_6 \cdot E_5 \cdot E_4 \cdot E_3 \cdot E_2 \cdot E_1 \cdot A.$$

We have

$\det(E_1) = -1$, $\det(E_2) = 1$, $\det(E_3) = 1$, $\det(E_4) = -1$, $\det(E_5) = 1/2$, $\det(E_6) = 1$, $\det(E_7) = 1/4$, $\det(E_8) = 1$.

Also we have the determinant of A

$$\det(A) = 8.$$

Thus, we have

$$\det(E_8 \cdot E_7 \cdot E_6 \cdot E_5 \cdot E_4 \cdot E_3 \cdot E_2 \cdot E_1 \cdot A) = \frac{1}{2} \cdot \frac{1}{4} \cdot 8 = 1.$$

In fact, we have

$$E_8 \cdot E_7 \cdot E_6 \cdot E_5 \cdot E_4 \cdot E_3 \cdot E_2 \cdot E_1 \cdot A = I_3$$

where I_3 is the identity matrix of size 3.

3.4.4 Checkmarks

- The properties of determinants.

- You can apply properties of determinants.

- You can apply these properties to a system of linear equations.

- With R, you can apply properties on the determinant of a square matrix using the det() function.

3.4.5 Conceptual Quizzes

Quiz 94 True or False: *The determinant of a square matrix A is equal to the determinant of a square matrix A^T.*

Quiz 95 True or False: *The determinant of a square matrix A is equal to the determinant of a square matrix A^{-1}.*

Quiz 96 True or False: *The determinant of an elementary matrix for the Scaling operation is equal to the 1.*

Quiz 97 True or False: *The determinant of an elementary matrix for the Replacement operation is equal to the 1.*

Quiz 98 True or False: *The determinant of an elementary matrix for the Exchange operation is equal to the 1.*

Quiz 99 True or False: *The determinant of the product of square matrices is the product of determinants of these matrices.*

Quiz 100 True or False: *The determinant of the inverse of an invertible matrix is the inverse of determinants of the invertible matrix.*

Quiz 101 True or False: *Suppose A is an $n \times n$ invertible matrix. Then $\det(A^{-1}) = 1/\det(A)$.*

Quiz 102 True or False: *A row replacement operation does not affect the determinant of a matrix.*

Quiz 103 True or False: *Suppose U is an $n \times n$ matrix such that*

$$U^T \cdot U = I_n.$$

Then $\det(U) = \pm 1$.

Quiz 104 True or False: *Suppose P and A are $n \times n$ matrices and P is invertible. Then* $\det(P^{-1} \cdot A \cdot P) = \det(A)$.

Quiz 105 Multiple Choice: *Suppose we have a matrix*

$$A = \begin{bmatrix} 3 & 0 & 0 \\ 0 & -1 & 0 \\ 0 & 0 & 1 \end{bmatrix}.$$

The determinant of A is

1. *3*

2. *−3*

3. *0*

4. *4*

Quiz 106 Multiple Choice: *Suppose we have a matrix*

$$A = \begin{bmatrix} 1 & -1 & 4 \\ 0 & -2 & 0 \\ 0 & 0 & 1 \end{bmatrix}.$$

The determinant of A is

1. *2*

2. *−2*

3. *0*

4. *4*

Quiz 107 Multiple Choice: *Suppose we have a matrix*

$$A = \begin{bmatrix} 1 & -1 & 0 \\ 0 & -2 & 0 \\ 1 & 0 & 3 \end{bmatrix}.$$

The determinant of A is

1. *−6*

2. *−2*

3. *0*

4. *6*

Quiz 108 Multiple Choice: *Suppose we have a matrix*

$$A = \begin{bmatrix} 1 & -1 & 4 \\ 0 & -2 & 0 \\ 0 & 0 & 1 \end{bmatrix}.$$

The determinant of A is

1. *2*

2. *−2*

3. *0*

4. *4*

Quiz 109 Multiple Choice: *Suppose we have a matrix*

$$A = \begin{bmatrix} 0 & 0 & 3 \\ 3 & 2 & 0 \\ -1 & 0 & 3 \end{bmatrix}.$$

The determinant of A is

1. *6*

2. *5*

3. *0*

4. *3*

Quiz 110 Multiple Choice: *Suppose we have a matrix*

$$A = \begin{bmatrix} 0 & 0 & 0 \\ 3 & 2 & 0 \\ -1 & 0 & 0 \end{bmatrix}.$$

The determinant of A is

1. *6*

2. *5*

3. *0*

4. *3*

3.4.6 Regular Quizzes

Exercise 3.32 *Consider the matrix in Exercise 3.1. Then verify Theorem 3.4.*

Exercise 3.33 *Consider the matrix in Exercise 3.2. Then verify Theorem 3.4.*

Exercise 3.34 *Consider the matrix in Exercise 3.3. Then verify Theorem 3.4.*

Exercise 3.35 *Consider the matrix in Exercise 3.4. Then verify Theorem 3.4.*

Exercise 3.36 *Consider the matrix in Exercise 3.5. Then verify Theorem 3.4.*

Exercise 3.37 *Consider the matrix in Exercise 3.1. Then compute the inverse of the coefficient matrix as the product of elementary matrices if it is invertible.*

Exercise 3.38 *Consider the matrix in Exercise 3.2. Then compute the inverse of the coefficient matrix as the product of elementary matrices if it is invertible.*

Exercise 3.39 *Consider the matrix in Exercise 3.3. Then compute the inverse of the coefficient matrix as the product of elementary matrices if it is invertible.*

Exercise 3.40 *Consider the matrix in Exercise 3.4. Then compute the inverse of the coefficient matrix as the product of elementary matrices if it is invertible.*

Exercise 3.41 *Consider the matrix in Exercise 3.5. Then compute the inverse of the coefficient matrix as the product of elementary matrices if it is invertible.*

Exercise 3.42 *Suppose A and B are 3×3 matrices such that $\det(A) = 4$ and $\det(B) = -3$. Using the properties of the determinant, compute:*

1. $\det(A \cdot B)$

2. $\det(5 \cdot A)$

3. $\det(B^T)$

4. $\det(A^{-1}$

5. $\det(A)$

Exercise 3.43 *Suppose A and B are 4×4 matrices such that $\det(A) = 2$ and $\det(B) = -2$. Using the properties of the determinant, compute:*

1. $\det(A \cdot B)$

2. $\det(B^3)$

3. $\det(2 \cdot A)$

4. $\det(A \cdot A^T$

5. $\det(B^{-1} \cdot A \cdot B)$

Exercise 3.44 *Suppose we have matrices*

$$A = \begin{bmatrix} -1 & 3 & 1 \\ 1 & 3 & -1 \\ -1 & 1 & 0 \end{bmatrix}, \ B = \begin{bmatrix} -1 & 1 & 3 \\ -1 & 2 & -2 \\ -2 & 1 & -3 \end{bmatrix}.$$

Then verify Theorem 3.6.

Exercise 3.45 *Suppose we have matrices*

$$A = \begin{bmatrix} 0 & 2 & -3 \\ 3 & -3 & -2 \\ 3 & -1 & -5 \end{bmatrix}, \ B = \begin{bmatrix} 1 & -1 & 3 \\ 1 & -3 & 0 \\ -3 & -1 & -2 \end{bmatrix}.$$

Then verify Theorem 3.6.

Exercise 3.46 *Suppose we have a matrix*

$$A = \begin{bmatrix} 0 & -1 & 0 \\ -1 & 2 & -1 \\ 3 & -2 & 0 \end{bmatrix}.$$

Then verify Theorem 3.7.

Exercise 3.47 *Suppose we have a matrix*

$$A = \begin{bmatrix} -2 & 2 & 1 & 1 \\ -2 & -3 & -3 & -2 \\ 3 & 2 & 1 & -2 \\ -3 & -2 & 0 & -2 \end{bmatrix}.$$

Then verify Theorem 3.7.

Exercise 3.48 *Suppose we have a matrix*

$$A = \begin{bmatrix} 2 & 3 & -2 \\ -2 & -1 & -2 \\ 3 & 1 & -1 \end{bmatrix}.$$

Then verify Theorem 3.8.

Exercise 3.49 *Suppose we have a matrix*

$$A = \begin{bmatrix} 0 & 2 & 1 & 2 \\ 3 & 1 & 3 & -3 \\ -2 & -1 & 0 & 3 \\ -3 & 3 & 0 & 1 \end{bmatrix}.$$

Then verify Theorem 3.8.

Exercise 3.50 *Construct a 4×4 matrix A such that*

$$\det(A^T \cdot A) = 1.$$

Exercise 3.51 *Suppose A is a square matrix and k is a scalar. Find the formula for*

$$\det(k \cdot A).$$

Exercise 3.52 *Suppose we have*

$$A = \begin{bmatrix} a_{11} & a_{12} & u_1 + v_1 \\ a_{21} & a_{22} & u_2 + v_2 \\ a_{31} & a_{32} & u_3 + v_3 \end{bmatrix}, B = \begin{bmatrix} a_{11} & a_{12} & u_1 \\ a_{21} & a_{22} & u_2 \\ a_{31} & a_{32} & u_3 \end{bmatrix}, C = \begin{bmatrix} a_{11} & a_{12} & v_1 \\ a_{21} & a_{22} & v_2 \\ a_{31} & a_{32} & v_3 \end{bmatrix}.$$

Verify

$$\det(A) = \det(B) + \det(C).$$

3.4.7 Lab Exercises

Lab Exercise 104 *Using the det() function in R, repeat Exercise 3.32.*

Lab Exercise 105 *Using the det() function in R, repeat Exercise 3.33.*

Lab Exercise 106 *Using the det() function in R, repeat Exercise 3.34.*

Lab Exercise 107 *Using the det() function in R, repeat Exercise 3.35.*

Lab Exercise 108 *Using the det() function in R, repeat Exercise 3.36.*

Lab Exercise 109 *Using the det() function in R, repeat Exercise 3.44.*

Lab Exercise 110 *Using the det() function in R, repeat Exercise 3.45.*

Lab Exercise 111 *Using the det() function in R, repeat Exercise 3.46.*

Lab Exercise 112 *Using the det() function in R, repeat Exercise 3.47.*

Lab Exercise 113 *Using the det() function in R, repeat Exercise 3.49.*

Lab Exercise 114 *First we construct a sequence of numbers such that*

$$(1, 1/2, \ldots, 1/100)$$

using the seq() function in R*. The seq() function has three arguments: seq(from = a, to = b, by = c) means that this function outputs a sequence of numbers from a to b by difference c. For example, if we want to create a sequence of numbers*

$$(0, 0.1, 0.2, \ldots, 0.9, 1)$$

then we type:

```
seq(0, 1, by = 0.1)
```

To create the sequence above, we have to type in R*:*

```
s <- seq(1, 100, by = 1)
k <- 1/s
```

Now, we consider a matrix A_k in R *such that*

$$A_k = \begin{bmatrix} k & k^2 & k^3 \\ k^2 & k^4 & k^6 \\ k^3 & k^6 & k^9 \end{bmatrix}.$$

Conduct experiments to see how $\det(A_k)$ *changes when k is getting smaller. In* R *we can use the following for loop:*

```
for(i in 2:100){
    Ak <- matrix(c(k[i], k[i]^2, k[i]^3, k[i]^2, 2*k[i]^4, 3*k[i]^6,
    k[i]^3, 2*k[i]^6, 3*k[i]^9), 3, 3)
    print(det(Ak))
}
```

How does $\det(A_k)$ *change when k is getting smaller?*

3.4.8   Practical Applications

Suppose we have a triangle with vertices

$$(x_1, y_1), \ (x_2, y_2), \ (x_3, y_3).$$

Then the area of the triangle is given by half of the absolute value of the determinant of the matrix

$$\begin{bmatrix} x_1 & y_1 & 1 \\ x_2 & y_2 & 1 \\ x_3 & y_3 & 1 \end{bmatrix}.$$

For example, suppose we have a triangle with vertices

$$(0,0), \ (3,1), \ (1,4).$$

Then, the determinant of the matrix

$$\begin{bmatrix} 0 & 0 & 1 \\ 3 & 1 & 1 \\ 1 & 4 & 1 \end{bmatrix}$$

is 11. Therefore, the area of the triangle is $11/2$.
Similarly, the volume of a tetrahedron with vertices

$$(x_1, y_1, z_1), \ (x_2, y_2, z_2), \ (x_3, y_3, z_3)$$

is one-sixth of the absolute value of the determinant of the matrix

$$\begin{bmatrix} x_1 & y_1 & z_1 & 1 \\ x_2 & y_2 & z_2 & 1 \\ x_3 & y_3 & z_3 & 1 \end{bmatrix}.$$

For our example we have a tetrahedron with vertices

$$(4,0,0) \quad (1,2,5) \quad (3,5,1) \quad (0,3,0).$$

Thus we consider the matrix

$$\begin{bmatrix} 4 & 0 & 0 & 1 \\ 1 & 2 & 5 & 1 \\ 3 & 5 & 1 & 1 \\ 0 & 3 & 0 & 1 \end{bmatrix}.$$

In R we define the matrix

```
A <- matrix(c(4, 0, 0, 1, 1, 2, 5, 1, 3, 5, 1, 1, 0, 3, 0, 1),
nrow=4, ncol=4, byrow=TRUE)
```

Then we use the det() function to compute the determinant of this matrix and R returns:

```
> det(A)
[1] -86
```

Therefore, the volume of the tetrahedron is $86/6 = 43/3$.

3.4.9 Supplements with python Code

In this section we did not use any new functions. Therefore we do not have any supplemental python code in this section.

3.5   Cramer's Rule

If we know a system of n linear equations with n variables has a unique solution, we can apply **Cramer's rule** to find the unique solution to the system of linear equations. This rule gives an explicit formula for the unique solution using determinants, thus it is very fast to compute. In this section we focus on Cramer's rule and discuss how we can apply it to systems of linear equations.

3.5.1   Task Completion Checklist

- During the Lecture:

    1. Read the definition of Cramer's rule.
    2. Learn how to apply Cramer's rule to solve a system of linear equations.
    3. Try some examples without the help of a computer.
    4. With R, learn how to perform computational examples in this section.
    5. Read the definition of the adjugate matrix of an invertible matrix.

- After the Lecture:

    1. Take conceptual quizzes to make sure you understand the materials in this section.
    2. Do some regular exercises.
    3. Conduct lab exercises with R.
    4. Conduct practical applications with R.
    5. If you are interested in `python`, read the supplement in this section and conduct lab exercises and practical applications with `python`.

3.5.2   Working Examples

We go back to the Practical Applications in Section 2.3 to see if the solution using Cramer's rule coincides with the solution using the reduced echelon form.

Recall that we worked on the Guinness data set with additional information collected from Guinness Ghana Ltd. on the supply of Malta Guinness from two production sites, Kaasi and Achimota, to nine key distributors geographically scattered in the regions of Ghana. Table 1.3 shows the demands from the distributors and supplies from the production sites.

Recall that here we have a system of linear equations

$$
\begin{aligned}
x_{11} + x_{12} + \ldots + x_{19} &= 1298 \\
x_{21} + x_{22} + \ldots + x_{29} &= 1948 \\
x_{11} + x_{21} &= 465 \\
x_{12} + x_{22} &= 605 \\
x_{13} + x_{23} &= 451 \\
x_{14} + x_{24} &= 338 \\
x_{15} + x_{25} &= 260 \\
x_{16} + x_{26} &= 183 \\
x_{17} + x_{27} &= 282 \\
x_{18} + x_{28} &= 127 \\
x_{19} + x_{29} &= 535,
\end{aligned}
$$

where $1 = \text{FTA}, 2 = \text{RICKY}, 3 = \text{OBIBAJK}, 4 = \text{KADOM}, 5 = \text{NAATO}, 6 = \text{LESK}, 7 = \text{DCEE}, 8 = \text{JOEMAN}, 9 = \text{KBOA}$ for the distributors and $1 = \text{Achimota}, 2 = \text{Kaasi}$ for the production sites. The coefficient matrix for this system is

$$
\begin{bmatrix}
1 & 1 & 1 & 1 & 1 & 1 & 1 & 1 & 1 & 0 & 0 & 0 & 0 & 0 & 0 & 0 & 0 & 0 \\
0 & 0 & 0 & 0 & 0 & 0 & 0 & 0 & 0 & 1 & 1 & 1 & 1 & 1 & 1 & 1 & 1 & 1 \\
1 & 0 & 0 & 0 & 0 & 0 & 0 & 0 & 0 & 1 & 0 & 0 & 0 & 0 & 0 & 0 & 0 & 0 \\
0 & 1 & 0 & 0 & 0 & 0 & 0 & 0 & 0 & 0 & 1 & 0 & 0 & 0 & 0 & 0 & 0 & 0 \\
0 & 0 & 1 & 0 & 0 & 0 & 0 & 0 & 0 & 0 & 0 & 1 & 0 & 0 & 0 & 0 & 0 & 0 \\
0 & 0 & 0 & 1 & 0 & 0 & 0 & 0 & 0 & 0 & 0 & 0 & 1 & 0 & 0 & 0 & 0 & 0 \\
0 & 0 & 0 & 0 & 1 & 0 & 0 & 0 & 0 & 0 & 0 & 0 & 0 & 1 & 0 & 0 & 0 & 0 \\
0 & 0 & 0 & 0 & 0 & 1 & 0 & 0 & 0 & 0 & 0 & 0 & 0 & 0 & 1 & 0 & 0 & 0 \\
0 & 0 & 0 & 0 & 0 & 0 & 1 & 0 & 0 & 0 & 0 & 0 & 0 & 0 & 0 & 1 & 0 & 0 \\
0 & 0 & 0 & 0 & 0 & 0 & 0 & 1 & 0 & 0 & 0 & 0 & 0 & 0 & 0 & 0 & 1 & 0 \\
0 & 0 & 0 & 0 & 0 & 0 & 0 & 0 & 1 & 0 & 0 & 0 & 0 & 0 & 0 & 0 & 0 & 1
\end{bmatrix},
$$

which is an 11×18 matrix and the vector for the right hand side of the system of linear equations is

$$
\begin{bmatrix}
1298 \\
1948 \\
465 \\
605 \\
451 \\
338 \\
260 \\
183 \\
282 \\
127 \\
535
\end{bmatrix},
$$

which is a 1×11 matrix. Then we generated additional information via sequential importance sampling as

$$x_{11} = 266, x_{12} = 223, x_{13} = 140, x_{14} = 264, x_{15} = 137, x_{16} = 67,$$
$$x_{17} = 130, x_{18} = 24.$$

In the Practical Application in Section 3.2, we found out that the determinant of the coefficient matrix for this system of linear equations is not zero. Therefore, by Theorem 3.4, we know that this system of linear equations has a unique solution.

In this section, we apply Cramer's rule to find the unique solution for this system of linear equations.

3.5.3 Cramer's Rule

In Section 1.4, we discussed that we can solve a system of linear equations using the reduced echelon form of the augmented matrix of the system. In addition, by Theorem 3.4 in the previous section, we can verify whether a system of n linear equations with n variables has a unique solution or not. If the system of n linear equations with n variables has a unique solution, then we can use Cramer's rule to find the solution. Here we define Cramer's rule and discuss how we can use the rule to find the solution if the system has a unique solution.

Theorem 3.10 (Cramer's rule) *Suppose that a system of n linear equations with n variables has a unique solution such that*

$$Ax = b$$

where A is the coefficient matrix, b is a vector for the right hand side, and x is a vector of variables such that

$$A = \begin{bmatrix} a_{11} & a_{12} & \cdots & a_{1n} \\ a_{21} & a_{22} & \cdots & a_{2n} \\ \vdots & \vdots & \vdots & \vdots \\ a_{n1} & a_{n2} & \cdots & a_{nn} \end{bmatrix}, \ b = \begin{bmatrix} b_1 \\ b_2 \\ \vdots \\ b_n \end{bmatrix}, \ x = \begin{bmatrix} x_1 \\ x_2 \\ \vdots \\ x_n \end{bmatrix}.$$

Let $A_i(b)$ be an $n \times n$ matrix created from A and b by replacing the ith column of A with b such that

$$A_1(b) = \begin{bmatrix} b_1 & a_{12} & \cdots & a_{1n} \\ b_2 & a_{22} & \cdots & a_{2n} \\ \vdots & \vdots & \vdots & \vdots \\ b_n & a_{n2} & \cdots & a_{nn} \end{bmatrix},$$

$$A_2(b) = \begin{bmatrix} a_{11} & b_1 & \cdots & a_{1n} \\ a_{21} & b_2 & \cdots & a_{2n} \\ \vdots & \vdots & \vdots & \vdots \\ a_{n1} & b_n & \cdots & a_{nn} \end{bmatrix},$$

$$\vdots,$$

$$A_n(b) = \begin{bmatrix} a_{11} & a_{12} & \cdots & b_1 \\ a_{21} & a_{22} & \cdots & b_2 \\ \vdots & \vdots & \vdots & \vdots \\ a_{n1} & a_{n2} & \cdots & b_n \end{bmatrix}.$$

Then the unique solution for the system of linear equations is

$$x_1 = \frac{\det(A_1(b))}{\det(A)}, \ x_2 = \frac{\det(A_2(b))}{\det(A)}, \ \cdots, x_n = \frac{\det(A_n(b))}{\det(A)}.$$

Example 89 *We use the system of linear equations from Example 44. Suppose we have the following system of linear equations:*

$$\begin{array}{rrrrrrrr} & x_2 & + & 3x_3 & - & x_4 & = & 1 \\ -x_1 & + & x_2 & - & 4x_3 & & & = & 1 \\ x_1 & & & + & 2x_3 & + & 4x_4 & = & 5 \\ & x_2 & & & & - & 4x_4 & = & -2. \end{array}$$

Its coefficient matrix and the vector for the right hand side are

$$A = \begin{bmatrix} 0 & 1 & 3 & -1 \\ -1 & 1 & -4 & 0 \\ 1 & 0 & 2 & 4 \\ 0 & 1 & 0 & -4 \end{bmatrix}, \ b = \begin{bmatrix} 1 \\ 1 \\ 5 \\ -2 \end{bmatrix}.$$

We showed in Example 84 that the determinant of the matrix A is 30. Therefore by Theorem 3.4, there exists a unique solution to the system.

Now we apply Cramer's rule to find the solution. Here we have

$$A_1(b) = \begin{bmatrix} 1 & 1 & 3 & -1 \\ 1 & 1 & -4 & 0 \\ 5 & 0 & 2 & 4 \\ -2 & 1 & 0 & -4 \end{bmatrix}, \ A_2(b) = \begin{bmatrix} 0 & 1 & 3 & -1 \\ -1 & 1 & -4 & 0 \\ 1 & 5 & 2 & 4 \\ 0 & -2 & 0 & -4 \end{bmatrix},$$

$$A_3(b) = \begin{bmatrix} 0 & 1 & 1 & -1 \\ -1 & 1 & 1 & 0 \\ 1 & 0 & 5 & 4 \\ 0 & 1 & -2 & -4 \end{bmatrix}, \ A_4(b) = \begin{bmatrix} 0 & 1 & 3 & 1 \\ -1 & 1 & -4 & 1 \\ 1 & 0 & 2 & 5 \\ 0 & 1 & 0 & -2 \end{bmatrix}.$$

We are going to use the det() function in R. *As usual, first we define the matrix A and a vector b:*

```
A <- matrix(c(0, 1, 3, -1, -1, 1, -4, 0, 1, 0, 2, 4, 0, 1, 0, -4),
nrow = 4, ncol = 4, byrow = TRUE)
b <- c(1, 1, 5, -2)
```

Then we define matrices $A_i(b)$ for $i = 1, 2, 3, 4$:

```
# Define A1(b)
A1 <- A
A1[, 1] <- b
# Define A2(b)
A2 <- A
A2[ ,2] <- b
# Define A3(b)
A3 <- A
A3[ ,3] <- b
# Define A4(b)
A4 <- A
A4[ ,4] <- b
```

Then we use the det() function to find the solution using Cramer's rule:

```
x1 <- det(A1)/det(A)
x2 <- det(A2)/det(A)
x3 <- det(A3)/det(A)
x4 <- det(A4)/det(A)
```

Then R *returns:*

```
> x1
[1] 1
> x2
[1] 2
> x3
[1] 7.401487e-17
> x4
[1] 1
```

Note that R *returns $x_3 = 7.401487e - 17$. This means $x_3 = 7.401487 \cdot 10^{-17}$, which is very small number and a very close to 0. This is caused by computational numerical errors.*

Now we check the solution by using the solve() function. Then R *returns:*

```
> solve(A,b)
[1] 1 2 0 1
```

Remark 3.11 *As you see from Example 89, Cramer's rule has larger numerical errors caused by divisions compared with the reduced echelon form used in the function solve(). Cramer's rule is nice in terms of mathematical theory but it is not stable in terms of numerical computations. Thus if we want to solve a system of linear equations we should use the reduced echelon form instead of Cramer's rule.*

Now we use Cramer's rule to compute the inverse of a square matrix if it is invertible. Recall that the (i,j)th cofactor of an invertible $n \times n$ square matrix A is

$$C_{ij} = (-1)^{i+j} \cdot \det(A_{ij}),$$

where A_{ij} is a $(n-1) \times (n-1)$ square matrix computed from A by deleting the ith row of A and the jth column of A. We can compute the cofactor C_{ji} using Cramer's rule.

Theorem 3.12 *Suppose A is an invertible $n \times n$ square matrix. Then the cofactor C_{ji} of A is*

$$C_{ji} = \det(A_i(e_j)),$$

where e_j is a vector whose ith entry equals one and otherwise all zero.

Definition 26 *Suppose A is an invertible $n \times n$ square matrix. Then the* **adjugate matrix** *of A is an $n \times n$ square matrix such that*

$$adj(A) = \begin{bmatrix} C_{11} & C_{12} & \dots & C_{1n} \\ C_{21} & C_{22} & \dots & C_{2n} \\ \vdots & \vdots & \vdots & \vdots \\ C_{n1} & C_{n2} & \dots & C_{nn} \end{bmatrix},$$

where C_{ij} is the (i,j)th cofactor of A.

With the notion of the adjugate matrix of a square matrix, we have the following theorem:

Theorem 3.13 *Suppose A is an invertible $n \times n$ square matrix. Then*

$$A^{-1} = adj(A)/\det(A).$$

Remark 3.14 *Again, computing the inverse of an invertible matrix with Cramer's rule is not numerically stable in a computer since we have to perform multiple divisions.*

3.5.4   Checkmarks

- The definition of Cramer's rule.

- You can apply Cramer's rule to solve a system of linear equations.

- With R, you can apply Cramer's rule to solve a system of linear equations using the det() function.

- The definition of the adjugate matrix of an invertible matrix.

3.5.5   Conceptual Quizzes

Quiz 111 True or False: *Cramer's rule is numerically stable.*

Quiz 112 True or False: *With Cramer's rule, a solution of a system of linear equations $Ax = b$ is $x_i = \det(A_i(b))/\det(A)$.*

Quiz 113 True or False: *The adjugate matrix of an invertible $n \times n$ matrix A is an $n \times n$ matrix whose (i,j)th element is the (i,j)th cofactor of A.*

Quiz 114 True or False: *With Cramer's rule, the (i,j)th cofactor of an invertible $n \times n$ matrix A can be computed as*

$$C_{ij} = \det(A_i(e_j)).$$

Quiz 115 True or False: *With the adjugate matrix $adj(A)$ of an invertible $n \times n$ matrix A, the inverse of A is*

$$A^{-1} = adj(A)/\det(A).$$

3.5.6   Regular Exercises

Exercise 3.53 *Consider the system of linear equations in Exercise 3.1. If there exists a unique solution to the system, find the solution using Cramer's rule.*

Exercise 3.54 *Consider the system of linear equations in Exercise 3.2. If there exists a unique solution to the system, find the solution using Cramer's rule.*

Exercise 3.55 *Consider the system of linear equations in Exercise 3.3. If there exists a unique solution to the system, find the solution using Cramer's rule.*

Exercise 3.56 *Consider the system of linear equations in Exercise 3.4. If there exists a unique solution to the system, find the solution using Cramer's rule.*

Exercise 3.57 *Consider the system of linear equations in Exercise 3.5. If there exists a unique solution to the system, find the solution using Cramer's rule.*

Exercise 3.58 *Consider the system of linear equations in Exercise 3.6. If there exists a unique solution to the system, find the solution using Cramer's rule.*

Exercise 3.59 *Consider the system of linear equations in Exercise 3.7. If there exists a unique solution to the system, find the solution using Cramer's rule.*

Exercise 3.60 *Consider the system of linear equations in Exercise 3.8. If there exists a unique solution to the system, find the solution using Cramer's rule.*

Exercise 3.61 *Consider the system of linear equations in Exercise 3.9. If there exists a unique solution to the system, find the solution using Cramer's rule.*

Exercise 3.62 *Consider the system of linear equations in Exercise 3.10. If there exists a unique solution to the system, find the solution using Cramer's rule.*

Exercise 3.63 *Consider the system of linear equations in Exercise 3.11. If there exists a unique solution to the system, find the solution using Cramer's rule.*

Exercise 3.64 *Consider the system of linear equations in Exercise 3.12. If there exists a unique solution to the system, find the solution using Cramer's rule.*

Exercise 3.65 *Consider the system of linear equations in Exercise 3.13. If there exists a unique solution to the system, find the solution using Cramer's rule.*

Exercise 3.66 *Consider the matrix in Exercise 3.1. If it is invertible, compute the inverse of A using the adjugate matrix of A.*

Exercise 3.67 *Consider the matrix in Exercise 3.2. If it is invertible, compute the inverse of A using the adjugate matrix of A.*

Exercise 3.68 *Consider the matrix in Exercise 3.3. If it is invertible, compute the inverse of A using the adjugate matrix of A.*

Exercise 3.69 *Consider the matrix in Exercise 3.4. If it is invertible, compute the inverse of A using the adjugate matrix of A.*

Exercise 3.70 *Consider the matrix in Exercise 3.5. If it is invertible, compute the inverse of A using the adjugate matrix of A.*

Exercise 3.71 *Consider the matrix in Exercise 3.6. If it is invertible, compute the inverse of A using the adjugate matrix of A.*

Exercise 3.72 *Consider the matrix in Exercise 3.7. If it is invertible, compute the inverse of A using the adjugate matrix of A.*

Exercise 3.73 *Consider the matrix in Exercise 3.8. If it is invertible, compute the inverse of A using the adjugate matrix of A.*

Exercise 3.74 *How will A^{-1} for an invertible matrix A be affected if*

1. we interchange the first row and second row of A?

2. we multiply a number k which is not equal to 0 to the first row of A?

Exercise 3.75 *Suppose we have three points (x_1, x_2), (y_1, y_2), (z_1, z_2), which are not co-linear, shown in Figure 3.4. Then, we consider a triangle whose vertices are these three points.*

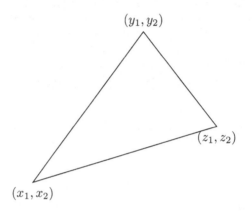

FIGURE 3.4
Triangle example in Exercise 3.75.

If we want to compute the area of the triangle, the formula of the area can be found as

$$\frac{1}{2} \cdot \left| \det \left(\begin{bmatrix} x_1 & x_2 & 1 \\ y_1 & y_2 & 1 \\ z_1 & z_2 & 1 \end{bmatrix} \right) \right|.$$

Compute the area of the triangles with the following vertices:

1. $(0,0)$, $(2,2)$, $(4,-1)$

2. $(1,-1)$, $(1,3)$, $(3,0)$

3. $(0,4)$, $(5,3)$, $(5,-2)$

Exercise 3.76 *Suppose we have five points $0,0)$, $(1,-1)$, $(2,2)$, $(5,3)$, $(4,-1)$ which are not co-linear shown in Figure 3.4. Then we consider a polygon whose vertices are these five points.*
    Using the formula of the area of the triangle

$$\frac{1}{2} \cdot \left| \det \left(\begin{bmatrix} x_1 & x_2 & 1 \\ y_1 & y_2 & 1 \\ z_1 & z_2 & 1 \end{bmatrix} \right) \right|,$$

compute the area of the polygon.

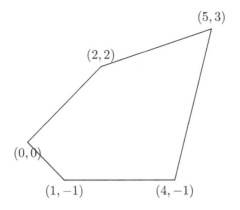

FIGURE 3.5
Polygon example in Exercise 3.76.

Exercise 3.77 *Suppose we have five points* $(0,0), (1,-1), (2,2), (5,3), (6,-2),$ $(4,-1)$ *which are not co-linear shown in Figure 3.4. Then we consider an object whose vertices are these five points.*

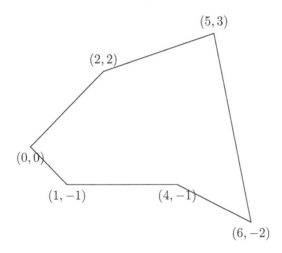

FIGURE 3.6
Object example in Exercise 3.77.

Using the formula of the area of the triangle

$$\frac{1}{2} \cdot \left| \det \left(\begin{bmatrix} x_1 & x_2 & 1 \\ y_1 & y_2 & 1 \\ z_1 & z_2 & 1 \end{bmatrix} \right) \right|,$$

compute the area of the object.

3.5.7   Lab Exercises

Lab Exercise 115 *Using the det() function in* R, *repeat Exercise 3.53 and then compare the solution from the solve() function.*

Lab Exercise 116 *Using the det() function in* R, *repeat Exercise 3.54 and then compare the solution from the solve() function.*

Lab Exercise 117 *Using the det() function in* R, *repeat Exercise 3.55 and then compare the solution from the solve() function.*

Lab Exercise 118 *Using the det() function in* R, *repeat Exercise 3.56 and then compare the solution from the solve() function.*

Lab Exercise 119 *Using the det() function in* R, *repeat Exercise 3.57 and then compare the solution from the solve() function.*

Lab Exercise 120 *Using the det() function in* R, *repeat Exercise 3.58 and then compare the solution from the solve() function.*

Lab Exercise 121 *Using the det() function in* R, *repeat Exercise 3.59 and then compare the solution from the solve() function.*

Lab Exercise 122 *Using the det() function in* R, *repeat Exercise 3.60 and then compare the solution from the solve() function.*

Lab Exercise 123 *Using the det() function in* R, *repeat Exercise 3.61 and then compare the solution from the solve() function.*

Lab Exercise 124 *Using the det() function in* R, *repeat Exercise 3.62 and then compare the solution from the solve() function.*

Lab Exercise 125 *Using the det() function in* R, *repeat Exercise 3.63 and then compare the solution from the solve() function.*

Lab Exercise 126 *Using the det() function in* R, *repeat Exercise 3.64 and then compare the solution from the solve() function.*

Lab Exercise 127 *Using the det() function in* R, *repeat Exercise 3.65 and then compare the solution from the solve() function.*

Lab Exercise 128 *Using the det() function in* R, *repeat Exercise 3.66 and then compare the solution from the inv() function from the* pracma *package.*

Lab Exercise 129 *Using the det() function in* R, *repeat Exercise 3.67 and then compare the solution from the inv() function from the* pracma *package.*

Lab Exercise 130 *Using the det() function in* R, *repeat Exercise 3.68 and then compare the solution from the inv() function from the* pracma *package.*

Lab Exercise 131 *Using the det() function in* R, *repeat Exercise 3.69 and then compare the solution from the inv() function from the* pracma *package.*

Lab Exercise 132 *Using the det() function in* R, *repeat Exercise 3.70 and then compare the solution from the inv() function from the* pracma *package.*

Lab Exercise 133 *Using the det() function in* R, *repeat Exercise 3.71 and then compare the solution from the inv() function from the* pracma *package.*

Lab Exercise 134 *Using the det() function in* R, *repeat Exercise 3.72 and then compare the solution from the inv() function from the* pracma *package.*

Lab Exercise 135 *Using the det() function in* R, *repeat Exercise 3.73 and then compare the solution from the inv() function from the* pracma*package.*

3.5.8 Practical Applications

Now we solve the system of linear equations from the Guinness data set using Cramer's rule and then compare the solution with the solve() function to see how they differ.

From Practical Applications we can compute the matrix M:

```
> M
                                                             b
 [1,] 1 0 0 0 0 0 0 0 0 0 -1 -1 -1 -1 -1 -1 -1 -1 -1483
 [2,] 0 1 0 0 0 0 0 0 0 0  1  0  0  0  0  0  0  0   605
 [3,] 0 0 1 0 0 0 0 0 0 0  0  1  0  0  0  0  0  0   451
 [4,] 0 0 0 1 0 0 0 0 0 0  0  0  1  0  0  0  0  0   338
 [5,] 0 0 0 0 1 0 0 0 0 0  0  0  0  1  0  0  0  0   260
 [6,] 0 0 0 0 0 1 0 0 0 0  0  0  0  0  1  0  0  0   183
 [7,] 0 0 0 0 0 0 1 0 0 0  0  0  0  0  0  1  0  0   282
 [8,] 0 0 0 0 0 0 0 1 0 0  0  0  0  0  0  0  1  0   127
 [9,] 0 0 0 0 0 0 0 0 1 0  0  0  0  0  0  0  0  1   535
[10,] 0 0 0 0 0 0 0 0 0 1  1  1  1  1  1  1  1  1  1948
[11,] 1 0 0 0 0 0 0 0 0 0  0  0  0  0  0  0  0  0   266
[12,] 0 1 0 0 0 0 0 0 0 0  0  0  0  0  0  0  0  0   223
[13,] 0 0 1 0 0 0 0 0 0 0  0  0  0  0  0  0  0  0   140
[14,] 0 0 0 1 0 0 0 0 0 0  0  0  0  0  0  0  0  0   264
[15,] 0 0 0 0 1 0 0 0 0 0  0  0  0  0  0  0  0  0   137
[16,] 0 0 0 0 0 1 0 0 0 0  0  0  0  0  0  0  0  0    67
[17,] 0 0 0 0 0 0 1 0 0 0  0  0  0  0  0  0  0  0   130
[18,] 0 0 0 0 0 0 0 1 0 0  0  0  0  0  0  0  0  0    24
```

The first 18 × 18 matrix is the coefficient matrix for the system of linear equations and the 19th column of the matrix M is the vector of the right hand side. So using the solve() function we have the solution:

```
> solve(M[, -19], M[, 19])
```

266 223 140 264 137 67 130 24 47 199 382 311 74 123 116 152 103 488

With Cramer's rule we have to define the matrix $A_i(b)$. In order to define them we will use the for() function and the array() function:

```
# Initialize the 18 x 18 array for storing the Ai(b) matrices
T <- array(dim=c(18,18,18))
# Create Ai(b) matrices for i = 1 to 18
for(i in 1:18)
    T[,,i] <- M[, -19]
    T[,i,i] <- M[, 19]
```

Then we compute the solution:

```
# Initialize the vector for the solution x,
# the zero vector with length 18
x <- rep(0, 18)
# Find the solution with Cramer's rule
for(i in 1:18)
    x[i] <- det(T[,,i])/det(M[, -19])
```

Then we have the solution:

```
> x
[1] 266 223 140 264 137  67 130  24  47 199 382 311  74 123 116 152 103
488
```

This solution is exactly the same as the solution from the solve() function via the reduced echelon form.

3.5.9   Supplements with python Code

In this section we did not use any new functions. Therefore we do not have any supplemental python code in this section.

3.6   Discussion

From the observed points, we are going to find the polynomial of the orbit for an ellipsoid given by

$$a \cdot x^2 + b \cdot x \cdot y + c \cdot y^2 + d \cdot x + e \cdot y + f = 0, \tag{3.2}$$

where a, b, c, d, e, f are constants so that we can draw the plot.

Suppose we observed the following five points:

$$\begin{bmatrix} -1 \\ 1 \end{bmatrix}, \begin{bmatrix} 1 \\ -1 \end{bmatrix}, \begin{bmatrix} 1 \\ 3 \end{bmatrix}, \begin{bmatrix} 3 \\ 1 \end{bmatrix}, \begin{bmatrix} 1+\sqrt{2} \\ 1+\sqrt{2} \end{bmatrix}.$$

From these points we are going to find constants a, b, c, d, e, f for the polynomial.

First, we consider the following matrix:

$$M = \begin{bmatrix} x^2 & x \cdot y & y^2 & x & y & 1 \\ x_1^2 & x_1 \cdot y_1 & y_1^2 & x_1 & y_1 & 1 \\ x_2^2 & x_2 \cdot y_2 & y_2^2 & x_2 & y_2 & 1 \\ x_3^2 & x_3 \cdot y_3 & y_3^2 & x_3 & y_3 & 1 \\ x_4^2 & x_4 \cdot y_4 & y_4^2 & x_4 & y_4 & 1 \\ x_5^2 & x_5 \cdot y_5 & y_5^2 & x_5 & y_5 & 1 \end{bmatrix},$$

where

$$\begin{bmatrix} x_1 \\ y_1 \end{bmatrix} = \begin{bmatrix} -1 \\ 1 \end{bmatrix}, \begin{bmatrix} x_2 \\ y_2 \end{bmatrix} = \begin{bmatrix} 1 \\ -1 \end{bmatrix}, \begin{bmatrix} x_3 \\ y_3 \end{bmatrix} = \begin{bmatrix} 1 \\ 3 \end{bmatrix}, \begin{bmatrix} x_4 \\ y_4 \end{bmatrix} = \begin{bmatrix} 3 \\ 1 \end{bmatrix},$$

$$\begin{bmatrix} x_5 \\ y_5 \end{bmatrix} = \begin{bmatrix} 1+\sqrt{2} \\ 1+\sqrt{2} \end{bmatrix}.$$

Then notice that

$$\det(M(x=-1,y=1)) = \det(M(x=1,y=-1)) = \det(M(x=1,y=3))$$
$$= \det(M(x=3,y=1)) = \det(M(x=1+\sqrt{2},y=1+\sqrt{2}))$$
$$= 0.$$

since

$$\begin{bmatrix} -1 \\ 1 \end{bmatrix}, \begin{bmatrix} 1 \\ -1 \end{bmatrix}, \begin{bmatrix} 1 \\ 3 \end{bmatrix}, \begin{bmatrix} 3 \\ 1 \end{bmatrix}, \begin{bmatrix} 1+\sqrt{2} \\ 1+\sqrt{2} \end{bmatrix}$$

are all roots of the polynomial in (3.2) and since there are only five roots of this polynomial, the polynomial in (3.2) can be written as

$$a \cdot x^2 + b \cdot x \cdot y + c \cdot y^2 + d \cdot x + e \cdot y + f = \det(M) = 0.$$

Therefore, we have

$$a = \det(M_{11}), b = -\det(M_{12}), c = \det(M_{13}), d = -\det(M_{14}),$$
$$e = \det(M_{15}), f = -\det(M_{16}).$$

With R we can do:

```
x <- c(-1, 1, 1, 3, 1+sqrt(2))
y <- c(1, -1, 3, 1, 1+sqrt(2))
```

This means that the ith entry of x is x_i and the ith entry of y is y_i for the matrix M above.

Now we create the 5×6 matrix M_1 created from M by deleting the first row.

```
## Initializing
M1 <- matrix(rep(0, 5*6), nrow = 5, ncol = 6)
## Make the last column of M1 have all 1.
M1[, 6] <- rep(1, 5)
## Creating the first column of M1
for(i in 1:5)
    M1[i, 1] <- x[i]*x[i]
## Creating the second column of M1
for(i in 1:5)
    M1[i, 2] <- x[i]*y[i]
## Creating the third column of M1
for(i in 1:5)
    M1[i, 3] <- y[i]*y[i]
## Creating the fourth column of M1
for(i in 1:5)
    M1[i, 4] <- x[i]
## Creating the fifth column of M1
for(i in 1:5)
    M1[i, 5] <- y[i]
```

Now we compute the coefficients of the polynomial in Equation (3.2).

```
a <- det(M1[, -1])
#determinant of the matrix deleting the first column of M1
b <- -det(M1[, -2])
#determinant of the matrix deleting the second column of M1
c <- det(M1[, -3])
#determinant of the matrix deleting the third column of M1
d <- -det(M1[, -4])
#determinant of the matrix deleting the fourth column of M1
e <- det(M1[, -5])
#determinant of the matrix deleting the fifth column of M1
f <- -det(M1[, -6])
#determinant of the matrix deleting the sixth column of M1
```

Now we draw the orbit using the `conics` package:

```
library(conics)
v <- c(a, b, c, d, e, f)
conicPlot(v)
```

The result is shown in Figure 3.7. In this case the ellipsoid is the circle centered at the point $(1,1)$ with the radius 2.

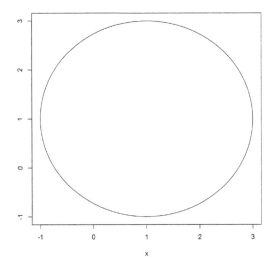

FIGURE 3.7
The result of the orbit of the ellipsoid from the conicPlot() function. In this case it is a circle around the point $(1, 1)$.

4

Vector Spaces

A vector space is a foundation in many areas. For example, in statistics, we assume properties of a vector space so that we can do a statistical analysis on a data set. In this chapter we focus on a vector space and its basic properties. First we define a vector space and then for its application, we assume "Euclidean spaces" as vector spaces.

4.1 Introductory Example from Data Science

In data science we have basically three sub-areas of statistical models: supervised learning, unsupervised learning, and reinforcement learning. In this chapter we focus on unsupervised learning, especially clustering. An unsupervised learning model is a descriptive statistical method unlike a supervised learning, which is an inferential statistical method to predict outcomes or parameters based on a data set. One of main areas in unsupervised learning is clustering. A clustering method is to group observations in a data set together by their "similarities". These "similarity measurements" are defined as a "dot product" of vectors in a vector space.

In this section we consider a crime data set from the United State, the "USArrests" data set from R. This data contains information about arrests made in the United States: average arrests per 100,000 residents for assault, murder, rape and the percent of the population living in urban areas in each of the 50 states in 1973.

Here we apply the hclust() function from R to see clusterings, i.e., groups by the similarity between each observations. First, we upload the "USArrests" data from R:

```
df <- USArrests
```

You can see the first 6 observations by using the head() function:

```
> head(df)
          Murder   Assault   UrbanPop       Rape
Alabama  1.24256408 0.7828393 -0.5209066 -0.003416473
Alaska   0.50786248 1.1068225 -1.2117642  2.484202941
```

DOI: 10.1201/9781003042259-4   199

```
Arizona     0.07163341 1.4788032  0.9989801  1.042878388
Arkansas    0.23234938 0.2308680 -1.0735927 -0.184916602
California  0.27826823 1.2628144  1.7589234  2.067820292
Colorado    0.02571456 0.3988593  0.8608085  1.864967207
```

Then we clean the data by deleting missing values using the na.omit() function:

```
df <- na.omit(df)
```

Then we use the scale() function to "standardize" each measurement. The reason why we apply this function is that some measurements are much bigger than the others. For example, one student takes ACT and SAT tests. SAT has a maximum score of 1600 and ACT has a maximum score of 36. If this student wants to know the test on which he or she did better, you cannot just compare these scores. If we compute how far away the student's score for SAT or ACT is from the mean, then we can compare the student's scores on ACT and SAT. The scale() function computes how far each observation locates from the mean.

```
df <- scale(df)
```

Then we use the "Euclidean" distance as a measurement of how far apart two observations are from each other. Here the "Euclidean" distance is a square root of the inner product of two vectors in $\mathbb{R}^4$ for the "USArrests" data (we will discuss the inner product in Chapter 5). We store the distance between each pair of observations in the "USArrests" data by the 50×50 matrix using the dist() function:

```
d <- dist(df, method = "euclidean")
```

Now we are ready to apply the hclust() function for clustering using a "hierarchical clustering" method:

```
hc <- hclust(d)
```

We can see how this method clusters the data together by a "dendrogram". There are several ways to see how many groups (clusters) exist in the data set. Here from the "dendrogram" we know there are basically four groups in the data set. Thus, we make the four groups together:

```
sub_grp <- cutree(hc, k = 4)
```

Then we use the **factoextra** package [26] to plot the clusters.

```
library(factoextra)
```

Now we plot the result. We use the fviz_cluster() function:

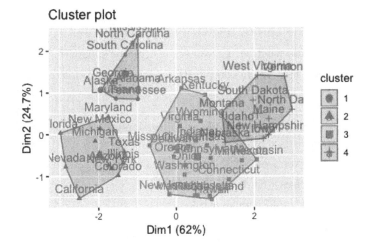

FIGURE 4.1
Clusters in the "USArrests" data via a hierarchical clustering method. A cluster has a unique color. The x-axis represents the first principal component and the y-axis represents the second principal component computed via the PCA.

```
fviz_cluster(list(data = df, cluster = sub_grp))
```

Then R outputs the plot shown in Figure 4.1.

4.2 Vector Spaces and Subspaces

A vector space is a foundation in many areas. However, a vector space is not intuitive to many students. Therefore, we define a vector space and a vector subspace with applications in data science. In Practical Applications, we work on **principal component analysis (PCA)** developed by K. Pearson in 1901 [32]. PCA projects data points in a high-dimensional vector space into a lower-dimensional vector subspace.

4.2.1 Task Completion Checklist

• During the Lecture:

1. Read the definition of a vector space.

2. Read the definition of a vector subspace.

3. Learn how to give an example of a vector space.

4. Learn how to give an example of a vector subspace.

5. Learn how to decide whether a given set is a vector space.

6. Learn how to decide whether a given set is a vector subspace.

7. With R, learn how to perform computational examples in this section.

- After the Lecture:

  1. Take conceptual quizzes to make sure you understand the materials in this section.

  2. Do some regular exercises.

  3. Conduct lab exercises with R.

  4. Conduct practical applications with R.

  5. If you are interested in python, read the supplement in this section and conduct lab exercises and practical applications with python.

4.2.2   Working Examples

In this section we are applying PCA to the "Auto" dataset from the ISLR package in R [24]. This contains gas mileage, horsepower, and other information for 392 vehicles.

This dataset was taken from the StatLib library maintained at Carnegie Mellon University. The dataset was used in the 1983 American Statistical Association Exposition [24].

First, we load the data set from the ISLR package.

```
library(ISLR)
data(Auto)
summary(Auto)
```

The summary() function shows the summary of the dataset including each variable's mean, etc.

```
> summary(Auto)
      mpg           cylinders       displacement      horsepower         weight
 Min.   : 9.00   Min.   :3.000   Min.   : 68.0   Min.   : 46.0   Min.   :1613
 1st Qu.:17.00   1st Qu.:4.000   1st Qu.:105.0   1st Qu.: 75.0   1st Qu.:2225
 Median :22.75   Median :4.000   Median :151.0   Median : 93.5   Median :2804
 Mean   :23.45   Mean   :5.472   Mean   :194.4   Mean   :104.5   Mean   :2978
 3rd Qu.:29.00   3rd Qu.:8.000   3rd Qu.:275.8   3rd Qu.:126.0   3rd Qu.:3615
 Max.   :46.60   Max.   :8.000   Max.   :455.0   Max.   :230.0   Max.   :5140

  acceleration        year          origin              name
 Min.   :13.78   Min.   :70.00   Min.   :1.000   amc matador       :  5
 1st Qu.:13.78   1st Qu.:73.00   1st Qu.:1.000   ford pinto        :  5
 Median :15.50   Median :76.00   Median :1.000   toyota corolla    :  5
 Mean   :15.54   Mean   :75.98   Mean   :1.577   amc gremlin       :  4
 3rd Qu.:17.02   3rd Qu.:79.00   3rd Qu.:2.000   amc hornet        :  4
 Max.   :24.80   Max.   :82.00   Max.   :3.000   chevrolet chevette:  4
                                                 (Other)           :365
```

Before doing anything, we will clean up the dataset so that we can use functions to apply PCA as well as visualize them.

```
auto <- Auto[,1:7]
summary(auto)
```

R outputs:

```
> summary(auto)
      mpg           cylinders      displacement      horsepower        weight
 Min.   : 9.00   Min.   :3.000   Min.   : 68.0    Min.   : 46.0    Min.   :1613
 1st Qu.:17.00   1st Qu.:4.000   1st Qu.:105.0    1st Qu.: 75.0    1st Qu.:2225
 Median :22.75   Median :4.000   Median :151.0    Median : 93.5    Median :2804
 Mean   :23.45   Mean   :5.472   Mean   :194.4    Mean   :104.5    Mean   :2978
 3rd Qu.:29.00   3rd Qu.:8.000   3rd Qu.:275.8    3rd Qu.:126.0    3rd Qu.:3615
 Max.   :46.60   Max.   :8.000   Max.   :455.0    Max.   :230.0    Max.   :5140
  acceleration         year
 Min.   : 8.00   Min.   :70.00
 1st Qu.:13.78   1st Qu.:73.00
 Median :15.50   Median :76.00
 Mean   :15.54   Mean   :75.98
 3rd Qu.:17.02   3rd Qu.:79.00
 Max.   :24.80   Max.   :82.00
```

Then, we will pick the first 7 variables and also make the "origin" variable as character class.

```
new.data <- Auto[, 1:8]
new.data$origin <- as.character(new.data$origin)
summary(new.data)
```

R outputs:

```
> summary(new.data)
      mpg           cylinders      displacement      horsepower        weight
 Min.   : 9.00   Min.   :3.000   Min.   : 68.0    Min.   : 46.0    Min.   :1613
 1st Qu.:17.00   1st Qu.:4.000   1st Qu.:105.0    1st Qu.: 75.0    1st Qu.:2225
 Median :22.75   Median :4.000   Median :151.0    Median : 93.5    Median :2804
 Mean   :23.45   Mean   :5.472   Mean   :194.4    Mean   :104.5    Mean   :2978
 3rd Qu.:29.00   3rd Qu.:8.000   3rd Qu.:275.8    3rd Qu.:126.0    3rd Qu.:3615
 Max.   :46.60   Max.   :8.000   Max.   :455.0    Max.   :230.0    Max.   :5140
  acceleration         year            origin
 Min.   : 8.00   Min.   :70.00    Length:392
 1st Qu.:13.78   1st Qu.:73.00    Class :character
 Median :15.50   Median :76.00    Mode  :character
 Mean   :15.54   Mean   :75.98
 3rd Qu.:17.02   3rd Qu.:79.00
 Max.   :24.80   Max.   :82.00
```

In Practical Applications, we will show how to pick the dimension to visualize this data set and how we can visualize the data in a lower-dimensional vector space.

4.2.3 Vector Spaces and Vector Subspaces

Definition 27 *A* **scalar multiplication** *is an operation such that*

$$c \cdot v$$

where c is a scalar and v is an element of a set V.

A vector space is a foundation in many areas of mathematics and it is a well-defined set with some nice properties. A vector space seems to be very intuitive, however, we have to be very careful with the definition.

Definition 28 *A* **vector space** *V* *is a non-empty set with two operations:* **addition** *and* **scalar multiplication** *such that the following are true:*

- $v + u$ *is in* V *if* v *and* u *are in* V.

- $v + u = u + v$ *for all* v *and* u *in* V.

- $(v + u) + w = v + (u + w)$ *for all* v, u, *and* w *in* V.

- *There exists a* **zero vector 0** *in* V *such that* $u + \mathbf{0} = u$ *for any* u *in* V.

- *For any* v *in* V *there exists a vector* $-v$ *in* V *such that* $v + (-v) = (-v) + v = \mathbf{0}$.

- *If* v *is in* V *and if* c *is a scalar, then* $c \cdot v$ *is in* V.

- $c \cdot (v + u) = c \cdot v + c \cdot u$ *for any* v, u *in* V *and for any scalar* c.

- $(c + d) \cdot v = c \cdot v + d \cdot v$ *for any scalar* c, d *and any* v *in* V.

- $c \cdot (d \cdot v) = (c \cdot d) \cdot v$ *for any scalar* c, d *and any* v *in* V.

- $1 \cdot v = v$ *for any* v *in* V.

From the definition of a vector space, we can show the following proposition:

Proposition 4.1 *For any vector* v *in a vector space* V,

- $0 \cdot v = \mathbf{0}$,

- $c \cdot \mathbf{0} = \mathbf{0}$ *for any scalar* c,

- $(-1) \cdot v = -v$.

Example 90 *Suppose* $V = \mathbb{R}^3 := \{(x, y, z) | x, y, z, \text{ are real numbers}\}$. *Now, we check all properties of a vector space so we can verify that* V *is a vector space.*

- *Want to verify:* $v + u$ *is in* V *if* v *and* u *are in* V. *Suppose we have* (x_1, y_1, z_1), (x_2, y_2, z_2) *in* V *for* $x_1, x_2, y_1, y_2, z_1, z_2$, *which are real numbers. Then* $(x_1, y_1, z_1) + (x_2, y_2, z_2) = (x_1 + x_2, y_1, +y_2, z_1 + z_2)$ *is in* V *since* $x_1 + x_2, y_1 + y_2, z_1 + z_2$ *are real numbers.*

- *Want to verify:* $v + u = u + v$ *for all* v *and* u *in* V. *Suppose we have* (x_1, y_1, z_1), (x_2, y_2, z_2) *in* V *for* $x_1, x_2, y_1, y_2, z_1, z_2$, *which are real numbers. Then* $(x_1, y_1, z_1) + (x_2, y_2, z_2) = (x_1 + x_2, y_1, +y_2, z_1 + z_2) = (x_2 + x_1, y_2, +y_1, z_2 + z_1) = (x_2, y_2, z_2) + (x_1, y_1, z_1)$.

- *Want to verify:* $(v + u) + w = v + (u + w)$ *for all* v, u, *and* w *in* V. *Suppose we have* (x_1, y_1, z_1), (x_2, y_2, z_2), (x_3, y_3, z_3) *in* V *for* $x_1, x_2, x_3, y_1, y_2, y_3, z_1, z_2, z_3$, *which are real numbers. Then*

$$((x_1, y_1, z_1) + (x_2, y_2, z_2)) + (x_3, y_3, z_3)$$
$$= ((x_1 + x_2) + x_3, (y_1, +y_2) + y_3, (z_1 + z_2) + z_3)$$
$$= (x_1 + (x_2 + x_3), y_1, +(y_2 + y_3), z_1 + (z_2 + z_3))$$
$$= (x_1, y_1, z_1) + ((x_2, y_2, z_2) + (x_3, y_3, z_3))$$

- *Want to verify: There exists a* **zero vector 0** *in* V *such that* $u + \mathbf{0} = u$ *for any* u *in* V. *In this example, we have* $\mathbf{0} = (0, 0, 0)$.

- *Want to verify: For any* v *in* V *there exists a vector* $-v$ *in* V *such that* $v + (-v) = (-v) + v = \mathbf{0}$. *Suppose we have* $v = (x_1, y_1, z_1)$ *in* V *for* x_1, y_1, z_1, *which are real numbers. Then* $-v = (-x_1, -y_1, -z_1)$.

- *Want to verify: If* v *is in* V *and if* c *is a scalar, then* $c \cdot v$ *is in* V. *Suppose we have* $v = (x_1, y_1, z_1)$ *in* V *for* x_1, y_1, z_1, *which are real numbers and a scalar* c *which is a real number. Then,* $c \cdot v = (c \cdot x_1, c \cdot x_2, c \cdot x_3)$ *is in* V *since* $c \cdot x_1, c \cdot y_1, c \cdot z_1$ *are real numbers.*

- *Want to verify:* $c \cdot (v + u) = c \cdot v + c \cdot u$ *for any* v, u *in* V *and for any scalar* c. *Suppose we have* $v = (x_1, y_1, z_1)$, $u = (x_2, y_2, z_2)$ *in* V *for* $x_1, x_2, y_1, y_2, z_1, z_2$, *which are real numbers and* c *is real number scalar. Then*

$$
\begin{aligned}
c \cdot (v + u) &= c \cdot (x_1 + x_2, y_1 + y_2, z_1 + z_2) \\
&= (c \cdot x_1 + c \cdot x_2, c \cdot y_1 + c \cdot y_2, c \cdot z_1 + c \cdot z_2) \\
&= c \cdot v + c \cdot u.
\end{aligned}
$$

- *Want to verify:* $(c + d) \cdot v = c \cdot v + d \cdot v$ *for any scalar* c, d *and any* v *in* V. *Suppose we have* $v = (x_1, y_1, z_1)$ *in* V *for* x_1, y_1, z_1, *which are real numbers and* c, d *are real number scalars. Then*

$$
\begin{aligned}
(c \cdot d) \cdot v &= ((c \cdot d) \cdot x_1, (c \cdot d) \cdot y_1, (c \cdot d) \cdot z_1) \\
&= (c \cdot x_1 + d \cdot x_1, c \cdot y_1 + d \cdot y_1, c \cdot z_1 + d \cdot z_1) \\
&= c \cdot v + d \cdot v.
\end{aligned}
$$

- *Want to verify:* $c \cdot (d \cdot v) = (c \cdot d) \cdot v$ *for any scalar* c, d *and any* v *in* V. *Suppose we have* $v = (x_1, y_1, z_1)$ *in* V *for* x_1, y_1, z_1 *are real numbers and* c, d *are real number scalars. Then*

$$
\begin{aligned}
c \cdot (d \cdot v) &= (c \cdot (d \cdot x_1), c \cdot (d \cdot y_1), c \cdot (d \cdot z_1)) \\
&= ((c \cdot d) \cdot x_1, (c \cdot d) \cdot y_1, (c \cdot d) \cdot z_1) \\
&= (c \cdot d) \cdot v.
\end{aligned}
$$

- *Want to verify:* $1 \cdot v = v$ *for any* v *in* V. *Suppose we have* $v = (x_1, y_1, z_1)$ *in* V *for* x_1, y_1, z_1, *which are real numbers. Then we have* $1 \cdot v = 1 \cdot (x_1, y_1, z_1) = (1 \cdot x_1, 1 \cdot y_1, 1 \cdot z_1) = (x_1, y_1, z_1) = v$.

Therefore, V is a vector space.

Example 91 *Let V be a set of all 2×2 matrices with real numbers. Then V is a vector space.*

Example 92 *Let V be a set of all symmetric 3×3 matrices with real numbers. Then V is a vector space.*

Example 93 *Let V be a set of all polynomials $a \cdot x^2 + b \cdot x \cdot y + c \cdot y^2 + d \cdot x + e \cdot y + f = 0$ for a, b, c, d, e, f, which are real numbers. Then V is a vector space.*

Now we have the definition of a vector subspace.

Definition 29 *A **vector subspace** W is a subset of a vector space V such that*

- *the zero vector $\mathbf{0}$ in V is also in W,*

- *W is closed under addition, namely, for any vectors v, u in W, $v + u$ is also in W, and*

- *W is closed under the scalar multiplication, namely, for any vector v in W and for any scalar c, $c \cdot v$ is also in W.*

Example 94 *This is from Example 90. Suppose $V = \{(x, y, z) | x, y, z \text{ are real numbers}\}$, i.e., the 3-dimensional real numbers. Then let $W = \{(x, y, 0) | x, y \text{ are real numbers}\}$. Now we want to show that W is a vector subspace.*

- *Want to verify: The zero vector $\mathbf{0}$ in V is also in W. When $x = y = 0$, then $\mathbf{0}$ is in W.*

- *Want to verify: W is closed under addition, namely, for any vectors v, u in W, $v + u$ is also in W. Suppose $v = (x_1, y_1, 0)$, $u = (x_2, y_2)$ for x_1, x_2, y_1, y_2, which are real numbers. Then $v + u = (x_1 + x_2, y_1 + y_2, 0)$, which is in W.*

- *Want to verify: W is closed under scalar multiplication, namely, for any vector v in W and for any scalar c, $c \cdot v$ is also in W. Suppose $v = (x_1, y_1, 0)$ for x_1, y_1, which are real numbers, and let c be a scalar of a real number. Then, $c \cdot v = (c \cdot x_1, c \cdot y_1, 0)$ is in W.*

All conditions are satisfied. Therefore, W is a vector subspace of V.

Example 95 *This is from Example 91. Let V be a set of all 2×2 matrices with real numbers and W be a set of all 2×2 symmetric matrices with real numbers. Then V is a vector space and W is a vector subspace of V.*

Example 96 *This is from Example 92. Let V be a set of all symmetric* 3×3 *matrices with real numbers and W be a set of all* 3×3 *diagonal matrices with real numbers. Then V is a vector space and W is a vector subspace of V.*

Example 97 *This is from Example 93. Let V be a set of all polynomials* $a \cdot x^2 + b \cdot x \cdot y + c \cdot y^2 + d \cdot x + e \cdot y + f = 0$ *for a, b, c, d, e, f and let W be a set of all polynomials* $d \cdot x + e \cdot y + f = 0$ *for e, f, which are real numbers. Then V is a vector space and W is a vector subspace of V.*

Example 98 *The vector space V is defined in Example 90. Suppose* $V = \{(x, y, z)|x, y, z$ *are real numbers*$\}$, *i.e., 3-dimensional real numbers. Then let* $W = \{(x, y, z)|z = 2 \cdot x - 3 \cdot y\}$. *Then W is a vector subspace.*

To visualize W from Example 98 in 3-dimensional space, we can use the `rgl` package in R [2].

```
library(rgl)
```

Then we initialize the plot:

```
# Create some dummy data
dat <- replicate(2, 1:3)
# Initialize the scene, no data plotted
plot3d(dat, type = 'n', xlim = c(-1, 8), ylim = c(-1, 8),
zlim = c(-10, 20), xlab = '', ylab = '', zlab = '')
```

Then we define the plane by a linear equation and define the origin:

```
# Define the linear plane
planes3d(2, 3, -1, 0, col = 'red', alpha = 0.6)
# Define the origin
points3d(x=0, y=0, z=0)
```

This will plot the linear plane in the 3-dimensional space and output the plot shown in Figure 4.2. If we want to see the plot from a different angle, we can hold the arrow and drag it.

Remark 4.2 *If we have a linear plane through the origin* $(0, 0, 0)$ *in the 3-dimensional space, then the set of all points on the linear plane becomes a vector subspace.*

Remark 4.3 *A vector subspace is also a vector space itself.*

From Example 98 we can represent the vector subspace as $W = \{(x, y, z)|z = 2 \cdot x - 3 \cdot y\}$. Also we can write this vector subspace as

$$W = \{(x, y, z) = \alpha_1(1, 1, -1) + \alpha_2(3, 1, 3)|\alpha_1, \alpha_2 \text{ are real numbers}\}.$$

Writing this form

$$W = \{\alpha_1 v_1 + \alpha_2 v_2 | \alpha_1, \alpha_2 \text{ are real numbers}\}$$

is called **spanned** by vectors v_1 and v_2.

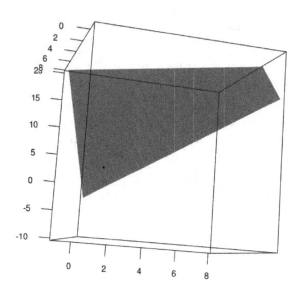

FIGURE 4.2
A vector subspace in the 3-dimensional space defined by a linear equation
$z = 2 \cdot x - 3 \cdot y$ plotted by the rgl package in R. The black point in the plot
is the origin $(0, 0, 0)$.

Example 99 *The 3-dimensional space* $\mathbb{R}^3$ *is spanned by vectors*

$$
\begin{aligned}
v_1 &= (1, 0, 0) \\
v_2 &= (0, 1, 0) \\
v_3 &= (0, 0, 1).
\end{aligned}
$$

Example 100 *This is from Example 95. Let* V *be a set of all* 2×2 *matrices
with real numbers and* W *be a set of all* 2×2 *symmetric matrices with real
numbers. Then* V *is a vector space and* W *is a vector subspace of* V. V *is
spanned by the matrices:*

$$
\begin{bmatrix} 1 & 0 \\ 0 & 0 \end{bmatrix}, \quad
\begin{bmatrix} 0 & 1 \\ 0 & 0 \end{bmatrix}, \quad
\begin{bmatrix} 0 & 0 \\ 1 & 0 \end{bmatrix}, \quad
\begin{bmatrix} 0 & 0 \\ 0 & 1 \end{bmatrix}.
$$

W is spanned by the matrices:

$$
\begin{bmatrix} 1 & 0 \\ 0 & 0 \end{bmatrix}, \quad
\begin{bmatrix} 0 & 1 \\ 1 & 0 \end{bmatrix}, \quad
\begin{bmatrix} 0 & 0 \\ 0 & 1 \end{bmatrix}.
$$

4.2.4 Checkmarks

- The definition of a vector space.

- The definition of a vector subspace.

- The definition of spanning.

- You can prove a set V is a vector space.

- You can prove a set V is a vector subspace.

- With R, you can plot a linear plane in the 3-dimensional space.

4.2.5 Conceptual Quizzes

Quiz 116 True or False: *A set $V = \{(0,0,0)\}$ is a vector space.*

Quiz 117 True or False: *A set of all 2×2 invertible matrices is a vector space.*

Quiz 118 True or False: *A set $V = \{(x,y,z)|x - 2 \cdot y + z = 10\}$ is a vector space.*

Quiz 119 True or False: *A set $V = \{(x,y,z)| - 5 \cdot x + 7 \cdot y - z = 0\}$ is a vector space.*

Quiz 120 True or False: *A set $V = \{(x,y,z)| - 5 \cdot x^2 + 7 \cdot y^2 - z = 0\}$ is a vector space.*

Quiz 121 True or False: *Let V be a set of all 4×4 matrices and let W be a set of all upper triangular matrices. Then V is a vector space and W is a vector subspace of V.*

Quiz 122 True or False: *Let V be a set of all 4×4 matrices and let W be a set of all lower triangular matrices. Then V is a vector space and W is a vector subspace of V.*

Quiz 123 True or False: *Let V be a set $V = \{(x,y,z)|x - 2 \cdot y + z = 0\}$. Let W be a set $W = \{(x,y,0)|x - 2 \cdot y + z = 0\}$. Then V is a vector space and W is a vector subspace of V.*

4.2.6 Regular Exercises

Exercise 4.1 *Consider Example 91. Verify V is a vector space.*

Exercise 4.2 *Consider Example 92. Verify V is a vector space.*

Exercise 4.3 *Consider Example 93. Verify V is a vector space.*

Exercise 4.4 *Consider Example 95. Verify W is a vector subspace of V.*

Exercise 4.5 *Consider Example 96. Verify W is a vector subspace of V.*

Exercise 4.6 *Consider Example 97. Verify W is a vector subspace of V.*

Exercise 4.7 *Consider Example 98. Verify W is a vector subspace of V.*

Exercise 4.8 *Consider Example 96. Find vectors spanning V and vectors spanning W.*

Exercise 4.9 *Consider Example 97. Find vectors spanning V and vectors spanning W.*

4.2.7   Lab Exercises

Lab Exercise 136 *Using the* `rgl` *package in* R, *plot the vector subspace in 3-dimensional space defined by* $\{(x, y, z)|x + y + z = 0\}$.

Lab Exercise 137 *Using the* `rgl` *package in* R, *plot the vector subspace in 3-dimensional space defined by* $\{(x, y, z)|2 \cdot x + y - z = 0\}$.

Lab Exercise 138 *Using the* `rgl` *package in* R, *plot the vector subspace in 3-dimensional space defined by* $\{(x, y, z)| - 3 \cdot x + 2 \cdot y + 5 \cdot z = 0\}$.

Lab Exercise 139 *Using the* `rgl` *package in* R, *plot the vector subspace in 3-dimensional space defined by* $\{(x, y, z)|x - y - z = 0\}$.

4.2.8   Practical Applications

To reduce the dimensionality of the "Auto" data from the ISLR package, we will use the prcomp() function for applying PCA to the dataset. For the prcomp() function, to scale the data we do not have to use the scale function. Just add the option "scale=TRUE".

```
pc <- prcomp(auto, scale=TRUE)
plot(pc)
```

This will display the plot shown in Figure 4.3. There are various rules of thumb for selecting the number of principal components to retain in an analysis of this type:

- Pick the number of components explaining 85% or greater of the variation.

To do so, we will use the cumsum() function. This function computes the cumulative sum.

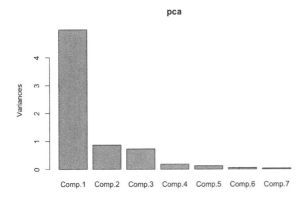

FIGURE 4.3
The picture shown here is the histogram for variances from each component.

```
plot(cumsum(pc$sdev^2/sum(pc$sdev^2)))
```

This will display the plot shown in Figure 4.4. From this plot, it seems

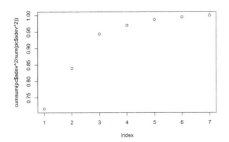

FIGURE 4.4
The cumulative variances.

that we need the first-order to the fourth-order principal components.

By the cumulative proportion of explained variance criterion, we keep as many as are needed to explain more than 80% of total variance. Thus, by looking at the plot, we will keep 2 variables of the data sets.

Now we finish using the autoplot() function from the **ggfortify** package [39]. Using this function we can easily see how the data points are clustered in terms of the different categories.

First, we would like to see how the data points are clustered in terms of the "origin" category.

```
autoplot(pc, data=new.data, colour = "origin")
```

R returns the plot shown in Figure 4.5. As you can see from Figure 4.5, the

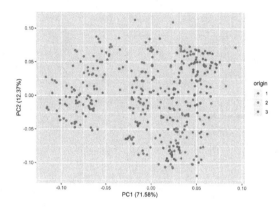

FIGURE 4.5
The "Auto" data projected onto the 2-dimensional linear plane found via PCA.

American cars (in red) are separated from cars made in other countries (in other colors). Using the dimensionality reduction via PCA, we can project all data points in a high-dimensional space onto the 2-dimensional space defined by a linear plane, which is a vector subspace in the higher-dimensional space, so that we can visualize how data points are distributed and form clusters.

4.2.9 Supplements with python Code

To visualize W from Example 98 in 3-dimensional space, we can use the matplotlib package [23] in python. We also use the numpy package in python.

```
import numpy as np
import matplotlib.pyplot as plt
from mpl_toolkits.mplot3d import Axes3D
```

Now we define the grid for x-axis and y-axis. Then we define the linear plane:

```
# create x,y
xx, yy = np.meshgrid(range(10), range(10))
# define the plane for a vector subspace
z = 2 * xx - 3 * yy
```

Now we set up the linear plane in the 3-dimensional space:

```
# plot the surface
plt3d = plt.figure().gca(projection='3d')
plt3d.plot_surface(xx, yy, z, alpha=0.2)
# Ensure that the next plot doesn't overwrite the first plot
ax = plt.gca()
ax.scatter(0, 0, 0, color='green')
# Show the plot
plt.show()
```

This creates the plot shown in Figure 4.6. If we want to see the plot from a different angle, we can hold the arrow and drag it.

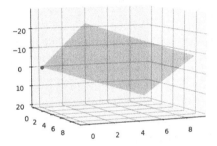

FIGURE 4.6
A vector subspace in the 3-dimensional space defined by a linear equation $z = 2 \cdot x - 3 \cdot y$ plotted by the `matplotlib` package in `python`. The green point in the plot is the origin $(0, 0, 0)$.

4.3 Null Space, Column Space, and Row Space

The null space, column space and row space of a matrix are related to the system of linear equations. In this section we will use a contingency table for the introductory example.

4.3.1 Task Completion Checklist

• During the Lecture:

1. Read the definition of the null space of a matrix.

2. Read the definition of the column space of a matrix.

3. Read the definition of the row space of a matrix.

4. Learn how to give an example of the null space of a matrix.

5. Learn how to give an example of the column space of a matrix.

6. Learn how to give an example of the row space of a matrix.

7. With R, learn how to perform computational examples in this section.

- After the Lecture:

  1. Take conceptual quizzes to make sure you understand the materials in this section.

  2. Do some regular exercises.

  3. Conduct lab exercises with R.

  4. Conduct practical applications with R.

  5. If you are interested in python, read the supplement in this section and conduct lab exercises and practical applications with python.

4.3.2   Working Examples

For the working example in this section we use a contingency table from [3]. Originally, this data was collected by Fisher. For this test, Fisher asked a British woman whether milk or tea was poured into the cup first. Fisher tested the woman eight times and he summarized the results as a contingency table. Here we have the null and alternative hypotheses as follows:

- H_0 : There is no association between the true order and the woman's guess, i.e., the woman's guess is random.

- H_1 : There is association between the true order and the woman's guess.

The data is organized as a contingency table as follows:

Guess/True	Milk	Tea	Total
Milk	3	1	4
Tea	1	3	4
Total	4	4	8

The column of the table is for the true first order and the row is the first pour of the woman's guess. For example, the $(1,1)$ entry is 3. This means that the true order of pouring into the cup was milk first and the woman guessed correctly three times. In the $(1,2)$ entry, the number in the first row and the second column, is 1. This means that the first pour into the cup was tea and the woman guessed wrong. The reader might notice that this contingency table is "sparse."

If an observed table is not sparse, we can simply apply the Chi-square distribution as we did to the problem of the birth-month and death-month being related to each other (maybe they are the same month) or not, as we discussed. However, in this case, we cannot simply apply the Chi-square distribution. In order to do this, Diaconis-Sturmfels in [16] proposed a Markov Chain Monte Carlo (MCMC) approach, a generalized Fisher's exact test. In order to do so, we have to set up the system of linear equations and inequalities as we saw in the practical applications in Section 1.1. Without going into too much detail, we will explain Fisher's exact test using the null space of the coefficient matrix of the system of linear equations for this example.

4.3.3 Null Space, Column Space, and Row Space

Suppose we have the following system of 3 linear equations with the variables x_1, x_2, x_3:

$$\begin{array}{rcrcrcr} x_1 & - & x_2 & + & 4x_3 & = & 1 \\ 2x_1 & & & - & x_3 & = & -1 \\ -x_1 & - & x_2 & + & 5x_3 & = & 2. \end{array}$$

Then we want to solve the system of linear equations. In addition to solving the system of linear equations, we would like to find **all** solutions if they exist.

We discussed how to solve a system of linear equations using the reduced echelon form of the augmented matrix in Chapter 1 and using Cramer's rule in Section 3.5. Here, we use the **null space** of the coefficient matrix to find all solutions of the system of linear equations.

Definition 30 *Suppose we have an $m \times n$ matrix A. The **null space** of the matrix A is the set of all solutions for the system of linear equations such that*

$$A \cdot x = \mathbf{0}.$$

Going back to the system of linear equations above, here we have

$$Ax = b$$

where

$$A = \begin{bmatrix} 1 & -1 & 4 \\ 2 & 0 & -1 \\ -1 & -1 & 5 \end{bmatrix}, b = \begin{bmatrix} 1 \\ -1 \\ 2 \end{bmatrix}.$$

In order to find the null space of the matrix A, we have to solve the system of linear equations:

$$A \cdot X = b$$

that is

$$\begin{array}{rcrcrcr} x_1 & - & x_2 & + & 4x_3 & = & 0 \\ 2x_1 & & & - & x_3 & = & 0 \\ -x_1 & - & x_2 & + & 5x_3 & = & 0. \end{array}$$

With the help of R, we solve this system of linear equations using the reduced echelon form of the matrix A with the rref() function from the pracma package:

```
library(pracma)
A <- matrix(c(1, -1, 4, 0, 2, 0, -1, 0, -1, -1, 5, 0),
nrow=3, ncol=4, byrow=TRUE)
rref(A)
```

R outputs as follows:

```
> rref(A)
     [,1] [,2] [,3] [,4]
[1,]    1    0 -0.5    0
[2,]    0    1 -4.5    0
[3,]    0    0  0.0    0
```

This means the reduced echelon form of the system is

$$
\begin{aligned}
x_1 \quad\quad\quad - \quad 0.5x_3 &= 0 \\
x_2 - \quad 4.5x_3 &= 0.
\end{aligned}
$$

The solutions of the system of linear equations are

$$
\begin{aligned}
x_1 &= 0.5t \\
x_2 &= 4.5t \\
x_3 &= t,
\end{aligned}
$$

where t is any real number. Therefore, the null space of the matrix A is spanned by the vector

$$
v = \begin{bmatrix} 0.5 \\ 4.5 \\ 1 \end{bmatrix}
$$

and the null space of the matrix A is the set

$$
\left\{ t \cdot \begin{bmatrix} 0.5 \\ 4.5 \\ 1 \end{bmatrix} \mid -\infty < t < \infty \right\}.
$$

Now we have the following theorem about the null space of a matrix.

Theorem 4.4 *The null space of a matrix A is a vector space.*

Then we have the following lemma:

Lemma 4.5 *Suppose we have an $m \times n$ matrix A. Then the null space of the matrix A is a vector subspace of $\mathbb{R}^n$.*

Let Null (A) be the null space of the matrix A. Here, we go through how the null space of the matrix A relates to the system of linear equations

$$A \cdot x = b.$$

Suppose y is a vector in the null space of the matrix A. Then y satisfies the system of linear equations

$$A \cdot y = \mathbf{0}.$$

Suppose x_0 is a solution which satisfies the system of linear equations above. Then, x_0 satisfies the system of linear equations

$$A \cdot x_0 = b.$$

Therefore, $x_0 + y$ satisfies the system of linear equations

$$A \cdot (x_0 + y) = A \cdot x_0 + A \cdot y = b + \mathbf{0} = b.$$

Thus, $x_0 + y$ is also another solution of the system of linear equations $A \cdot x = b$. Therefore, if the system of linear equations $A \cdot x = b$ has a unique solution, then the null space of the matrix A consists of the zero vector, i.e., Null $(A) = \{\mathbf{0}\}$.

Theorem 4.6 *Suppose we have a system of linear equations such that*

$$A \cdot x = b.$$

Then the system of linear equations has a unique solution if and only if $Null(A) = \{\mathbf{0}\}$.

Going back to the example in the beginning, we have the null space of the coefficient matrix A of the system is

$$\text{Null}\,(A) = \left\{ t \cdot \begin{bmatrix} 0.5 \\ 4.5 \\ 1 \end{bmatrix} \,\middle|\, -\infty < t < \infty \right\}.$$

Therefore, the system of linear equations has infinitely many solutions since Null $(A) \neq \{\mathbf{0}\}$. Also by solving a system of linear equations, note that the vector

$$x_0 = \begin{bmatrix} -1/2 \\ -3/2 \\ 0 \end{bmatrix}$$

is a solution to the system of linear equations. Therefore the solution set of the system is

$$\left\{ \begin{bmatrix} -1/2 \\ -3/2 \\ 0 \end{bmatrix} + t \cdot \begin{bmatrix} 0.5 \\ 4.5 \\ 1 \end{bmatrix} \,\middle|\, -\infty < t < \infty \right\}.$$

In R we can use the Null() function from the MASS package to compute the vectors spanning the null space of a matrix A. We will use the example in the beginning to demonstrate how to use the function in R. First we load the package and define the matrix:

```
library(MASS)
A <- matrix(c(1, -1, 4, 2, 0, -1, -1, -1, 5),
nrow=3, ncol=3, byrow=TRUE)
```

Then we use the NULL() function to the **transpose** of the matrix A:

```
Null(t(A))
```

Example 101 *Then* R *returns as follows:*

```
> Null(t(A))
          [,1]
[1,]  0.1078328
[2,]  0.9704950
[3,]  0.2156655
```

Suppose we have the following system of 3 linear equations with the variables x_1, x_2, x_3, x_4:

$$
\begin{array}{rcrcrcrcr}
x_1 & - & x_2 & + & 4x_3 & + & x_4 & = & 1 \\
2x_1 & & & - & x_3 & - & 2x_4 & = & -1 \\
-x_1 & - & x_2 & + & 5x_3 & + & 3x_4 & = & 2.
\end{array}
$$

Then we can re-write this system as

$$Ax = b$$

where

$$
A = \begin{bmatrix} 1 & -1 & 4 & 1 \\ 2 & 0 & -1 & -2 \\ -1 & -1 & 5 & 3 \end{bmatrix}, b = \begin{bmatrix} 1 \\ -1 \\ 2 \end{bmatrix}.
$$

First we will find the null space of the matrix A and then find the set of all solutions to the system.

Using the rref() function from the **pracma** *package:*

```
library(pracma)
A <- matrix(c(1, -1, 4, 1, 0, 2, 0, -1, -2, 0, -1, -1, 5, 3, 0),
nrow=3, ncol=5, byrow=TRUE)
rref(A)
```

R *outputs as follows:*

```
> rref(A)
     [,1] [,2] [,3] [,4] [,5]
[1,]    1    0 -0.5   -1    0
[2,]    0    1 -4.5   -2    0
[3,]    0    0  0.0    0    0
```

This means the reduced echelon form of the system is

$$\begin{aligned} x_1 \quad - \quad 0.5x_3 \quad - \quad x_4 &= \quad 0 \\ x_2 \quad - \quad 4.5x_3 \quad - \quad 2x_2 &= \quad 0. \end{aligned}$$

This means the solutions of the system of linear equations are

$$\begin{aligned} x_1 &= \quad 0.5t + s \\ x_2 &= \quad 4.5t + 2s \\ x_3 &= \quad t \\ x_4 &= \quad s \end{aligned}$$

where t and s are any real numbers. The null space of the coefficient matrix A is

$$Null(A) = \left\{ t \cdot \begin{bmatrix} 0.5 \\ 4.5 \\ 1 \\ 0 \end{bmatrix} + s \cdot \begin{bmatrix} 1 \\ 2 \\ 0 \\ 1 \end{bmatrix} \mid -\infty < t, s < \infty \right\}.$$

Therefore $Null(A)$ is a vector subspace of $\mathbb{R}^4$ spanned by two vectors

$$\begin{bmatrix} 0.5 \\ 4.5 \\ 1 \\ 0 \end{bmatrix}, \begin{bmatrix} 1 \\ 2 \\ 0 \\ 1 \end{bmatrix}.$$

Now we notice that the vector

$$x_0 = \begin{bmatrix} -1/2 \\ -3/2 \\ 0 \\ 0 \end{bmatrix}$$

is a solution to the system of linear equations. Thus, the set of all solutions for the system of linear equations is:

$$\left\{ \begin{bmatrix} -1/2 \\ -3/2 \\ 0 \\ 0 \end{bmatrix} + t \cdot \begin{bmatrix} 0.5 \\ 4.5 \\ 1 \\ 0 \end{bmatrix} + s \cdot \begin{bmatrix} 1 \\ 2 \\ 0 \\ 1 \end{bmatrix} \mid -\infty < t, s < \infty \right\}.$$

Now we shift our focus to the **column space** of the matrix A. The column space of the matrix A is also a vector space and if the right hand side vector b is in the column space, then the system of linear equations $A \cdot x = b$ is feasible, i.e., there exists a solution.

Definition 31 *Suppose we have an $m \times n$ matrix A and let a_i be the ith column of the matrix A. Then the* **column space** *of the matrix A, denoted by $Col(A)$, is a vector space spanned by $a_1, a_2, \ldots, a_n$, i.e.,*

$$Col(A) = \{\alpha_1 \cdot a_1 + \alpha_2 \cdot a_2 + \ldots + \alpha_n \cdot a_n | -\infty < \alpha_1, \alpha_2, \ldots \alpha_n < \infty\}.$$

Theorem 4.7 *The system of linear equations $A \cdot x = b$ has a solution if and only if the vector b is in the column space of the matrix A, $Col(A)$.*

Going back to the example in the beginning, recall that we have the following system of linear equations:

$$
\begin{array}{rrrrr}
x_1 & - & x_2 & + & 4x_3 & = & 1 \\
2x_1 & & & - & x_3 & = & -1 \\
-x_1 & - & x_2 & + & 5x_3 & = & 2.
\end{array}
$$

For this example, the column space of the coefficient matrix A is

$$\text{Col}(A) = \left\{ \alpha_1 \cdot \begin{bmatrix} 1 \\ 2 \\ -1 \end{bmatrix} + \alpha_2 \cdot \begin{bmatrix} -1 \\ 0 \\ -1 \end{bmatrix} + \alpha_3 \cdot \begin{bmatrix} 4 \\ -1 \\ 5 \end{bmatrix} \,\middle|\, -\infty < \alpha_1, \alpha_2, \alpha_3 < \infty \right\}.$$

If $\alpha_1 = -1/2$, $\alpha_2 = -3/2$, $\alpha_3 = 0$, we have

$$(-1/2) \cdot \begin{bmatrix} 1 \\ 2 \\ -1 \end{bmatrix} + (-3/2) \cdot \begin{bmatrix} -1 \\ 0 \\ -1 \end{bmatrix} + 0 \cdot \begin{bmatrix} 4 \\ -1 \\ 5 \end{bmatrix} = \begin{bmatrix} 1 \\ -1 \\ 2 \end{bmatrix}.$$

Therefore, this system has a solution.

Finally, we introduce the **row space** of the matrix A.

Definition 32 *Suppose we have an $m \times n$ matrix A and let a^i be the ith row of the matrix A. Then the* **row space** *of the matrix A, denoted by $Row(A)$, is a vector space spanned by $a^1, a^2, \ldots, a^m$, i.e.,*

$$Col(A) = \{\beta_1 \cdot a^1 + \beta_2 \cdot a^2 + \ldots + \beta_m \cdot a^m | -\infty < \beta_1, \beta_2, \ldots \beta_m < \infty\}.$$

Theorem 4.8 *Suppose we have a system of linear equations $A \cdot x = b$. Let B be the augmented matrix of the system of linear equations, i.e.,*

$$B = [A|b].$$

Suppose v is in the row space of B, $Row(B)$, and let B' be the matrix created by adding v to B as its row. Then the system of linear equations whose augmented matrix is B' has the same solutions as the system of linear equations $A \cdot x = b$.

Going back to the example in the beginning, recall that we have the following system of linear equations:

$$
\begin{array}{rcrcrcr}
x_1 & - & x_2 & + & 4x_3 & = & 1 \\
2x_1 & & & - & x_3 & = & -1 \\
-x_1 & - & x_2 & + & 5x_3 & = & 2.
\end{array}
$$

Recall that the solution set of the system is

$$
\left\{ \begin{bmatrix} -1/2 \\ -3/2 \\ 0 \end{bmatrix} + t \cdot \begin{bmatrix} 0.5 \\ 4.5 \\ 1 \end{bmatrix} \,\middle|\, -\infty < t < \infty \right\}.
$$

For this example, the row space of the augmented matrix B is

Row $(B) =$
$\{\beta_1 \cdot [1,-1,4,1] + \beta_2 \cdot [2,0,-1,-1] + \beta_3 \cdot [-1,-1,5,2] \,|\, -\infty < \beta_1, \beta_2, \beta_3 < \infty\}.$

Let $\beta_1 = \beta_2 = \beta_3 = 1$. Then we have

$$
[1,-1,4,1] + [2,0,-1,-1] + [-1,-1,5,2] = [2,-2,8,2].
$$

The new matrix B' is

$$
B' = \begin{bmatrix} 1 & -1 & 4 & 1 \\ 2 & 0 & -1 & -1 \\ -1 & -1 & 5 & 2 \\ 2 & -2 & 8 & 2 \end{bmatrix}.
$$

The new system of linear equations is

$$
\begin{array}{rcrcrcr}
x_1 & - & x_2 & + & 4x_3 & = & 1 \\
2x_1 & & & - & x_3 & = & -1 \\
-x_1 & - & x_2 & + & 5x_3 & = & 2 \\
2x_1 & - & 2x_2 & + & 8x_3 & = & 2.
\end{array}
$$

Let

$$
v = \begin{bmatrix} -1/2 \\ -3/2 \\ 0 \end{bmatrix} + t \cdot \begin{bmatrix} 0.5 \\ 4.5 \\ 1 \end{bmatrix}
$$

for t is a real number. Then we plug v into the new system of linear equations:

$$
\begin{array}{rcrcrcr}
(-1/2 + 0.5t) & - & (-3/2 + 4.5t) & + & 4t & = & 1 \\
2(-1/2 + 0.5t) & & & - & t & = & -1 \\
-(-1/2 + 0.5t) & - & (-3/2 + 4.5t) & + & 5t & = & 2 \\
2(-1/2 + 0.5t) & - & 2(-3/2 + 4.5t) & + & 8t & = & 2.
\end{array}
$$

Therefore the solution set for this new system of linear equations is also

$$
\left\{ \begin{bmatrix} -1/2 \\ -3/2 \\ 0 \end{bmatrix} + t \cdot \begin{bmatrix} 0.5 \\ 4.5 \\ 1 \end{bmatrix} \,\middle|\, -\infty < t < \infty \right\}.
$$

222 *Linear Algebra and Its Applications with R*

4.3.4 Checkmarks

- The definition of the null space of a matrix.

- The definition of the column space of a matrix.

- The definition of the row space of a matrix.

- You can give an example of the null space of a matrix.

- You can give an example of the column space of a matrix.

- You can give an example of the row space of a matrix.

- With the Null() function from the MASS package in R, you can perform computational examples in this section.

4.3.5 Conceptual Quizzes

Quiz 124 True or False: *Suppose A is an $m \times n$ matrix. Then the null space is a vector subspace of $\mathbb{R}^n$.*

Quiz 125 True or False: *Suppose A is an $m \times n$ matrix. Then the null space is a vector subspace of $\mathbb{R}^m$.*

Quiz 126 True or False: *Suppose A is an $m \times n$ matrix. Then the column space is a vector subspace of $\mathbb{R}^n$.*

Quiz 127 True or False: *Suppose A is an $m \times n$ matrix. Then the column space is a vector subspace of $\mathbb{R}^m$.*

Quiz 128 True or False: *Suppose A is an $m \times n$ matrix. Then the row space is a vector subspace of $\mathbb{R}^n$.*

Quiz 129 True or False: *Suppose A is an $m \times n$ matrix. Then the row space is a vector subspace of $\mathbb{R}^m$.*

Quiz 130 True or False: *Suppose A is an $m \times n$ matrix and b is an m-dimensional vector. Then the system $A \cdot x = b$ has a unique solution if the null space of the matrix A is empty.*

Quiz 131 True or False: *Suppose A is an $m \times n$ matrix and b is an m-dimensional vector. Then the system $A \cdot x = b$ has a unique solution if the system of linear equations $A \cdot x = 0$ has no solution.*

Quiz 132 True or False: *Suppose A is an $m \times n$ matrix and b is an m-dimensional vector. Then the system $A \cdot x = b$ has a unique solution if the system of linear equations $A \cdot x = 0$ has a unique solution.*

Quiz 133 True or False: *Suppose A is an $m \times n$ matrix and b is an m-dimensional vector. Then the system $A \cdot x = b$ has a unique solution if the system of linear equations $A \cdot x = 0$ has infinitely many solutions.*

4.3.6 Regular Exercises

Exercise 4.10 *Verify Theorem 4.4.*

Exercise 4.11 *Verify Lemma 4.5.*

Exercise 4.12 *Suppose we have a matrix*

$$A = \begin{bmatrix} 0 & 0 & 3 \\ -3 & -1 & -2 \\ 2 & 1 & 3 \end{bmatrix}.$$

1. *Compute the null space of A.*

2. *Compute the column space of A.*

3. *Compute the row space of A.*

Exercise 4.13 *Suppose we have a matrix*

$$A = \begin{bmatrix} 3 & -1 & -3 \\ 0 & 3 & 1 \\ 3 & 2 & -2 \end{bmatrix}.$$

1. *Compute the null space of A.*

2. *Compute the column space of A.*

3. *Compute the row space of A.*

Exercise 4.14 *Suppose we have a matrix*

$$A = \begin{bmatrix} -2 & 0 & 0 & 0 \\ 3 & 0 & 1 & 1 \\ 2 & 0 & -2 & -1 \end{bmatrix}.$$

1. *Compute the null space of A.*

2. *Compute the column space of A.*

3. *Compute the row space of A.*

Exercise 4.15 *Suppose we have a matrix*

$$A = \begin{bmatrix} -3 & 3 & 2 & 2 \\ 2 & -1 & -2 & -3 \\ 1 & 2 & -2 & -1 \end{bmatrix}.$$

1. *Compute the null space of A.*

2. *Compute the column space of A.*

3. *Compute the row space of A.*

Exercise 4.16 *Suppose we have a matrix*

$$A = \begin{bmatrix} 1 & -3 & 2 & -2 \\ 1 & -2 & 0 & -3 \\ 0 & 3 & -3 & -3 \\ 0 & -1 & 1 & 1 \end{bmatrix}.$$

1. *Compute the null space of A.*

2. *Compute the column space of A.*

3. *Compute the row space of A.*

Exercise 4.17 *Suppose we have a matrix*

$$A = \begin{bmatrix} 2 & -2 & -3 & -3 \\ 1 & 3 & 3 & 0 \\ -3 & -2 & -1 & 3 \\ -2 & 2 & -3 & 0 \end{bmatrix}.$$

1. *Compute the null space of A.*

2. *Compute the column space of A.*

3. *Compute the row space of A.*

Exercise 4.18 *Suppose we have a matrix*

$$A = \begin{bmatrix} 1 & -1 & 1 & 1 & -2 \\ -2 & 2 & -1 & 2 & 1 \\ 3 & 2 & -3 & 0 & -2 \\ 1 & 4 & -4 & 2 & -1 \end{bmatrix}.$$

1. *Compute the null space of A.*

2. *Compute the column space of A.*

3. *Compute the row space of A.*

Exercise 4.19 *Consider the system of linear equations in Exercise 3.1. Using the null space of A, compute the set of all solutions to the system.*

Exercise 4.20 *Consider the system of linear equations in Exercise 3.2. Using the null space of A, compute the set of all solutions to the system.*

Exercise 4.21 *Consider the system of linear equations in Exercise 3.3. Using the null space of A, compute the set of all solutions to the system.*

Exercise 4.22 *Consider the system of linear equations in Exercise 3.4. Using the null space of A, compute the set of all solutions to the system.*

Exercise 4.23 *Consider the system of linear equations in Exercise 3.5. Using the null space of A, compute the set of all solutions to the system.*

Exercise 4.24 *Consider the system of linear equations in Exercise 3.6. Using the null space of A, compute the set of all solutions to the system.*

Exercise 4.25 *Consider the system of linear equations in Exercise 3.7. Using the null space of A, compute the set of all solutions to the system.*

Exercise 4.26 *Consider the system of linear equations in Exercise 3.8. Using the null space of A, compute the set of all solutions to the system.*

Exercise 4.27 *Consider the system of linear equations in Exercise 3.9. Using the null space of A, compute the set of all solutions to the system.*

Exercise 4.28 *Consider the system of linear equations in Exercise 3.10. Using the null space of A, compute the set of all solutions to the system.*

4.3.7 Lab Exercises

Lab Exercise 140 *Using the Null() function in* R, *compute the null space of the matrix in Exercise 4.12.*

Lab Exercise 141 *Using the Null() function in* R, *compute the null space of the matrix in Exercise 4.13.*

Lab Exercise 142 *Using the Null() function in* R, *compute the null space of the matrix in Exercise 4.14.*

Lab Exercise 143 *Using the Null() function in* R, *compute the null space of the matrix in Exercise 4.15.*

Lab Exercise 144 *Using the Null() function in* R, *compute the null space of the matrix in Exercise 4.16.*

Lab Exercise 145 *Using the Null() function in* R, *compute the null space of the matrix in Exercise 4.17.*

Lab Exercise 146 *Using the Null() function in* R, *compute the null space of the matrix in Exercise 4.18.*

4.3.8 Practical Applications

Now we go back to the introductory example, the British woman's tea problem. Let 1 = Milk and 2 = Tea for notation. As we discussed in Section 1.1, in order to test the hypotheses, we fix the sum of each row and the sum of each column. Then we assign a variable x_{ij} for each (i,j)th entry in the 2×2 contingency table.

Guess/True	Milk	Tea	Total
Milk	x_{11}	x_{12}	4
Tea	x_{21}	x_{22}	4
Total	4	4	8

This can be written as a system of linear equations such that:

$$
\begin{array}{ccccccc}
x_{11} & + & x_{12} & & & = & 4 \\
 & & & x_{21} & + & x_{22} & = & 4 \\
x_{11} & & & + & x_{21} & & = & 4 \\
 & & x_{12} & & & + & x_{22} & = & 4.
\end{array}
$$

Therefore we have the system of linear equations

$$A \cdot x = b$$

where

$$
A = \begin{bmatrix} 1 & 1 & 0 & 0 \\ 0 & 0 & 1 & 1 \\ 1 & 0 & 1 & 0 \\ 0 & 1 & 0 & 1 \end{bmatrix}, b = \begin{bmatrix} 4 \\ 4 \\ 4 \\ 4 \end{bmatrix}.
$$

Here we compute the null space of the matrix A with the Null function from the MASS package in R. First, we upload the MASS library and then define the coefficient matrix A and the right hand side vector b:

```
library(MASS)
A <- matrix(c(1,1,0,0,0,0,1,1,1,0,1,0,0,1,0,1), 4, 4, byrow=TRUE)
b <- c(4, 4, 4, 4)
```

Then we use the Null() function to compute the null space of the matrix A:

```
v <- Null(t(A))
```

R outputs the vector spanning the null space of A:

```
> v
      [,1]
[1,]  0.5
[2,] -0.5
[3,] -0.5
[4,]  0.5
```

Therefore, the vector

$$v = \begin{bmatrix} 1 \\ -1 \\ -1 \\ 1 \end{bmatrix}$$

spans the null space of A. Note that the observed table is a feasible solution. Thus, we set

$$x_0 = \begin{bmatrix} 3 \\ 1 \\ 1 \\ 3 \end{bmatrix}$$

and we know that

$$x_0 + t \cdot v \tag{4.1}$$

For any real numbers, t is a feasible solution for the system of linear equations. However, note that the number in each cell in the table must be a non-negative integer. Thus, we have

$$x_0 + t \cdot v \geq 0$$

for some integer t.

Fisher's exact test is to enumerate all possible contingency tables satisfying the given row sums and column sums. Thus we enumerate all tables satisfying the given row and column sums using the conditions in (4.1). For $t = 1$ we have

$$x_1 = x_0 + v = \begin{bmatrix} 4 \\ 0 \\ 0 \\ 4 \end{bmatrix}$$

which represents the table

Guess/True	Milk	Tea	Total
Milk	4	0	4
Tea	0	4	4
Total	4	4	8

For $t = 10$ we have the observed table

$$x_2 = x_0 + 0 \cdot v = \begin{bmatrix} 3 \\ 1 \\ 1 \\ 3 \end{bmatrix}$$

which represents the table

Guess/True	Milk	Tea	Total
Milk	3	1	4
Tea	1	3	4
Total	4	4	8

For $t = -1$ we have

$$x_3 = x_0 - v = \begin{bmatrix} 2 \\ 2 \\ 2 \\ 2 \end{bmatrix}$$

which represents the table

Guess/True	Milk	Tea	Total
Milk	2	2	4
Tea	2	2	4
Total	4	4	8

For $t = -2$ we have

$$x_4 = x_0 - 2 \cdot v = \begin{bmatrix} 1 \\ 3 \\ 3 \\ 1 \end{bmatrix}$$

which represents the table

Guess/True	Milk	Tea	Total
Milk	1	3	4
Tea	3	1	4
Total	4	4	8

For $t = -3$ we have

$$x_5 = x_0 + v = \begin{bmatrix} 0 \\ 4 \\ 4 \\ 0 \end{bmatrix}$$

which represents the table

Guess/True	Milk	Tea	Total
Milk	0	4	4
Tea	4	0	4
Total	4	4	8

Now we compute the probability to observing each table with given row sums and column sums under the **hypergeometric distribution**. We can use the dhyper() function in R. For $t = 1$, we have the probability to observing the table x_1 is

```
> dhyper(4, 4, 4, 4)
[1] 0.01428571
```

For $t = 0$, we have the probability to observing the table x_2 is

```
> dhyper(3, 4, 4, 4)
[1] 0.2285714
```

The p-value for the hypotheses is $P(x_1) + P(x_2) = 0.01428571 + 0.2285714 = 0.2428571$.

If we use the fisher.test() function, we can conduct Fisher's exact test. For this we create the table as a matrix and call the fisher.test() function:

```
Tea <- matrix(c(3, 1, 1, 3), 2, 2, byrow = TRUE)
fisher.test(Tea, alternative = "greater")
```

Then R outputs as follows:

```
> fisher.test(Tea, alternative = "greater")

Fisher's Exact Test for Count Data

data:  Tea
p-value = 0.2429
alternative hypothesis: true odds ratio is greater than 1
95 percent confidence interval:
 0.3135693      Inf
sample estimates:
odds ratio
  6.408309
```

Thus we have the same p-value. If we set the significance level of 0.05, then the p-value is bigger than 0.05. Thus, we fail to reject the null hypothesis and this means that her guess cannot be differentiated from a random guess.

4.3.9 Supplements with python Code

We use the nullspace() function from the sympy package in python [28]. Suppose we have a matrix A such that

$$A = \begin{bmatrix} 1 & -1 & 4 \\ 2 & 0 & -1 \\ -1 & -1 & 5 \end{bmatrix}.$$

We will compute the vector(s) spanning the null space of the matrix A.

First, we upload the sympy package:

```
# Sympy is a library in python for
# symbolic Mathematics
from sympy import Matrix
```

Then we define the matrix A:

```
A = Matrix([[1, -1, 4],[2, 0, -1],[-1, -1, 5]])
```

To compute the vector(s) to span the null space of the matrix A we call the nullspace() function:

```
# Null Space of A
NullSpace = A.nullspace()
```

Then **python** outputs:

```
>>> NullSpace
[Matrix([
[1/2],
[9/2],
[  1]])]
```

This means the vector v such that

$$v = \begin{bmatrix} 1/2 \\ 9/2 \\ 1 \end{bmatrix}$$

spans the null space of the matrix A.

4.4   Spanning Sets and Bases

In Practical Applications in Section 4.3, we focused on a sparse 2×2 contingency table for the British woman's claims on teas. To conduct a hypothesis test on the sparse table we used Fisher's exact test. In order to do so, we had to enumerate all possible 2×2 contingency tables with the same row and column sums of the observed table. To enumerate them all, we computed the vector which spans the null space of the coefficient matrix for the system of linear equations defined from the row and column sums. However, how do we know that the vector we computed in Practical Applications in Section 4.3 is enough to span the null space of the coefficient matrix? Do we need more? In this section, we will discuss a **spanning set** and also discuss a **basis** for a vector space.

4.4.1   Task Completion Checklist

- During the Lecture:

    1. Read the definition of a spanning set of a vector space.

    2. Read the definition of an independent set of vectors in a vector space.

3. Read the definition of a basis of a vector space.

4. Learn how to give an example of a spanning set of a vector space.

5. Learn how to give an example of an independent set of vectors in a vector space.

6. Read how to give an example of a basis of a vector space.

7. Read the definition of the rank of a matrix.

8. With R, learn how to perform computational examples in this section.

• After the Lecture:

1. Take conceptual quizzes to make sure you understand the materials in this section.

2. Do some regular exercises.

3. Conduct lab exercises with R.

4. Conduct practical applications with R.

5. If you are interested in python, read the supplement in this section and conduct lab exercises and practical applications with python.

4.4.2 Working Examples

For this section we work an example from [14]. Descamps et.al. worked on king penguins (Aptenodytes patagonicus) and how their nesting structures affect their survival. For each of the following three nesting areas (lower, middle, and upper) on Possession Island in the Crozet Archipelago, they marked 50 penguins. Then they checked these penguins a year later to learn if they survived or not. The results are shown in Table 4.1.

Area	Alive	Dead	Total
Lower nesting area	43	7	50
Middle nesting area	44	6	50
Upper nesting area	49	1	50
Total	136	14	150

TABLE 4.1
Contingency table for king penguins in Possession Island in the Crozet Archipelago collected by [14].

Descamps et.al. had these hypotheses:

• H_0 : The location of the nest and death rate among king penguins in Possession Island are not correlated.

• H_1 : The location of the nest and death rate among king penguins in Possession Island are correlated.

They used the Chi-square distribution to conduct a hypothesis test with these hypotheses and also used Fisher's exact test. The results from the Chi-square distribution and Fisher's exact test disagreed about the conclusion. Thus, in Practical Applications, we will apply the Chi-square distribution and Fisher's exact test to see how they are different. In order to apply Fisher's exact test, we need to know a **spanning set** and a **basis** of the null space of the coefficient matrix for the system of linear equations to define the row and column sums for this hypothesis test. Therefore, in Practical Applications, we will compute them with the help of R.

4.4.3   Spanning Sets and Bases

Since a vector space usually has infinitely many vectors (except the trivial vector space which consists of only the zero vector **0**), we would like to have a way to describe a vector space explicitly. One way is a **spanning set** of a vector space. First, it is useful to define **linear combinations**.

Definition 33 *Suppose we have a set of vectors* $\{v_1, v_2, \ldots, v_k\}$ *which is a subset of a vector space* V. *Suppose we also have a vector* x *from* V. *Then we say* x *is a* **linear combination** *of* $v_1, v_2, \ldots, v_k$ *if* x *can be written as*

$$x = \alpha_1 \cdot v_1 + \alpha_2 \cdot v_2 + \ldots + \alpha_k \cdot v_k$$

where $\alpha_1, \ldots \alpha_k$ *are real numbers.*

Definition 34 *Suppose we have a set of vectors* $\{v_1, v_2, \ldots, v_k\}$ *which is a subset of a vector space* V. *Then we say* $v_1, v_2, \ldots, v_k$ **span** V *if any vector* x *in* V *can be written as a linear combination of* $v_1, v_2, \ldots, v_k$.

Now we define **spanning sets** of a vector space:

Definition 35 *Suppose a set of vectors* $\{v_1, v_2, \ldots, v_k\}$ *is a subset of a vector space* V. *Then we call* $\{v_1, v_2, \ldots, v_k\}$ *a* **spanning set** *of* V *if* $v_1, v_2, \ldots, v_k$ *span* V.

Example 102 *Let* $V = \mathbb{R}^3$. *We take vectors*

$$v_1 = \begin{bmatrix} 1 \\ 0 \\ 0 \end{bmatrix}, v_2 = \begin{bmatrix} 0 \\ 1 \\ 0 \end{bmatrix}, v_3 = \begin{bmatrix} 0 \\ 0 \\ 1 \end{bmatrix}.$$

Let

$$x = \begin{bmatrix} x_1 \\ x_2 \\ x_3 \end{bmatrix}.$$

Then we set

$$\begin{bmatrix} x_1 \\ x_2 \\ x_3 \end{bmatrix} = \alpha_1 \cdot \begin{bmatrix} 1 \\ 0 \\ 0 \end{bmatrix} + \alpha_2 \cdot \begin{bmatrix} 0 \\ 1 \\ 0 \end{bmatrix} + \alpha_3 \cdot \begin{bmatrix} 0 \\ 0 \\ 1 \end{bmatrix}.$$

This forms a system of linear equations such that

$$
\begin{array}{rcl}
\alpha_1 & = & x_1 \\
\alpha_2 & = & x_2 \\
\alpha_3 & = & x_3.
\end{array}
$$

Then we have

$$x_1 = \alpha_1, \ x_2 = \alpha_2, \ x_3 = \alpha_3.$$

Since $\alpha_1, \alpha_2, \alpha_3$ are all real numbers, $\{v_1, v_2, v_3\}$ is a spanning set of V.

Example 103 *Let $V = \mathbb{R}^3$. We take vectors*

$$
v_1 = \begin{bmatrix} 1 \\ 0 \\ 0 \end{bmatrix}, \ v_2 = \begin{bmatrix} 0 \\ 1 \\ 0 \end{bmatrix}, \ v_3 = \begin{bmatrix} 0 \\ 0 \\ 1 \end{bmatrix}, \ v_4 = \begin{bmatrix} 1 \\ 1 \\ 1 \end{bmatrix}.
$$

Let

$$
x = \begin{bmatrix} x_1 \\ x_2 \\ x_3 \end{bmatrix}.
$$

Then we set

$$
\begin{bmatrix} x_1 \\ x_2 \\ x_3 \end{bmatrix} = \alpha_1 \cdot \begin{bmatrix} 1 \\ 0 \\ 0 \end{bmatrix} + \alpha_2 \cdot \begin{bmatrix} 0 \\ 1 \\ 0 \end{bmatrix} + \alpha_3 \cdot \begin{bmatrix} 0 \\ 0 \\ 1 \end{bmatrix} + \alpha_4 \cdot \begin{bmatrix} 1 \\ 1 \\ 1 \end{bmatrix}.
$$

This forms a system of linear equations such that

$$
\begin{array}{rcccl}
\alpha_1 & & + & \alpha_4 & = & x_1 \\
\alpha_2 & & + & \alpha_4 & = & x_2 \\
& \alpha_3 & + & \alpha_4 & = & x_3.
\end{array}
$$

Then we have

$$x_1 = \alpha_1 + \alpha_4, \ x_2 = \alpha_2 + \alpha_4, \ x_3 = \alpha_3 + \alpha_4.$$

Since $\alpha_1, \alpha_2, \alpha_3, \alpha_4$ are all real numbers, $\{v_1, v_2, v_3, v_4\}$ is a spanning set of V.

Example 104 *Let $V = \mathbb{R}^3$. We take vectors*

$$
v_1 = \begin{bmatrix} 1 \\ 0 \\ 0 \end{bmatrix}, \ v_2 = \begin{bmatrix} 1 \\ 1 \\ 1 \end{bmatrix}.
$$

Let

$$
x = \begin{bmatrix} x_1 \\ x_2 \\ x_3 \end{bmatrix}.
$$

Then we set

$$\begin{bmatrix} x_1 \\ x_2 \\ x_3 \end{bmatrix} = \alpha_1 \cdot \begin{bmatrix} 1 \\ 0 \\ 0 \end{bmatrix} + \alpha_2 \cdot \begin{bmatrix} 1 \\ 1 \\ 1 \end{bmatrix}.$$

This forms a system of linear equations such that

$$\begin{aligned} \alpha_1 + \alpha_2 &= x_1 \\ \alpha_2 &= x_2 \\ \alpha_2 &= x_3. \end{aligned}$$

Then we have

$$x_1 = \alpha_1 + \alpha_2, \; x_2 = \alpha_2, \; x_3 = \alpha_2.$$

This means that $x_2 = x_3$ always. Therefore we cannot write all vectors in V as a linear combination of v_1, v_2. For example, a vector

$$\begin{bmatrix} 1 \\ 2 \\ 1 \end{bmatrix}$$

cannot be written as a linear combination of v_1, v_2. Therefore v_1, v_2 do not span V.

We notice from Example 102 and Example 103, $\{v_1, v_2, v_3, v_4\}$ is a bigger set than $\{v_1, v_2, v_3\}$ but both of them are spanning sets of V. This means that $\{v_1, v_2, v_3, v_4\}$ is not the smallest spanning set for V. In order to have an explicit description of a vector space we would like to have the minimum spanning set. To do so, we introduce the notion of **linearly independent**.

Definition 36 *Suppose we have a set of vectors $\{v_1, v_2, \ldots v_k\}$ from a vector space V. Then if there exists a vector v_i such that*

$$v_i = \alpha_1 \cdot v_1 + \alpha_2 \cdot v_2 + \ldots + \alpha_{i-1} \cdot v_{i-1} + \alpha_{i+1} \cdot v_{i+1} + \ldots + \alpha_k \cdot v_k$$

*with $\alpha_j \neq 0$ for some j, then we say $v_1, v_2, \ldots v_k$ are **linearly dependent**. If $v_1, v_2, \ldots v_k$ are not linearly dependent, then we say they are **linearly independent**.*

Theorem 4.9 *Vectors $v_1, v_2, \ldots v_k$ from a vector space V are linearly independent if and only if the system of vector equations*

$$\alpha_1 \cdot v_1 + \alpha_2 \cdot v_2 + \ldots + \alpha_k \cdot v_k = 0$$

has a unique solution

$$\alpha_1 = \alpha_2 = \ldots = \alpha_k = 0.$$

Example 105 *From Example 103, we set*

$$\begin{bmatrix} 0 \\ 0 \\ 0 \end{bmatrix} = \alpha_1 \cdot \begin{bmatrix} 1 \\ 0 \\ 0 \end{bmatrix} + \alpha_2 \cdot \begin{bmatrix} 0 \\ 1 \\ 0 \end{bmatrix} + \alpha_3 \cdot \begin{bmatrix} 0 \\ 0 \\ 1 \end{bmatrix} + \alpha_4 \cdot \begin{bmatrix} 1 \\ 1 \\ 1 \end{bmatrix}.$$

This forms a system of linear equations such that

$$\begin{array}{ccccc} \alpha_1 & & + & \alpha_4 & = & 0 \\ & \alpha_2 & + & \alpha_4 & = & 0 \\ & & \alpha_3 & + & \alpha_4 & = & 0. \end{array}$$

Then we have only one solution for the system of linear equations

$$\alpha_1 = -\alpha_4,\ \alpha_2 = -\alpha_4,\ \alpha_3 = -\alpha_4.$$

Let $\alpha_4 = -1$. Then we have

$$\begin{bmatrix} 1 \\ 1 \\ 1 \end{bmatrix} = \begin{bmatrix} 1 \\ 0 \\ 0 \end{bmatrix} + \begin{bmatrix} 0 \\ 1 \\ 0 \end{bmatrix} + \begin{bmatrix} 0 \\ 0 \\ 1 \end{bmatrix}.$$

Thus, v_1, v_2, v_3, v_4 are linearly dependent.

Example 106 *From Example 102, we set*

$$\begin{bmatrix} 0 \\ 0 \\ 0 \end{bmatrix} = \alpha_1 \cdot \begin{bmatrix} 1 \\ 0 \\ 0 \end{bmatrix} + \alpha_2 \cdot \begin{bmatrix} 0 \\ 1 \\ 0 \end{bmatrix} + \alpha_3 \cdot \begin{bmatrix} 0 \\ 0 \\ 1 \end{bmatrix}.$$

This forms a system of linear equations such that

$$\begin{array}{ccc} \alpha_1 & = & 0 \\ \alpha_2 & = & 0 \\ \alpha_3 & = & 0. \end{array}$$

Then we have a solution for the system of linear equations

$$\alpha_1 = 0,\ \alpha_2 = 0,\ \alpha_3 = 0.$$

Therefore

$$\begin{bmatrix} 1 \\ 0 \\ 0 \end{bmatrix}, \begin{bmatrix} 0 \\ 1 \\ 0 \end{bmatrix}, \begin{bmatrix} 0 \\ 0 \\ 1 \end{bmatrix}$$

are linearly independent.

Definition 37 *A **basis** of a vector space V is a subset $\{v_1, v_2, \ldots, v_k\}$ from V such that*

1. $v_1, v_2, \ldots v_k$ are linearly independent; and

2. $\{v_1, v_2, \ldots, v_k\}$ spans V.

Example 107 *From Example 102,*

$$v_1 = \begin{bmatrix} 1 \\ 0 \\ 0 \end{bmatrix}, v_2 = \begin{bmatrix} 0 \\ 1 \\ 0 \end{bmatrix}, v_3 = \begin{bmatrix} 0 \\ 0 \\ 1 \end{bmatrix}$$

are linearly independent and they span $V = \mathbb{R}^3$, thus $\{v_1, v_2, v_3\}$ is a basis of V.

Example 108 *From Example 103, since*

$$\begin{bmatrix} 1 \\ 1 \\ 1 \end{bmatrix} = \begin{bmatrix} 1 \\ 0 \\ 0 \end{bmatrix} + \begin{bmatrix} 0 \\ 1 \\ 0 \end{bmatrix} + \begin{bmatrix} 0 \\ 0 \\ 1 \end{bmatrix}.$$

Thus, v_1, v_2, v_3, v_4 are linearly dependent and therefore, $\{v_1, v_2, v_3, v_4\}$ is not a basis of $V = \mathbb{R}^3$.

Example 109 *Let $V = \mathbb{R}^3$. Let*

$$v_1 = \begin{bmatrix} 1 \\ 1 \\ 1 \end{bmatrix}, v_2 = \begin{bmatrix} 0 \\ 1 \\ 1 \end{bmatrix}, v_3 = \begin{bmatrix} 0 \\ 0 \\ 1 \end{bmatrix}.$$

Then we want to know whether v_1, v_2, v_2 form a basis for V. First we want to know if v_1, v_2, v_2 span V. We set

$$\begin{bmatrix} x_1 \\ x_2 \\ x_3 \end{bmatrix} = \alpha_1 \cdot \begin{bmatrix} 1 \\ 1 \\ 1 \end{bmatrix} + \alpha_2 \cdot \begin{bmatrix} 0 \\ 1 \\ 1 \end{bmatrix} + \alpha_3 \cdot \begin{bmatrix} 0 \\ 0 \\ 1 \end{bmatrix}.$$

This forms a system of linear equations such that

$$\begin{aligned} \alpha_1 & & & & & = & x_1 \\ \alpha_1 & + & \alpha_2 & & & = & x_2 \\ \alpha_1 & + & \alpha_2 & + & \alpha_3 & = & x_3. \end{aligned}$$

Thus we have

$$x_1 = \alpha_1, \ x_2 = x_1 + \alpha_2, \ x_3 = x_2 + \alpha_3.$$

Since $\alpha_1, \alpha_2, \alpha_3$ are any real numbers, x_1, x_2, x_3 can be any real numbers. Thus v_1, v_2, v_3 span V.

Now we want to show v_1, v_2, v_3 are linearly independent. We set

$$\begin{bmatrix} 0 \\ 0 \\ 0 \end{bmatrix} = \alpha_1 \cdot \begin{bmatrix} 1 \\ 1 \\ 1 \end{bmatrix} + \alpha_2 \cdot \begin{bmatrix} 0 \\ 1 \\ 1 \end{bmatrix} + \alpha_3 \cdot \begin{bmatrix} 0 \\ 0 \\ 1 \end{bmatrix}.$$

This forms a system of linear equations such that

$$
\begin{aligned}
\alpha_1 &&&&&=& 0 \\
\alpha_1 &+& \alpha_2 &&&=& 0 \\
\alpha_1 &+& \alpha_2 &+& \alpha_3 &=& 0.
\end{aligned}
$$

The unique solution for the system of linear equations is

$$\alpha_1 = \alpha_2 = \alpha_3 = 0.$$

Thus, v_1, v_2, v_3 are linearly independent and therefore, they form a basis for V.

Remark 4.10 *The Null() function from the MASS package in R outputs a basis for the null space of a matrix. Also a method that we used to computing a spanning set of the null space of a matrix in Section 4 outputs a basis of the null space of the matrix.*

Example 110 *Suppose we have the following system of 3 linear equations with the variables x_1, x_2, x_3:*

$$
\begin{aligned}
x_1 &-& x_2 &+& 4x_3 &=& 1 \\
2x_1 &&&-& x_3 &=& -1 \\
-x_1 &-& x_2 &+& 5x_3 &=& 2.
\end{aligned}
$$

Here we have

$$Ax = b$$

where

$$A = \begin{bmatrix} 1 & -1 & 4 \\ 2 & 0 & -1 \\ -1 & -1 & 5 \end{bmatrix}, b = \begin{bmatrix} 1 \\ -1 \\ 2 \end{bmatrix}.$$

As we saw in Section 4.3, the null space of the matrix A is spanned by the vector

$$v = \begin{bmatrix} 0.5 \\ 4.5 \\ 1 \end{bmatrix}.$$

Therefore, the null space of the matrix A has a basis $\{v\}$.

Remark 4.11 *To compute a basis for the row space of a matrix, we compute the reduced echelon form of the matrix. The non-zero rows in the reduced echelon form of the matrix form a basis for the row space of the matrix.*

Example 111 *We have the matrix:*

$$A = \begin{bmatrix} 1 & -1 & 4 \\ 2 & 0 & -1 \\ -1 & -1 & 5 \end{bmatrix}.$$

Using the rref() function from the pracma package in R, we have

```
library(pracma)
A <- matrix(c(1, -1, 4, 2, 0, -1, -1, -1, 5), nrow=3, ncol=3,
byrow=TRUE)
rref(A)
```

Then, R *returns:*

```
> rref(A)
     [,1] [,2] [,3]
[1,]   1    0 -0.5
[2,]   0    1 -4.5
[3,]   0    0  0.0
```

Therefore,

$$[1, 0, -0.5], [0, 1, -4.5]$$

form a basis for the row space of the matrix A.

Remark 4.12 *To compute a basis for the column space of a matrix, we compute the reduced echelon form of the matrix. The columns in the reduced echelon form of the matrix, which contains a leading non-zero entry, form a basis of the column space of the matrix.*

Example 112 *From Example 111, we want to compute a basis for the column space of the matrix*

$$A = \begin{bmatrix} 1 & -1 & 4 \\ 2 & 0 & -1 \\ -1 & -1 & 5 \end{bmatrix}.$$

Using the rref() function from the **pracma** *package in* R, *we have the reduced echelon form of the matrix* A

$$\begin{bmatrix} 1 & 0 & -0.5 \\ 0 & 1 & -4.5 \\ 0 & 0 & 0 \end{bmatrix}.$$

Thus,

$$\begin{bmatrix} 1 \\ 0 \\ 0 \end{bmatrix}, \begin{bmatrix} 0 \\ 1 \\ 0 \end{bmatrix}$$

form a basis for the column space of the matrix A.

Definition 38 *The* **dimension** *of a vector space is the number of vectors in a basis for the vector space.*

Example 113 *From Example 111, the dimension of the null space of the matrix*

$$A = \begin{bmatrix} 1 & -1 & 4 \\ 2 & 0 & -1 \\ -1 & -1 & 5 \end{bmatrix}$$

is 1 *since there is only one vector in the basis. Also the dimension of the row space of the matrix A is* 2 *since there are two vectors in the basis. The dimension of the column space of A is also* 2 *since there are two vectors in the basis.*

Theorem 4.13 *Suppose we have an* $m \times n$ *matrix A. Then the dimension of the column space of A is equal to the dimension of the row space of A.*

This theorem leads to the following definition:

Definition 39 *The* **rank** *of the matrix A is the dimension of the row space and the column space of the matrix A.*

Example 114 *From Example 111, suppose we have*

$$A = \begin{bmatrix} 1 & -1 & 4 \\ 2 & 0 & -1 \\ -1 & -1 & 5 \end{bmatrix}.$$

The rank of A is 2.

Theorem 4.14 *Suppose we have an* $m \times n$ *matrix A. Let* n_1 *be the dimension of the null space of A and* n_2 *be the rank of A. Then*

$$n = n_1 + n_2.$$

Example 115 *From Example 111, suppose we have*

$$A = \begin{bmatrix} 1 & -1 & 4 \\ 2 & 0 & -1 \\ -1 & -1 & 5 \end{bmatrix}.$$

The rank of A is 2 *and the dimension of the null space of A is* 1. *So we have* $1 + 2 = 3$.

The following theorem is useful to check the dimension of the null space of a matrix A:

Theorem 4.15 *Suppose we have an* $m \times n$ *matrix A where* $m < n$. *Then the dimension of the null space of A is at least* $n - m$.

4.4.4 Checkmarks

- The definition of a spanning set of a vector space.
- The definition of linearly independent.
- The definition of a basis of a vector space.
- The definition of the rank of a matrix.

- You can give an example of a spanning set of a vector space.

- You can give an example of a set of vectors which are linearly independent.

- You can give an example of a set of vectors which are linearly dependent.

- You can compute a basis of a vector space.

- You can compute the rank of a matrix.

4.4.5   Conceptual Quizzes

Quiz 134 True or False: *Let $V = \mathbb{R}^3$. Then*

$$\begin{bmatrix} 1 \\ 0 \\ 0 \end{bmatrix}, \begin{bmatrix} 0 \\ 1 \\ 0 \end{bmatrix}$$

form a basis for V.

Quiz 135 True or False: *Let $V = \mathbb{R}^3$. Then*

$$\begin{bmatrix} 1 \\ 0 \\ 1 \end{bmatrix}, \begin{bmatrix} 0 \\ 1 \\ 0 \end{bmatrix}, \begin{bmatrix} 1 \\ 1 \\ 0 \end{bmatrix}$$

form a basis for V.

Quiz 136 True or False: *Let $V = \mathbb{R}^3$. Then*

$$\begin{bmatrix} 1 \\ 0 \\ 1 \end{bmatrix}, \begin{bmatrix} 0 \\ 1 \\ 0 \end{bmatrix}, \begin{bmatrix} 1 \\ 1 \\ 0 \end{bmatrix}, \begin{bmatrix} 1 \\ 1 \\ 1 \end{bmatrix}$$

form a basis for V.

Quiz 137 True or False: *Suppose*

$$A = \begin{bmatrix} 1 & -1 & 3 \\ -1 & 0 & -1 \\ 0 & -1 & 2 \end{bmatrix}.$$

Then,

$$\begin{bmatrix} 1 \\ -1 \\ 0 \end{bmatrix}, \begin{bmatrix} -1 \\ 0 \\ -1 \end{bmatrix}, \begin{bmatrix} 3 \\ -1 \\ 2 \end{bmatrix}$$

span the column space of A.

Quiz 138 True or False: *Suppose*

$$A = \begin{bmatrix} 1 & -1 & 3 \\ -1 & 0 & -1 \\ 0 & -1 & 2 \end{bmatrix}.$$

Then,

$$\begin{bmatrix} 1 \\ -1 \\ 0 \end{bmatrix}, \begin{bmatrix} -1 \\ 0 \\ -1 \end{bmatrix}, \begin{bmatrix} 3 \\ -1 \\ 2 \end{bmatrix}$$

span the row space of A.

Quiz 139 True or False: *Suppose*

$$A = \begin{bmatrix} 1 & -1 & 3 \\ -1 & 0 & -1 \\ 0 & -1 & 2 \end{bmatrix}.$$

Then,

$$\begin{bmatrix} 1 \\ -1 \\ 3 \end{bmatrix}, \begin{bmatrix} -1 \\ 0 \\ -1 \end{bmatrix}, \begin{bmatrix} 0 \\ -1 \\ 2 \end{bmatrix}$$

span the row space of A.

Quiz 140 True or False: *Suppose*

$$A = \begin{bmatrix} 1 & -1 & 3 \\ -1 & 0 & -1 \\ 0 & -1 & 2 \end{bmatrix}.$$

Then,

$$\begin{bmatrix} 1 \\ -1 \\ 3 \end{bmatrix}, \begin{bmatrix} -1 \\ 0 \\ -1 \end{bmatrix}, \begin{bmatrix} 0 \\ -1 \\ 2 \end{bmatrix}$$

span the column space of A.

Quiz 141 True or False: *Suppose*

$$A = \begin{bmatrix} 1 & -1 & 3 \\ -1 & 0 & -1 \\ 0 & -1 & 2 \end{bmatrix}.$$

The rank of A is 3.

Quiz 142 True or False: *Suppose*

$$A = \begin{bmatrix} 1 & -1 & 3 \\ -1 & 0 & -1 \\ 0 & -1 & 2 \end{bmatrix}.$$

The dimension of the null space of A is 1.

Quiz 143 True or False: *Suppose*

$$A = \begin{bmatrix} 1 & -1 & 3 & 1 \\ -1 & 0 & -1 & 0 \\ 0 & -1 & 2 & 1 \end{bmatrix}.$$

Then,

$$\begin{bmatrix} 1 \\ -1 \\ 3 \\ 1 \end{bmatrix}, \begin{bmatrix} -1 \\ 0 \\ -1 \\ 0 \end{bmatrix}, \begin{bmatrix} 0 \\ -1 \\ 2 \\ 1 \end{bmatrix}$$

span the column space of A.

Quiz 144 True or False: *Suppose*

$$A = \begin{bmatrix} 1 & -1 & 3 & 1 \\ -1 & 0 & -1 & 0 \\ 0 & -1 & 2 & 1 \end{bmatrix}.$$

The rank of A is 2.

Quiz 145 True or False: *Suppose*

$$A = \begin{bmatrix} 1 & -1 & 3 & 1 \\ -1 & 0 & -1 & 0 \\ 0 & -1 & 2 & 1 \end{bmatrix}.$$

The dimension of the null space of A is 2.

4.4.6   Regular Exercises

Exercise 4.29 *Suppose we have a matrix*

$$A = \begin{bmatrix} 3 & -1 & -3 \\ 0 & 3 & 1 \\ 3 & 2 & -2 \end{bmatrix}.$$

1. *Compute a basis for the null space of A.*

2. *Compute a basis for the column space of A.*

3. *Compute a basis for the row space of A.*

4. *What is the rank of A?*

5. *What is the dimension of the null space of A?*

Exercise 4.30 *Suppose we have a matrix*

$$A = \begin{bmatrix} 1 & 0 & 3 \\ 0 & -2 & 2 \\ -1 & -1 & -2 \end{bmatrix}.$$

1. *Compute a basis for the null space of A.*

2. *Compute a basis for the column space of A.*

3. *Compute a basis for the row space of A.*

4. *What is the rank of A?*

5. *What is the dimension of the null space of A?*

Exercise 4.31 *Suppose we have a matrix*

$$A = \begin{bmatrix} 0 & 1 & 2 \\ 2 & -2 & 2 \\ -1 & 1 & -1 \end{bmatrix}.$$

1. *Compute a basis for the null space of A.*

2. *Compute a basis for the column space of A.*

3. *Compute a basis for the row space of A.*

4. *What is the rank of A?*

5. *What is the dimension of the null space of A?*

Exercise 4.32 *Suppose we have a matrix*

$$A = \begin{bmatrix} -3 & 0 & 3 & -1 \\ -1 & 1 & -3 & 1 \\ -3 & -3 & 3 & 2 \end{bmatrix}.$$

1. *Compute a basis for the null space of A.*

2. *Compute a basis for the column space of A.*

3. *Compute a basis for the row space of A.*

4. *What is the rank of A?*

5. *What is the dimension of the null space of A?*

Exercise 4.33 *Suppose we have a matrix*

$$A = \begin{bmatrix} -2 & 0 & 0 & 0 \\ 3 & 0 & 1 & 1 \\ 2 & 0 & -2 & -1 \end{bmatrix}.$$

1. *Compute a basis for the null space of A.*

2. *Compute a basis for the column space of A.*

3. *Compute a basis for the row space of A.*

4. *What is the rank of A?*

5. *What is the dimension of the null space of A?*

Exercise 4.34 *Suppose we have a matrix*

$$A = \begin{bmatrix} 0 & 0 & 3 & -1 & 2 \\ 1 & 3 & -3 & 2 & -3 \\ 3 & -1 & 0 & 0 & -3 \end{bmatrix}.$$

1. *Compute a basis for the null space of A.*

2. *Compute a basis for the column space of A.*

3. *Compute a basis for the row space of A.*

4. *What is the rank of A?*

5. *What is the dimension of the null space of A?*

Exercise 4.35 *Suppose we have a matrix*

$$A = \begin{bmatrix} 3 & 0 & -1 & 0 & 3 \\ -1 & 2 & -3 & -3 & -1 \\ -3 & -2 & -2 & 0 & 0 \end{bmatrix}.$$

1. *Compute a basis for the null space of A.*

2. *Compute a basis for the column space of A.*

3. *Compute a basis for the row space of A.*

4. *What is the rank of A?*

5. *What is the dimension of the null space of A?*

Exercise 4.36 *From Exercise 1.38, suppose we have a system of linear equations such that*

$$\begin{aligned} x_1 \ - \ 6x_2 &= \ 3 \\ -x_1 \ + \ 2x_2 &= \ -1. \end{aligned}$$

1. *Compute a basis for the null space of the coefficient matrix.*

2. *Compute a basis for the column space of the coefficient matrix.*

3. *Compute a basis for the row space of the coefficient matrix.*

4. *What is the rank of the coefficient matrix?*

5. *What is the dimension of the null space of the coefficient matrix?*

6. Does this system of linear equations have a unique solution?

Exercise 4.37 *From Exercise 1.39, suppose we have a system of linear equations such that*

$$-x_1 \quad - \quad 9x_2 \quad = \quad 1$$
$$-10x_1 \quad + \quad 9x_2 \quad = \quad -10.$$

1. Compute a basis for the null space of the coefficient matrix.

2. Compute a basis for the column space of the coefficient matrix.

3. Compute a basis for the row space of the coefficient matrix.

4. What is the rank of the coefficient matrix?

5. What is the dimension of the null space of the coefficient matrix?

6. Does this system of linear equations have a unique solution?

Exercise 4.38 *From Exercise 1.40, suppose we have a system of linear equations such that*

$$-5x_1 \quad + \quad 5x_2 \quad - \quad x_3 \quad = \quad 57$$
$$-7x_1 \quad - \quad 2x_2 \quad - \quad 4x_3 \quad = \quad 21$$
$$x_1 \quad + \quad 3x_2 \quad + \quad 4x_3 \quad = \quad 3.$$

1. Compute a basis for the null space of the coefficient matrix.

2. Compute a basis for the column space of the coefficient matrix.

3. Compute a basis for the row space of the coefficient matrix.

4. What is the rank of the coefficient matrix?

5. What is the dimension of the null space of the coefficient matrix?

6. Does this system of linear equations have a unique solution?

Exercise 4.39 *From Exercise 1.41, suppose we have a system of linear equations such that*

$$4x_1 \quad - \quad 10x_2 \quad + \quad 4x_3 \quad = \quad 66$$
$$-4x_1 \quad + \quad 3x_2 \quad \quad \quad = \quad -27$$
$$9x_1 \quad + \quad 6x_2 \quad + \quad 2x_3 \quad = \quad -1.$$

1. Compute a basis for the null space of the coefficient matrix.

2. Compute a basis for the column space of the coefficient matrix.

3. Compute a basis for the row space of the coefficient matrix.

4. What is the rank of the coefficient matrix?

5. What is the dimension of the null space of the coefficient matrix?

6. *Does this system of linear equations have a unique solution?*

Exercise 4.40 *From Exercise 1.42, suppose we have a system of linear equations such that*

$$-5x_1 + 5x_2 - x_3 = 57$$
$$-7x_1 - 2x_2 - 4x_3 = 21$$
$$x_1 + 3x_2 + 4x_3 = 3.$$

1. *Compute a basis for the null space of the coefficient matrix.*

2. *Compute a basis for the column space of the coefficient matrix.*

3. *Compute a basis for the row space of the coefficient matrix.*

4. *What is the rank of the coefficient matrix?*

5. *What is the dimension of the null space of the coefficient matrix?*

6. *Does this system of linear equations have a unique solution?*

4.4.7   Lab Exercises

Lab Exercise 147 *Using the Null() function from the MASS package and the rref() from pracma package in R, repeat Exercise 4.29.*

Lab Exercise 148 *Using the Null() function from the MASS package and the rref() from pracma package in R, repeat Exercise 4.30.*

Lab Exercise 149 *Using the Null() function from the MASS package and the rref() from pracma package in R, repeat Exercise 4.31.*

Lab Exercise 150 *Using the Null() function from the MASS package and the rref() from pracma package in R, repeat Exercise 4.32.*

Lab Exercise 151 *Using the Null() function from the MASS package and the rref() from pracma package in R, repeat Exercise 4.33.*

Lab Exercise 152 *Using the Null() function from the MASS package and the rref() from pracma package in R, repeat Exercise 4.34.*

Lab Exercise 153 *Using the Null() function from the MASS package and the rref() from pracma package in R, repeat Exercise 4.35.*

Lab Exercise 154 *Using the Null() function from the MASS package and the rref() from pracma package in R, repeat Exercise 4.36.*

Lab Exercise 155 *Using the Null() function from the MASS package and the rref() from pracma package in R, repeat Exercise 4.37.*

Lab Exercise 156 *Using the Null() function from the* MASS *package and the rref() from* pracma *package in* R, *repeat Exercise 4.38.*

Lab Exercise 157 *Using the Null() function from the* MASS *package and the rref() from* pracma *package in* R, *repeat Exercise 4.39.*

Lab Exercise 158 *Using the Null() function from the* MASS *package and the rref() from* pracma *package in* R, *repeat Exercise 4.40.*

4.4.8 Practical Applications

Going back to the dataset from [14], recall that Descamps et.al. worked on king penguins (Aptenodytes patagonicus) and how their nesting structures affect their survival. Descamps et.al. wish to know if the location of a nest might have some effect on babies' survival. In order to study this we have to know some concepts from probability, namely **independence** between two events.

First, to make it easier to analyze, we assign some variables in each cell of the contingency table in Table 4.1. For each of the following three nesting areas, we let 1 = lower, 2 = middle, and 3 = upper on Possession Island in the Crozet Archipelago. Also, for the survival, we let 1 = Alive, and 2 = Dead keeping the row and column sums from Table 4.1 and assigning the variables x_{ij} for nesting areas i and survival status j:

Area	Alive	Dead	Total
Lower nesting area	x_{11}	x_{12}	50
Middle nesting area	x_{21}	x_{22}	50
Upper nesting area	x_{31}	x_{32}	50
Total	136	14	150

Now we review the concept of **independence** between two events from probability. The idea of probability is based in the idea of a **random experiment**. A random experiment is a process whose outcomes have some uncertainty and we are interested in these outcomes. For example, tossing a coin twice is a random experiment. An **event** is a subset of outcomes from all possible outcomes. Going back to the example of tossing a coin twice, an event can be the outcome that the first toss is heads which includes the cases that the first toss is heads and the second toss is heads and that the first toss is heads and the second toss is tails. Suppose we have two events E_1 and E_2. Then we say E_1 and E_2 are independent if the probability of the event that E_1 and E_2 both happen is the multiplication of the probability of E_1 and the probability of E_2. For example, E_1 is the event that the first toss is heads and E_2 is the event that the second toss is tails. Then the probability of E_1 is $1/2$ and the probability of E_2 is $1/2$. Thus, the probability of the event E_1 and E_2 both happen, which is the case that the first toss is heads and the second toss is tails is $1/4 \times 1/4 = 1/2$. Thus, E_1 and E_2 are independent.

A contingency table can be seen as an estimation of the probability of two random variables. In this example, we consider random variables which can be "Lower nesting area", "Middle nesting area", and "Upper nesting area". The second random variable is "Alive" or "Dead". From the table above, the estimated probability of "Lower nesting area" equals $1/3$, the estimated probability of "Middle nesting area" equals $1/3$, and the estimated probability of "Upper nesting area" equals $1/3$. In addition, the estimated probability of "Alive" is $136/150$ and the estimated probability of "Dead" is $14/150$. If these are independent, then the probability of "Lower nesting area" and "Alive" is $1/3 \times 136/150$ and so on. From the contingency table in Table 4.1, we can verify that $1/3 \times 136/150$ is not equal to $43/150$. This is because the contingency table in Table 4.1 is computed from a **sample** of the entire king penguins. So we do not know the true probabilities. Therefore we have to consider some variability of sampling. Thus, we use hypothesis testing to allow "benefit of doubt" by sampling errors.

To set up hypotheses for hypothesis testing, Descamps et.al. had the following hypotheses:

- H_0 : The location of the nest and death rate among king penguins in Possession Island are not "related".

- H_1 : The location of the nest and death rate among king penguins in Possession Island are "related".

Here H_0 is denoted as the **null hypothesis** and H_1 is denoted as the **alternative hypothesis**. This can be rewritten as the following:

- H_0 : The location of the nest and death rate among king penguins in Possession Island are independent.

- H_1 : The location of the nest and death rate among king penguins in Possession Island are not independent.

When conducting a hypothesis test, the first step is to assume that the null hypothesis is true. Often we compare a hypothesis test with an American court system. In a court, we first assume that the defendant is innocent. In this case the null hypothesis is that the defendant is innocent and the alternative hypothesis is that the defendant is guilty. A prosecutor tries to collect evidence to reject the null hypothesis, i.e., to prove that the defendant is guilty. If we have enough evidence, then we reject the null hypothesis (the defendant is guilty). If we do not have enough evidence, then we cannot reject the null hypothesis (we cannot prove that the defendant is guilty). Here we cannot use a hypothesis to prove that the null hypothesis is true (i.e., the defendant is innocent). Therefore, the first step is that we assume the null hypothesis is true.

From Table 4.1, we have the following system of 5 linear equations with 6 unknown variables:

$$
\begin{aligned}
x_{11} + x_{12} &&&&&&= 50 \\
x_{21} + x_{22} &&&&&= 50 \\
&& x_{31} + x_{32} &= 50 \\
x_{11} + x_{21} + x_{31} &&&= 136 \\
x_{12} + x_{22} + x_{32} &&&= 14.
\end{aligned}
$$

Also note that a contingency table can have only non-negative integers in each cell. For example, we cannot have 1.5 king penguins in a contingency table in Table 4.1. So x_{ij} for $i = 1, 2, 3$ and $j = 1, 2$ is a non-negative integer. A solution for this system of linear equations with non-negative integer constraints represents a contingency table which satisfies the same row and column sums of the contingency table in Table 4.1.

It is also known that the set of all possible solutions for such linear equations with non-negative integer constraints is finite. So if we can enumerate all of the solutions of this system of linear equations, then we can enumerate all contingency tables satisfying the row and column sums of the contingency table in Table 4.1. Then if we can compute the probability of each contingency table under the independent assumption, then we can compute the probability of how rare it is to observe the contingency table in Table 4.1 if we assume independence. The probability of each contingency table given the row and column sums is assigned according to the **hypergeometric distribution**. Conducting a hypothesis test of independence by enumerating all contingency tables with the given row and column sums is called **Fisher's** exact test.

Example 116 *Suppose we interview 20 people with the following questions:*

- *What is your political affiliation?*

- *Are you introvert or extroverts?*

Then suppose we have an observed 2×2 contingency table such that

	Dem	Rep	Total
Introvert	3	7	10
Extrovert	6	4	10
Total	9	11	20

Now we form the following hypotheses:

H_0 : *Political affiliation and Introvert/Extrovert are not "related".*
H_0 : *Political affiliation and Introvert/Extrovert are "related".*

In order to find out which hypothesis is more likely from the observed data set, we conduct a hypothesis testing via Fisher's exact test.

In order to conduct Fisher's exact test, we need to enumerate all contingency tables with the same row and column sums of the observed table. There

are 10 tables with the same row and column sums of this observed table in the
following list of all contingency tables with the same row and column sums of
the observed table:

prob	.0001	.0027	.0322	.1500	.3151
Table	0 10 9 1	1 9 8 2	2 8 7 3	3 7 6 4	4 6 5 5

prob	.3151	.1500	.0322	.0027	.0001
Table	5 5 4 6	6 4 3 7	7 3 2 8	8 2 1 9	9 1 0 10

The table entries in red in the list is the observed table. Each table is assigned
its probability according to the hypergeometric distribution shown in the list
above. The observed table has the probability 0.15. *Now we compute the p-value*
for the hypothesis test. This can be computed by summing the probabilities of
the tables with smaller probabilities:

$$.0001 + .0027 + .0322 + .1500 + .1500 + .0322 + .0027 + .0001 = .3698.$$

Therefore, the p-value is 0.3698 *and if the significance level* $\alpha = 0.05$, *then*
we do not reject the null hypothesis. This means that Political affiliation and
Introvert/Extrovert are not "related".

To enumerate all contingency tables with the given row and column sums
is a very hard problem. This is called the #-P hard problem. So in general it
is infeasible to enumerate all solutions of the system of linear equations with
non-negative constraints for contingency tables. Therefore, Diaconis-Sturmfels
suggested **sampling** contingency tables from the set of all possible tables with
the given row and column sums [16]. Since a solution of the system of linear
equations with the non-negative integer constraint represents a contingency
table with the given row and column sums, sampling a contingency table from
the set of all possible tables with the given row and column sums is the same
as sampling a non-negative integer solution of the system of linear equations.

How can we sample a non-negative solution of the system of linear equations? We can run a **random walk** over the set of all possible solutions!

A random walk is a walk between some states (objects) randomly with
some probability. One can find an example of a random walk in Figure 4.7.
The number associated with each arrow in the figure is a probability from one
state to the other. For example, the probability from the state S_1 to S_2 is 1
and the probability from the state S_2 to S_4 is 1/3. If there is no arrow from
one state to another, then the probability is 0. For example, there is no arrow
from the state S_1 to the state S_3, so the probability from the state S_1 to the
state S_3 is 0.

In order to walk randomly from one solution to another solution, we have
to consider a **move**. A move is an arrow in Figure 4.7 for example. In the case

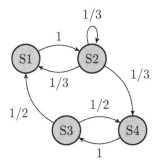

FIGURE 4.7
Example of a random walk on 4 states S_1, S_2, S_3, S_4. The number associated with each arrow is a probability from one state to the other.

of a random walk on the set of all non-negative integer solutions of a system of linear equations, we can use integer vectors in the kernel of the coefficient matrix. For example, suppose we have 3×2 contingency tables with the row and column sums such that

			total
	x_{11}	x_{12}	2
	x_{21}	x_{22}	2
	x_{31}	x_{32}	2
Total	3	3	6

Also note that a contingency table

$$
\begin{array}{cc}
1 & -1 \\
-1 & 1 \\
0 & 0
\end{array}
$$

is in the kernel of the system of linear equations since its row and column sums equal zero.

Also consider the contingency table

			total
	1	1	2
	1	1	2
	1	1	2
Total	3	3	6

which satisfies the given row and column sums. Then notice that

$$
\begin{bmatrix} 1 & 1 \\ 1 & 1 \\ 1 & 1 \end{bmatrix} + \begin{bmatrix} 1 & -1 \\ -1 & 1 \\ 0 & 0 \end{bmatrix} = \begin{bmatrix} 2 & 0 \\ 0 & 2 \\ 1 & 1 \end{bmatrix},
$$

which also satisfies the given row and column sums. If we keep adding integer vectors in the kernel of the coefficient matrix, we can run a random walk in the set of all possible contingency tables with the given row and column sums!

Thus, now let us compute the kernel of the coefficient matrix of the system of linear equations. We have the coefficient matrix of the system of linear equations:

$$A = \begin{bmatrix} 1 & 1 & 0 & 0 & 0 & 0 \\ 0 & 0 & 1 & 1 & 0 & 0 \\ 0 & 0 & 0 & 0 & 1 & 1 \\ 1 & 0 & 1 & 0 & 1 & 0 \\ 0 & 1 & 0 & 1 & 0 & 1 \end{bmatrix}.$$

Clearly, since the number of row vectors is smaller than the number of column vectors of A, the null space has at least $6 - 5 = 1$ dimension.

With the Null() function from the MASS package in R, we compute a basis for the null space of A:

```
library(MASS)
A <- matrix(c(1, 1, 0, 0, 0, 0,
              0, 0, 1, 1, 0, 0,
              0, 0, 0, 0, 1, 1,
              1, 0, 1, 0, 1, 0,
              0, 1, 0, 1, 0, 1), 5, 6, byrow=TRUE)
V <- Null(t(A))
```

Then R outputs a basis for the null space of A:

```
> V
            [,1]        [,2]
[1,] -0.1999668  0.54161482
[2,]  0.1999668 -0.54161482
[3,] -0.3690688 -0.44398374
[4,]  0.3690688  0.44398374
[5,]  0.5690356 -0.09763107
[6,] -0.5690356  0.09763107
```

Therefore, since we have two vectors in a basis, the dimension of the null space of A is 2.

If for all i and j, x_{ij} are real numbers, this basis spans all vectors in the null space of A. But for this problem, x_{ij} are non-negative integers for all i and j, not real numbers. Therefore we need more vectors to span all tables with non-negative integers. In fact, a basis consisting of all integer vectors to span the null space can be obtained by:

$$\left\{ \begin{bmatrix} 1 \\ -1 \\ -1 \\ 1 \\ 0 \\ 0 \end{bmatrix}, \begin{bmatrix} 0 \\ 0 \\ 1 \\ -1 \\ -1 \\ 1 \end{bmatrix}, \begin{bmatrix} 1 \\ -1 \\ 0 \\ 0 \\ -1 \\ 1 \end{bmatrix} \right\}.$$

Diaconis and Sturmfels called such a set a **Markov basis** [16] and in terms of 3×2 tables, they look like

$$
\begin{bmatrix} 1 & -1 \\ -1 & 1 \\ 0 & 0 \end{bmatrix}, \begin{bmatrix} 0 & 0 \\ 1 & 1 \\ -1 & 1 \end{bmatrix}, \begin{bmatrix} 1 & -1 \\ 0 & 0 \\ -1 & 1 \end{bmatrix}.
$$

One can compute a Markov basis using the R package latte [25]. To install the package, we can use the install.packages() function:

```
install_version("latte")
library(latte)
```

This is not the end of installation in order to run. You have to install the LattE software [13]. This is software written in C++ and you can also download binaries as well as source codes to install the code. To download binaries, you have to go to the website https://www.math.ucdavis.edu/~latte/software/packages/binary/ and install LattE in a directory at your computer. If you open the zip file, you can find a folder named "bin". Find the path to the folder and type:

```
set_4ti2_path("PATH TO THE FOLDER")
```

After setting up the folder, use the markov() function to find these vectors:

```
> markov(A)
     [,1] [,2] [,3]
[1,]    0    1    1
[2,]    0   -1   -1
[3,]    1   -1    0
[4,]   -1    1    0
[5,]   -1    0   -1
[6,]    1    0    1
```

In the Fisher's exact test procedure on two-way contingency tables, each time, we randomly select two distinct rows and two distinct columns and we add and subtract by one. This is exactly the same as spanning the set of all integers in the null space of the matrix A.

Then, we add and subtract one of these three vectors in the Markov basis of A to enumerate all possible tables with the row sums and the column sums given in Table 4.1 as we did for the British woman's tea test example in Practical Applications in Section 4.3.

In R, we have the fisher.test() function to conduct Fisher's exact test on the two-way contingency tables under such hypotheses. For our example, we do:

```
X <- matrix(c(43, 7, 44, 6, 49, 1), 3, 2, byrow = TRUE)
fisher.test(X, alternative = "greater")
```

We set "alternative = "greater"" in our case. Then, R outputs:

```
> fisher.test(X, alternative = "greater")

Fisher's Exact Test for Count Data

data:  X
p-value = 0.08963
alternative hypothesis: greater
```

Therefore, we have the p-value equal 0.08963 and with the significance level 0.05, then we fail to reject the null hypothesis, i.e., the location of the nest does not matter for their survival, since the p-value is bigger than the significance level. Now we conduct the Chi-square test where we assume that the observed table is not sparse. We can use the chisq.test() function in R:

```
chisq.test(X)
```

Then we have:

```
> chisq.test(X)

Pearson's Chi-squared test

data:  X
X-squared = 4.8845, df = 2, p-value = 0.08697

Warning message:
In chisq.test(X) : Chi-squared approximation may be incorrect
```

The p-value from the Chi-square test is 0.08697, which is close to the p-value from Fisher's exact test.

There is another method, called the G-test, for the null hypothesis on two-way contingency tables. To conduct the G-test we use the G.test() function from the **RVAideMemoire** package [21]:

```
library(RVAideMemoire)
G.test(X)
```

Then R outputs:

```
> G.test(X)

G-test

data:  X
G = 6.0621, df = 2, p-value = 0.04827
```

The p-value with the G-test is 0.04827, which is smaller than the significance level. Thus, under the G-test with the null hypothesis we reject the null hypothesis, i.e., the location of the nest matters for their survival.

Since this table is sparse, Fisher's exact test is more appropriate to conduct the hypothesis test. Thus, it is more likely that the location of the nest does not matter for their survival according to the observed table.

4.4.9 Supplements with python Code

In this section we did not use any new functions. Therefore we do not have any supplemental python codes in this section.

4.5 Coordinates Systems and Change of Basis

Principal Component Analysis (PCA) is a statistical method used to reduce the dimensionality and for visualization. One of the most important applications is to change a coordinate system so that a new system of coordinates are uncorrelated to each other. In statistics and data science, if variables are correlated, it is not easy to conduct statistical analysis on a data set. Therefore, for the working example in this section, we will apply PCA on the "Default" data set in the ISLR package.

4.5.1 Task Completion Checklist

- During the Lecture:

  1. Read the definition of the basis coordinate of a vector.

  2. Read a graphical interpretation of a change of coordinate.

  3. Read the definition of a linear transformation.

  4. Learn how to compute a coordinate in terms of a basis for a vector space.

  5. Learn how to compute a linear transformation.

  6. With R, learn how to perform computational examples in this section.

- After the Lecture:

  1. Take conceptual quizzes to make sure you understand the materials in this section.

  2. Do some regular exercises.

  3. Conduct lab exercises with R.

  4. Conduct practical applications with R.

5. If you are interested in python, read the supplement in this section and conduct lab exercises and practical applications with python.

4.5.2   Working Examples

For the working example in this section we use the "Default" data set from the ISLR package [24]. The "Default" data set is simulated and it is aimed at predicting which customers will default on their credit card debt. It contains 10,000 observations.

There are four variables: "default", "student", "balance", and "income". The "default" is a binary variable: It states "Yes" if the person defaulted on the credit card and it states "No" if the person did not. The "student" is also a binary variable: It states "Yes" if the person is a student and it states "No" if the person is not a student. The "balance" is a numerical variable which is the person's balance on the credit card in terms of dollars. The "income" is also a numerical variable which reflects the person's annual income in terms of dollars.

Figure 4.8 shows the plot for the "Default" data set. The "default" status is colored red if it is "Yes" and blue if it is "No". Note that there are no negative "balance" values so it is cut off at 0.

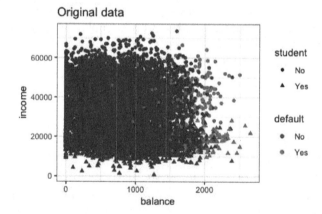

FIGURE 4.8
This plot shows the "Default" data. The x-axis shows the "balance" and the y-axis shows "income". The "default" status is colored red if it is "Yes" and blue if it is "No". We created this plot using the ggplot() function from the ggplot2 package.

The "balance" variable and the "income" variable are correlated. Thus, we have to know that these variables are correlated so that we cannot just analyze the "balance" variable or the "income" variable separately. They must be analyzed together. For example, in order to find outliers, we cannot just use the "empirical rule" on each variable as it requires uncorrelated variables. So in Practical Applications we will find a new coordinate system so we can use the new coordinate to find outliers.

4.5.3 Coordinate Systems and Change of Basis

In the previous section we discussed how vectors in a basis span a vector space. We also discussed there are many ways to form a basis in a vector space and, in fact, there are infinitely many ways to form a basis in a vector space. In this section we discuss a coordinate in terms of a basis for a vector space. Then we discuss how we can change the basis for a coordinate in a vector space.

Theorem 4.16 *Suppose we have a d-dimensional vector space V and suppose $\mathcal{B} = \{b_1, b_2, \ldots, b_d\}$ is a basis for V. Then, for any vector v in V, there is a unique set of scalars $c_1, c_2, \ldots, c_d$ such that*

$$v = c_1 \cdot b_1 + c_2 \cdot v_2 + \ldots + c_d \cdot b_d.$$

Example 117 *Suppose we have $V = \mathbb{R}^3$ and suppose*

$$\mathcal{B} = \left\{ b_1 = \begin{bmatrix} 1 \\ 0 \\ 0 \end{bmatrix}, b_2 = \begin{bmatrix} 0 \\ 1 \\ 0 \end{bmatrix}, b_3 = \begin{bmatrix} 0 \\ 0 \\ 1 \end{bmatrix} \right\}.$$

Let

$$v = \begin{bmatrix} 2 \\ 4 \\ 0 \end{bmatrix}.$$

Then we have

$$v = 2 \cdot b_1 + 4 \cdot b_2 + 0 \cdot b_3.$$

Example 118 *Suppose we have $V = \mathbb{R}^3$ and suppose*

$$\mathcal{B} = \left\{ b_1 = \begin{bmatrix} 1 \\ 1 \\ 1 \end{bmatrix}, b_2 = \begin{bmatrix} 0 \\ 1 \\ 1 \end{bmatrix}, b_3 = \begin{bmatrix} 0 \\ 0 \\ 1 \end{bmatrix} \right\}.$$

Let

$$v = \begin{bmatrix} 2 \\ 4 \\ 0 \end{bmatrix}.$$

Then we have

$$v = 2 \cdot b_1 + 2 \cdot b_2 - 4 \cdot b_3.$$

In order to represent a vector v in V we simply write

$$v = (c_1, c_2, \ldots, c_d)$$

in terms of a basis $\mathcal{B} = \{b_1, b_2, \ldots, b_d\}$. This representation is called the $\mathcal{B}$-**coordinates** of v. We notate the $\mathcal{B}$-coordinates of v as

$$[v]_\mathcal{B} = \begin{bmatrix} c_1 \\ c_2 \\ \vdots \\ c_d \end{bmatrix}.$$

Example 119 *Suppose we have $V = \mathbb{R}^3$ and suppose*

$$\mathcal{B} = \left\{ b_1 = \begin{bmatrix} 1 \\ 0 \\ 0 \end{bmatrix}, b_2 = \begin{bmatrix} 0 \\ 1 \\ 0 \end{bmatrix}, b_3 = \begin{bmatrix} 0 \\ 0 \\ 1 \end{bmatrix} \right\}.$$

Let

$$v = \begin{bmatrix} 2 \\ 4 \\ 0 \end{bmatrix}.$$

Then the $\mathcal{B}$-coordinates of v are

$$[v]_\mathcal{B} = \begin{bmatrix} 2 \\ 4 \\ 0 \end{bmatrix}.$$

Example 120 *Suppose we have $V = \mathbb{R}^3$ and suppose*

$$\mathcal{B} = \left\{ b_1 = \begin{bmatrix} 1 \\ 1 \\ 1 \end{bmatrix}, b_2 = \begin{bmatrix} 0 \\ 1 \\ 1 \end{bmatrix}, b_3 = \begin{bmatrix} 0 \\ 0 \\ 1 \end{bmatrix} \right\}.$$

Let

$$v = \begin{bmatrix} 2 \\ 4 \\ 0 \end{bmatrix}.$$

Then we have

$$[v]_\mathcal{B} = \begin{bmatrix} 2 \\ 2 \\ -4 \end{bmatrix}.$$

If a vector space $V = \mathbb{R}^d$ and a basis consists of vectors such that

$$\mathcal{B} = \left\{ e_1 = \begin{bmatrix} 1 \\ 0 \\ 0 \\ \vdots \\ 0 \end{bmatrix}, e_2 = \begin{bmatrix} 0 \\ 1 \\ 0 \\ \vdots \\ 0 \end{bmatrix}, \cdots, e_d = \begin{bmatrix} 0 \\ 0 \\ 0 \\ \vdots \\ 1 \end{bmatrix} \right\},$$

then $\mathcal{B}$ is called the *standard basis* for $\mathbb{R}^d$.

Example 121 *Let $V = \mathbb{R}^2$. Let*

$$\mathcal{B} = \left\{ e_1 = \begin{bmatrix} 1 \\ 0 \end{bmatrix}, e_2 = \begin{bmatrix} 0 \\ 1 \end{bmatrix} \right\}.$$

Then $\mathcal{B}$ is the standard basis for $\mathbb{R}^2$.

Let $V = \mathbb{R}^2$. Let

$$\mathcal{B} = \left\{ e_1 = \begin{bmatrix} 1 \\ 0 \end{bmatrix}, e_2 = \begin{bmatrix} 0 \\ 1 \end{bmatrix} \right\}$$

which is the standard basis for $\mathbb{R}^2$. Then the $\mathcal{B}$-coordinates system in $\mathbb{R}^2$ is shown in Figure 4.9.

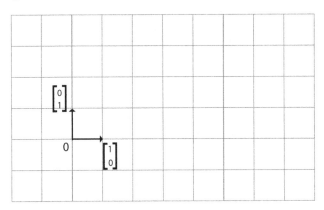

FIGURE 4.9
The vector space $\mathbb{R}^2$ spanned by vectors e_1 and e_2 in the standard basis.

If we have a basis

$$\mathcal{B} = \left\{ b_1 = \begin{bmatrix} 1 \\ 0 \end{bmatrix}, b_2 = \begin{bmatrix} 1 \\ 1 \end{bmatrix} \right\},$$

then the $\mathcal{B}$-coordinates system in $\mathbb{R}^2$ is shown in Figure 4.10.
    In order to find the $\mathcal{B}$-coordinates of a vector in a vector space, we can set up a system of linear equations and find a solution to the system. In order to demonstrate we will use Example 120. Suppose we have $V = \mathbb{R}^3$ and suppose

$$\mathcal{B} = \left\{ b_1 = \begin{bmatrix} 1 \\ 1 \\ 1 \end{bmatrix}, b_2 = \begin{bmatrix} 0 \\ 1 \\ 1 \end{bmatrix}, b_3 = \begin{bmatrix} 0 \\ 0 \\ 1 \end{bmatrix} \right\}.$$

Let

$$v = \begin{bmatrix} 2 \\ 4 \\ 0 \end{bmatrix}.$$

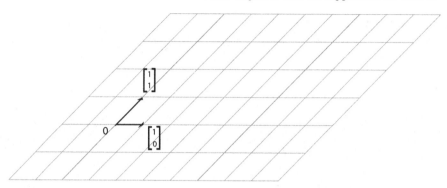

FIGURE 4.10
The vector space $\mathbb{R}^2$ spanned by vectors b_1 and b_2 in $\mathcal{B}$.

Then we want to find the $\mathcal{B}$-coordinates of v. Let

$$[v]_{\mathcal{B}} = \begin{bmatrix} c_1 \\ c_2 \\ c_3 \end{bmatrix}.$$

Then by the definition of the $\mathcal{B}$-coordinates of v, we have

$$v = c_1 \cdot b_1 + c_2 \cdot b_2 + c_3 \cdot b_3.$$

By substituting v, b_1, b_2, b_3, we have

$$\begin{bmatrix} 2 \\ 4 \\ 0 \end{bmatrix} = c_1 \cdot \begin{bmatrix} 1 \\ 1 \\ 1 \end{bmatrix} + c_2 \cdot \begin{bmatrix} 0 \\ 1 \\ 1 \end{bmatrix} + c_3 \cdot \begin{bmatrix} 0 \\ 0 \\ 1 \end{bmatrix}.$$

We can restate this as a system of linear equations:

$$
\begin{array}{ccccccc}
c_1 & & & & & = & 2 \\
c_1 & + & c_2 & & & = & 4 \\
c_1 & + & c_2 & + & c_3 & = & 0.
\end{array}
$$

Solving the system of linear equations gives a unique solution, which is

$$c_1 = 2,\ c_2 = 2,\ c_4 = -4.$$

Therefore we have

$$[v]_{\mathcal{B}} = \begin{bmatrix} 2 \\ 2 \\ -4 \end{bmatrix}.$$

This system of linear equations can be written as a matrix form:

$$A \cdot x = b$$

where

$$A = [b_1|b_2|b_3] = \begin{bmatrix} 1 & 0 & 0 \\ 1 & 1 & 0 \\ 1 & 1 & 1 \end{bmatrix}, x = \begin{bmatrix} c_1 \\ c_2 \\ c_3 \end{bmatrix}, b = \begin{bmatrix} 2 \\ 4 \\ 0 \end{bmatrix}.$$

In general, suppose we have a d-dimensional vector space and a basis $\mathcal{B} = \{b_1, b_2, \ldots, b_d\}$, then

$$v = B_{\mathcal{B}} \cdot [v]_{\mathcal{B}},$$

where

$$B_{\mathcal{B}} = [b_1|b_2|\ldots|b_d].$$

Thus, if we want to compute the $\mathcal{B}$-coordinates of v, then we simply compute

$$[v]_{\mathcal{B}} = B_{\mathcal{B}}^{-1} \cdot v$$

since $B_{\mathcal{B}}$ is invertible.

The change of the coordinates by the matrix $B_{\mathcal{B}}$ is called a **linear transformation**.

We can use the inv() function from the **pracma** package. From Example 120, we will demonstrate how to find the $\mathcal{B}$-coordinates of v. Suppose we have $V = \mathbb{R}^3$ and suppose

$$\mathcal{B} = \left\{ b_1 = \begin{bmatrix} 1 \\ 1 \\ 1 \end{bmatrix}, b_2 = \begin{bmatrix} 0 \\ 1 \\ 1 \end{bmatrix}, b_3 = \begin{bmatrix} 0 \\ 0 \\ 1 \end{bmatrix} \right\}.$$

Let

$$v = \begin{bmatrix} 2 \\ 4 \\ 0 \end{bmatrix}.$$

In R, first we upload the **pracma** package:

```
library(pracma)
```

Then we define $B_{\mathcal{B}}$ and v:

```
B <- matrix(c(1, 0, 0, 1, 1, 0, 1, 1, 1), 3, 3, byrow = TRUE)
v <- c(2, 4, 0)
```

Note that in the function matrix(), we can skip statement "nrows" and "ncol" in its argument. Then we compute $B_{\mathcal{B}}^{-1} \cdot v$ using the inv() function:

```
inv(B) %*% v
```

R outputs:

```
> inv(B) %*% v
      [,1]
[1,]     2
[2,]     2
[3,]    -4
```

Example 122 *Suppose we have* $V = \mathbb{R}^3$ *and suppose*

$$B = \left\{ b_1 = \begin{bmatrix} 0 \\ 1 \\ 1 \end{bmatrix}, b_2 = \begin{bmatrix} 1 \\ 0 \\ 1 \end{bmatrix}, b_3 = \begin{bmatrix} 1 \\ 1 \\ 0 \end{bmatrix} \right\}.$$

Suppose we have

$$v = \begin{bmatrix} 2 \\ 4 \\ 0 \end{bmatrix}.$$

We want to write the B-coordinates of v in terms of the basis B. Now we have

$$B_{\mathcal{B}} = \begin{bmatrix} 0 & 1 & 1 \\ 1 & 0 & 1 \\ 1 & 1 & 0 \end{bmatrix}$$

Thus,

$$B_{\mathcal{B}}^{-1} = \begin{bmatrix} -1/2 & 1/2 & 1/2 \\ 1/2 & -1/2 & 1/2 \\ 1/2 & 1/2 & -1/2 \end{bmatrix}.$$

So

$$[v]_{\mathcal{B}} = B_{\mathcal{B}}^{-1} \cdot v = \begin{bmatrix} 1 \\ -1 \\ 3 \end{bmatrix}.$$

4.5.4   Checkmarks

1. The definition of the basis coordinate of a vector.

2. A graphical interpretation of a change of coordinate.

3. The definition of a linear transformation.

4. You can compute a coordinate in terms of a basis for a vector space.

5. You can compute a linear transformation.

6. With R, you can perform computational examples in this section.

4.5.5 Conceptual Quizzes

Quiz 146 True or False: *Let $V = \mathbb{R}^3$. Then suppose we have*

$$\mathcal{B} = \left\{ \begin{bmatrix} 1 \\ 1 \\ 1 \end{bmatrix}, \begin{bmatrix} 0 \\ 1 \\ 1 \end{bmatrix}, \begin{bmatrix} 0 \\ 1 \\ 0 \end{bmatrix} \right\}$$

and

$$v = \begin{bmatrix} 0 \\ 0 \\ 0 \end{bmatrix}.$$

Then

$$[v]_{\mathcal{B}} = \begin{bmatrix} 0 \\ 0 \\ 0 \end{bmatrix}.$$

Quiz 147 True or False: *We have a d-dimensional vector space and $\mathcal{B} = \{b_1, b_2, \ldots, b_d\}$ is a basis. Let*

$$B_{\mathcal{B}} = [b_1 | b_2 | \ldots | b_d].$$

Then $B_{\mathcal{B}}$ is invertible.

Quiz 148 True or False: *We have a d-dimensional vector space V and $\mathcal{B} = \{b_1, b_2, \ldots, b_d\}$ is a basis. Then there exists v in V such that there are two different sets of the $\mathcal{B}$-coordinates of v.*

Quiz 149 True or False: *Let $V = \mathbb{R}^2$. Then suppose we have*

$$\mathcal{B} = \left\{ \begin{bmatrix} 1 \\ 1 \end{bmatrix}, \begin{bmatrix} 0 \\ 1 \end{bmatrix} \right\}$$

and

$$v = \begin{bmatrix} 2 \\ 4 \end{bmatrix}.$$

Then

$$[v]_{\mathcal{B}} = \begin{bmatrix} 2 \\ 2 \end{bmatrix}.$$

Quiz 150 True or False: *Let $V = \mathbb{R}^3$. Then suppose we have*

$$\mathcal{B} = \left\{ \begin{bmatrix} 1 \\ 1 \\ 1 \end{bmatrix}, \begin{bmatrix} 0 \\ 1 \\ 1 \end{bmatrix}, \begin{bmatrix} 0 \\ 1 \\ 0 \end{bmatrix} \right\}$$

and

$$v = \begin{bmatrix} 2 \\ 4 \\ 0 \end{bmatrix}.$$

Then

$$[v]_\mathcal{B} = \begin{bmatrix} 2 \\ -2 \\ 4 \end{bmatrix}.$$

4.5.6   Regular Exercises

Exercise 4.41 *Let $V = \mathbb{R}^2$. Then suppose we have*

$$\mathcal{B} = \left\{ \begin{bmatrix} 1 \\ 1 \end{bmatrix}, \begin{bmatrix} 0 \\ 1 \end{bmatrix} \right\}$$

and

$$v = \begin{bmatrix} 3 \\ 1 \end{bmatrix}.$$

Compute

$$[v]_\mathcal{B}.$$

Exercise 4.42 *Let $V = \mathbb{R}^2$. Then suppose we have*

$$\mathcal{B} = \left\{ \begin{bmatrix} 2 \\ 1 \end{bmatrix}, \begin{bmatrix} 1 \\ 2 \end{bmatrix} \right\}$$

and

$$v = \begin{bmatrix} 3 \\ 5 \end{bmatrix}.$$

Compute

$$[v]_\mathcal{B}.$$

Exercise 4.43 *Let $V = \mathbb{R}^3$. Then suppose we have*

$$\mathcal{B} = \left\{ \begin{bmatrix} 1 \\ 0 \\ 1 \end{bmatrix}, \begin{bmatrix} 2 \\ 1 \\ 0 \end{bmatrix}, \begin{bmatrix} 0 \\ 1 \\ 1 \end{bmatrix} \right\}$$

and

$$v = \begin{bmatrix} 3 \\ 1 \\ 1 \end{bmatrix}.$$

Compute

$$[v]_\mathcal{B}.$$

Exercise 4.44 *Let $V = \mathbb{R}^3$. Then suppose we have*

$$\mathcal{B} = \left\{ \begin{bmatrix} 0 \\ 1 \\ 0 \end{bmatrix}, \begin{bmatrix} 3 \\ -1 \\ 1 \end{bmatrix}, \begin{bmatrix} 1 \\ 1 \\ 1 \end{bmatrix} \right\}$$

and

$$v = \begin{bmatrix} 0 \\ 3 \\ 3 \end{bmatrix}.$$

Compute

$$[v]_{\mathcal{B}}.$$

Exercise 4.45 *Let $V = \mathbb{R}^4$. Then suppose we have*

$$\mathcal{B} = \left\{ \begin{bmatrix} 1 \\ 1 \\ 0 \\ 0 \end{bmatrix}, \begin{bmatrix} 1 \\ 0 \\ 1 \\ 0 \end{bmatrix}, \begin{bmatrix} 0 \\ 1 \\ 0 \\ 1 \end{bmatrix}, \begin{bmatrix} 0 \\ 0 \\ 1 \\ 1 \end{bmatrix} \right\}$$

and

$$v = \begin{bmatrix} 0 \\ 3 \\ 3 \\ 1 \end{bmatrix}.$$

Compute

$$[v]_{\mathcal{B}}.$$

Exercise 4.46 *Let $V = \mathbb{R}^4$. Then suppose we have*

$$\mathcal{B} = \left\{ \begin{bmatrix} 0 \\ 1 \\ 1 \\ 1 \end{bmatrix}, \begin{bmatrix} 1 \\ 0 \\ 1 \\ 1 \end{bmatrix}, \begin{bmatrix} 1 \\ 1 \\ 0 \\ 1 \end{bmatrix}, \begin{bmatrix} 1 \\ 1 \\ 1 \\ 0 \end{bmatrix} \right\}$$

and

$$v = \begin{bmatrix} 0 \\ 3 \\ 3 \\ 1 \end{bmatrix}.$$

Compute

$$[v]_{\mathcal{B}}.$$

Exercise 4.47 *Let $V = \mathbb{R}^5$. Then suppose we have*

$$\mathcal{B} = \left\{ \begin{bmatrix} 0 \\ 1 \\ 1 \\ 1 \\ 1 \end{bmatrix}, \begin{bmatrix} 1 \\ 0 \\ 1 \\ 1 \\ 1 \end{bmatrix}, \begin{bmatrix} 1 \\ 1 \\ 0 \\ 1 \\ 1 \end{bmatrix}, \begin{bmatrix} 1 \\ 1 \\ 1 \\ 0 \\ 1 \end{bmatrix}, \begin{bmatrix} 1 \\ 1 \\ 1 \\ 1 \\ 0 \end{bmatrix} \right\}$$

and

$$v = \begin{bmatrix} 0 \\ 3 \\ 3 \\ 1 \\ 0 \end{bmatrix}.$$

Compute

$$[v]_{\mathcal{B}}.$$

4.5.7 Lab Exercises

Lab Exercise 159 *Using the inv() function from the* pracma *package in* R, *repeat Exercise 4.41.*

Lab Exercise 160 *Using the inv() function from the* pracma *package in* R, *repeat Exercise 4.42.*

Lab Exercise 161 *Using the inv() function from the* pracma *package in* R, *repeat Exercise 4.43.*

Lab Exercise 162 *Using the inv() function from the* pracma *package in* R, *repeat Exercise 4.44.*

Lab Exercise 163 *Using the inv() function from the* pracma *package in* R, *repeat Exercise 4.45.*

Lab Exercise 164 *Using the inv() function from the* pracma *package in* R, *repeat Exercise 4.46.*

Lab Exercise 165 *Using the inv() function from the* pracma *package in* R, *repeat Exercise 4.47.*

4.5.8 Practical Applications

[PCA related to change of coordinates] Principal component analysis (PCA) can be seen as a visualization using the change of coordinates. Suppose we have a basis of d many vectors for the d-dimensional vector space $\mathbb{R}^d$ such that

$$\left\{ \begin{bmatrix} 1 \\ 0 \\ \vdots \\ 0 \end{bmatrix}, \begin{bmatrix} 0 \\ 1 \\ \vdots \\ 0 \end{bmatrix}, \cdots \begin{bmatrix} 0 \\ 0 \\ \vdots \\ 1 \end{bmatrix} \right\}.$$

This basis is called the **standard basis** of the d-dimensional vector space $\mathbb{R}^d$. PCA is to find a basis of the d-dimensional vector space $\mathbb{R}^d$, which "explains the given data set at most". We discuss more details in Chapter 6. In order to see an intuitive view of the process, we walk through a simple example. Suppose we have a set of 500 random points generated from a bivariate (two-variables) Gaussian distribution with the mean $(0,0)$ and the "covariant" matrix

$$\begin{bmatrix} 4 & 2 \\ 2 & 3 \end{bmatrix}$$

using the `mvtnorm` package:

```
## library
library(mvtnorm)
## setting a seed to reproduce the same result
set.seed(123)
sigma <- matrix(c(4,2,2,3), ncol=2)
x <- rmvnorm(n=500, mean=c(0,0), sigma=sigma)
```

Then, using the plot() function, they are distributed shown in Figure 4.11. Then, using PCA, we can **rotate** these 500 data points shown in Figure 4.12.

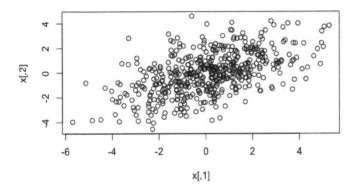

FIGURE 4.11
500 data points generated from the bivariate Gaussian distribution with the mean $(0,0)$ and the "covariant" matrix shown above.

For this simple example, via PCA we find the basis for the 2-dimensional vector space which explains the 500 data points most as

$$\mathcal{B} = \left\{ \begin{bmatrix} 0.781 \\ 0.625 \end{bmatrix}, \begin{bmatrix} 0.625 \\ -0.781 \end{bmatrix} \right\}.$$

We can compute this basis using the princomp() function:

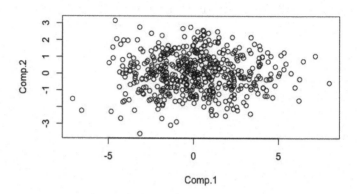

FIGURE 4.12
Rotated 500 data points generated from the bivariate Gaussian distribution
with the mean $(0,0)$ and the "covariant" matrix shown above so that these
two variables are decorrelated.

```
pca<- princomp(x)
```

It outputs the following:

```
> pca
Call:
princomp(x = x)

Standard deviations:
  Comp.1   Comp.2
2.241521 1.231119


 2  variables and  500 observations.
```

In order to see the basis of $\mathbb{R}^2$ via PCA based on the data set, we can use
the loading function:

```
> pca$loadings

Loadings:
     Comp.1 Comp.2
[1,]  0.781  0.625
[2,]  0.625 -0.781


               Comp.1 Comp.2
SS loadings       1.0    1.0
```

```
Proportion Var    0.5    0.5
Cumulative Var    0.5    1.0
```

If a basis which we found via PCA is not the standard basis for the d-dimensional vector space, then these variables are "correlated", which means they are related to each other. Since the basis which we found via PCA is not the standard basis for $\mathbb{R}^2$, the variable X and variable Y are correlated (shown in Figure 4.11). This means that if we change one variable, then other variables also change according to some relationship. If this is the case, we cannot just compute the **95 percent confidence interval** for each variable since they are correlated. Therefore, using the change of bases from the coordinate with the current basis to the coordinate with the standard basis, we will **decorrelate** these variables. For our simple example, after we change the basis $\mathcal{B}$ to the standard basis, all data points are shown in Figure 4.12. After the change of bases, we can finally compute the **confidence interval** for each variable. To change bases to the standard basis, we can do the following in R using the function t() for the transpose of a matrix and the inv() function for computing the inverse of a matrix:

```
T <- pca$loadings
D <- inv(T) %*% t(x)
y <- t(D)
plot(y)
```

Now we go back to the example in the beginning of this section. Note that in the "Default" data set, each observation is recorded as a row of the matrix, so to change the coordinates of each observation in the data set we need to transpose them. In order for R to read the data set to plot them, we have to make sure that the data frame is back to the original coordinates. In R, the data frame is formatted so that each row represents an observation in the data set and each column represents as a variable (or one coordinate of a vector).

First we upload all necessary packages:

```
## Library for the Credit data set
library(ISLR)

## Libraries for Plotting our Results
library(ggplot2)
library(ggfortify)
library(gridExtra)

## for inv() function
library(pracma)
```

Then we upload the "Default" data set:

```
## Loading the data
data(Credit)
```

The first column represents the "default" status and the second column represents the "student" status. So we take only the third and fourth columns:

```
credit.data <- Default[,3:4]
```

First we translate the mean of all the data points in the first column and the mean of all data points in the second column to the origin $(0,0)$ using the mean() function:

```
m1 <- mean(credit.data[,1])
m2 <- mean(credit.data[,2])
credit.data2 <- credit.data - c(m1, m2)
```

Then we **standardize** all data points. Basically, in order to compare the same scales, we compute the **z-score** of each measurement in the data set for the first column and for the second column. If we do, then we do not have to worry about scales of measurements. In order to do so, we apply the scale() function:

```
credit.data3 <- scale(credit.data2)
```

Then we apply the princomp() function to compute the first and second principal components, which form a basis for $\mathbb{R}^2$:

```
pca<- princomp(data.frame(scale(credit.data3)))
pca$loadings
```

This will output the first and second principal components:

```
> pca$loadings

Loadings:
        Comp.1 Comp.2
balance -0.707 -0.707
income   0.707 -0.707

        Comp.1 Comp.2
SS loadings       1.0    1.0
Proportion Var    0.5    0.5
Cumulative Var    0.5    1.0
```

Therefore, the basis computed from the PCA is

$$\mathcal{B} = \left\{ \begin{bmatrix} -0.707 \\ 0.707 \end{bmatrix}, \begin{bmatrix} -0.707 \\ -0.707 \end{bmatrix} \right\}$$

Thus we have the matrix:

$$B_{\mathcal{B}} = \begin{bmatrix} -0.707 & -0.707 \\ 0.707 & -0.707 \end{bmatrix}.$$

We store this as a matrix:

```
T <- pca$loadings
```

credit.data3 is a 1000×2 matrix so we have to apply transpose. Thus we have

```
D <- inv(T) %*% t(scale(credit.data3))
```

Then we apply the t() function to get back to the data frame format for R to plot.

```
credit.data4 <- t(D)
```

We want to eliminate outliers from our computation, so we first calculate values that will include 95% of the observations:

```
s <- c(-1.96, 1.96)
Non_Outlier_v1 <- mean(credit.data4[,1]) + s * sd(credit.data4[,1])
Non_Outlier_v2 <- mean(credit.data4[,2]) + s * sd(credit.data4[,2])
```

The outputs are:

```
> Non_Outlier_v1
[1] -2.608577  2.608577
> Non_Outlier_v2
[1] -0.9372965  0.9372965
```

This still does not look useful. In order to be useful, we have to bring these values back to the original coordinate. Since a z-scare is a measurement in terms of standard deviations, i.e., it measures standard deviations away from the mean, we rotate the confidence intervals that we computed with the coordinates in terms of principal components back to the original coordinates using the inverse of the matrix T, and then we multiply by the standard deviation of the data set using the sd() function:

```
M <- rbind(Non_Outlier_v1, Non_Outlier_v2)
mean(credit.data[,1]) + sd(credit.data[,1])*(T %*% M)[1,]
mean(credit.data[,2]) + sd(credit.data[,2])*(T %*% M)[2,]
```

The output looks like:

```
> M <- rbind(Non_Outlier_v1, Non_Outlier_v2)
> mean(credit.data[,1]) + sd(credit.data[,1])*(T %*% M)[1,]
[1] -377.4492 2048.1990
> mean(credit.data[,2]) + sd(credit.data[,2])*(T %*% M)[2,]
[1] 17756.09 49277.88
```

Therefore, if an observation has a balance which is less than -377.4492 or more than 2048.1990 and if the observation has income which is less than 17756.09 or more than 49277.88, then the observation is an outlier. The summary() function outputs the summary of the data:

```
> summary(credit.data)
     balance             income
 Min.    :   0.0   Min.    :  772
 1st Qu.: 481.7    1st Qu.:21340
 Median : 823.6    Median :34553
 Mean    : 835.4   Mean    :33517
 3rd Qu.:1166.3    3rd Qu.:43808
 Max.    :2654.3   Max.    :73554
```

Since it is not obvious from the summary, we will use the sum() function and Boolean functions to see how many observations are outside of these confidence intervals:

```
sum((credit.data[,1] < -377.4492) && (credit.data[,1] > 2048.1990)
&& (credit.data[,2] < 17756.09) && (credit.data[,2] > 49277.88))
```

Then R returns zero. Therefore There are no outliers in this data set.

To plot all observations after the linear transformation, we can use the autoplot() function:

```
pc <- prcomp(credit.data, scale=TRUE)
autoplot(pc, data=Default, colour = "default")
```

This will print the plot shown in Figure 4.13.

4.5.9 Supplements with python Code

In this section we did not use any new functions. Therefore we do not have any supplemental python codes in this section.

4.6 Discussion

We return to the "USArrests" data set. For this example, we use the "Euclidean distance", namely the root square of the inner product of the difference between two vectors. We will discuss more detail on inner products in the next chapter. Here we discuss how we determine that there are four clusters in this data set.

In order to do so we use a "dendrogram". A dendrogram is a graphical representation of which observations are close to each other. This "closeness"

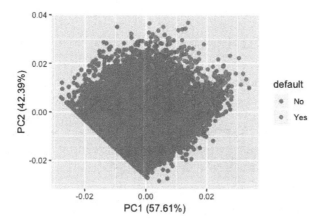

FIGURE 4.13
This plot shows the "Default" data after the linear transformation. The x-axis shows the first principal component and the y-axis shows the second principal component. The "default" status is colored red for "No" and blue for "Yes". We created this plot using the autoplot() function from the **ggplot2** package.

is measured by the "Euclidean distance" in the vector space $\mathbb{R}^4$. When we draw a dendrogram, we first look at all pairwise distances, the Euclidean distance between any two observations. Then we pair the two vectors which have the smallest distance. Then we pair the vectors with the second smallest distance, and so on. When we pair vectors, we draw an edge between them and record their distance. When we finish pairing all of them we have a dendrogram.

In R, we use the plot() function to plot the dendrogram of the data set.

```
plot(hc, cex = 0.6)
```

Then R outputs the plot shown in Figure 4.14. In the bottom part of the dendrogram, there are very short edges and in the top part of the dendrogram, there are long edges. When edge lengths shift from short to long, we just cut the dendrogram shown as in Figure 4.15. In Figure 4.15, the green line indicates the cut-off where the edge lengths shift. If we look at the dendrogram below the green line, there are four groups. Thus we group these four clusters together as shown in Figure 4.16. The plot shown in Figure 4.16 can be created via:

```
plot(hc, cex = 0.6)
rect.hclust(hc, k = 4, border = 2:5)
```

Cluster Dendrogram

d
hclust (*, "complete")

FIGURE 4.14
Dendrogram of the "USArrests" data via a hierarchical clustering method.

Cluster Dendrogram

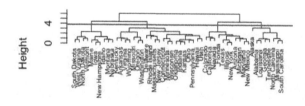

d
hclust (*, "complete")

FIGURE 4.15
Dendrogram of the "USArrests" data via a hierarchical clustering method.
The green line is the cut-off where edge lengths shift from small to large.

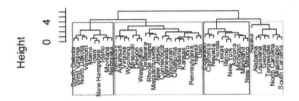

FIGURE 4.16
Dendrogram of the "USArrests" data via a hierarchical clustering method.
We determined there are four clusters in this data set.

5

Inner Product Space

An inner product is an operation which defines a vector space. This operation is key to conduct any analysis in a vector space including a Euclidean space, such as $\mathbb{R}^d$. In this chapter we define an inner product and discuss how we can apply such an operation to geometrical objects.

5.1   Introductory Example from Statistics

In an introductory statistics course, we learn "least squares" method and linear regression. If we have a data set and we would like to fit this data set to a curve or linear plane in order to predict an outcome, we use the least squares method. If we are fitting the data set to a linear plane, this is called a linear regression. least squares methods and linear regressions were first used to analyze a data set around 1720. In this chapter, we look at the "Hitters" data set from the ISLR package [24]. This dataset was collected from the StatLib library at Carnegie Mellon University and it contains 322 observations of major league players with 20 variables including their "Salary", "AtBat", the number of times at-bats in 1986, and "Hits", the number of hits in 1986 for each player.

Here we build a model so that we can predict the salary of a player based on their number of at-bats and hits in a year.

First, to see how observations in the data are distributed in the vector space, we draw two plots: Salary vs. AtBat and Salary vs. Hits. The plot for Salary vs. AtBat is shown in the left side of Figure 5.1 and the plot for Salary vs Hits is shown in the right side of Figure 5.1.

5.2   Inner Products

An inner product is a fundamental operation to define a vector space. This defines a magnitude or length of a vector in a vector space as well as the angle between two vectors in a vector space. This can also be applied to measure

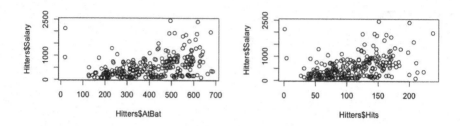

FIGURE 5.1
Two-dimensional plot of observations for Salary vs AtBat in the "Hitters" data from the ISLR package. LEFT: The x-axis represents the AtBat variable and the y-axis represents the Salary variable. RIGHT: The x-axis represents the Hits variable and the y-axis represents the Salary variable.

a distance between two points in a vector space, which is useful to cluster observations in a data set, for example, as we did in the previous chapter. For the working example in this section we again look at clustering methods from data science to see how inner products play their role in clustering.

5.2.1   Task Completion Checklist

- During the Lecture:

    1. Read the definition of an inner product in a vector space.
    2. Read the definition of the length of a vector in a vector space.
    3. Read the definition of the distance between two vectors in a vector space.
    4. Learn how to give an example of an inner product in a vector space.
    5. Learn how to compute the length of a vector in a vector space.
    6. Learn how to compute the distance between two vectors in a vector space.
    7. With R, learn how to perform computational examples in this section.

- After the Lecture:

    1. Take conceptual quizzes to make sure you understand the materials in this section.
    2. Do some regular exercises.
    3. Conduct lab exercises with R.
    4. Conduct practical applications with R.
    5. If you are interested in python, read the supplement in this section and conduct lab exercises and practical applications with python.

5.2.2   Working Examples

For this section we use the "USArrests" data set which we used as the Introductory Example for Chapter 4. In Chapter 4, we used hierarchical clustering to find clusters in the data set. In this section we will use the **k-means clustering** method to find clusters in the data set using the kmeans() function in R.

The k-means clustering is an unsupervised method which uses means of clusters to find clusters. Basically this method minimizes the sum of **distances** between data points to a mean for each cluster. In the Discussion for this section, we will discuss how the procedure of k-means clustering works. First, however, we will show how to compute the clusters via k-means clustering in R with the "USArrests" data set.

First we upload the data set and clean the data as we did in Chapter 4:

```
df <- USArrests
df <- na.omit(df)
df <- scale(df)
```

Now we use the kmeans() function to find clusters in the data set:

```
# nstart is the number of random starting points
clusters <- kmeans(df, 4, nstart = 10)
```

Here "nstart" is set for the number of initial random points. The first argument of the kmeans() function is the name of the data set, and the second argument of the kmeans() function is the number of clusters in the data set.

One disadvantage of the k-means clustering method is that we have to know the number of clusters in a given data set. There are several ways to find the number of clusters in a given data set. For now, we will use the number of clusters found in Chapter 4, which is 4.

Now we plot the result from the kmeans() function. As we did in Chapter 4, we will use the fviz_cluster() function from the **factoextra** package. First we upload the package:

```
library(factoextra)
```

Then, we plot the result:

```
fviz_cluster(clusters, df)
```

The first argument of the fviz_cluster() function is the result from the function and the second argument of the fviz_cluster() function is the name of the data set. The result of this function is shown in Figure 5.2.

5.2.3   Inner Products

An inner product is an operation on vectors in a vector space and it will define geometric properties of the vector space. A vector space paired with an inner

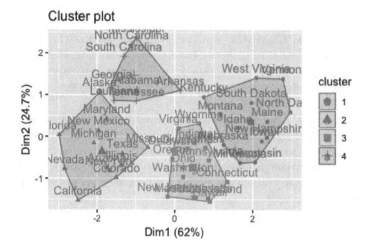

FIGURE 5.2
The clusters computed via the k-means clustering method. Each cluster has
a distinct color. The x-axis represents the first principal component and the
y-axis represents the second principal component computed via the PCA.

product is called an **inner product space**. With an inner product we can
define the length of a vector and also the distance between two vectors. These
properties are important for conducting data analysis, such as computing the
mean and standard deviation, for example.

Definition 40 *An* **inner product** $\langle\,,\rangle$ *is a function sending a pair of vectors
in a vector space to a real number such that: for all vectors u, v, w in a vector
space V and a scale c,*

- $\langle u, v \rangle = \langle v, u \rangle$,

- $\langle u + v, w \rangle = \langle u, w \rangle + \langle v, w \rangle$,

- $\langle c \cdot u, v \rangle = c \cdot \langle u, v \rangle$,

- $\langle u, u \rangle \geq 0$ *for all u in V, and*

- $\langle u, u \rangle = 0$ *if and only if* $u = \mathbf{0}$.

Example 123 *Suppose we have* $V = \mathbb{R}^3$. *Then we let*

$$\langle u, v \rangle = u_1 \cdot v_1 + u_2 \cdot v_2 + u_3 \cdot v_3,$$

where

$$u = \begin{bmatrix} u_1 \\ u_2 \\ u_3 \end{bmatrix}, \ v = \begin{bmatrix} v_1 \\ v_2 \\ v_3 \end{bmatrix}$$

are vectors in V. Now we want to show $\langle\,,\rangle$ is an inner product. We have to check all conditions in Definition 40.

- *Show $\langle u, v\rangle = \langle v, u\rangle$:*

$$\langle u, v\rangle = u_1 \cdot v_1 + u_2 \cdot v_2 + u_3 \cdot v_3$$
$$= v_1 \cdot u_1 + v_2 \cdot u_2 + v_3 \cdot u_3 = \langle v, u\rangle$$

- *Show $\langle u + v, w\rangle = \langle u, w\rangle + \langle v, w\rangle$: Let*

$$w = \begin{bmatrix} w_1 \\ w_2 \\ w_3 \end{bmatrix}.$$

Then,

$$\langle u + v, w\rangle = (u_1 + v_1) \cdot w_1 + (u_2 + v_2) \cdot w_2 + (u_3 + v_3) \cdot w_3$$
$$= u_1 \cdot w_1 + v_1 \cdot w_1 + u_2 \cdot w_2 + v_2 \cdot w_2 + u_3 \cdot w_3 + v_3 \cdot w_3$$
$$= (u_1 \cdot w_1 + u_2 \cdot w_2 + u_3 \cdot w_3) + (v_1 \cdot w_1 + v_2 \cdot w_2 + v_3 \cdot w_3)$$
$$= \langle u, w\rangle + \langle v, w\rangle.$$

- *Show $\langle c \cdot u, v\rangle = c \cdot \langle u, v\rangle$ for a scalar c.*

$$\langle c \cdot u, v\rangle = c \cdot u_1 \cdot v_1 + c \cdot u_2 \cdot v_2 + c \cdot u_3 \cdot v_3$$
$$= c(\cdot u_1 \cdot v_1 + \cdot u_2 \cdot v_2 + \cdot u_3 \cdot v_3) = c \cdot \langle u, v\rangle.$$

- *Show $\langle u, u\rangle \geq 0$.*

$$\langle u, u\rangle = u_1 \cdot u_1 + u_2 \cdot u_2 + u_3 \cdot u_3$$
$$= u_1^2 + u_2^2 + u_3^2 \geq 0$$

since u_1, u_2, u_3 are all real numbers.

- *Now show $\langle u, u\rangle = 0$ if and only if $u = \mathbf{0}$.*

$$\langle u, u\rangle = u_1 \cdot u_1 + u_2 \cdot u_2 + u_3 \cdot u_3$$
$$= u_1^2 + u_2^2 + u_3^2 = 0$$

if and only if $u_1 = u_2 = u_3 = 0$.

In general, if we have $V = \mathbb{R}^d$, then we let

$$\langle u, v\rangle = u_1 \cdot v_1 + u_2 \cdot v_2 + \ldots + u_d \cdot v_d,$$

where

$$u = \begin{bmatrix} u_1 \\ u_2 \\ \vdots \\ u_d \end{bmatrix}, v = \begin{bmatrix} v_1 \\ v_2 \\ \vdots \\ v_d \end{bmatrix}$$

are vectors in V. Then $\langle\,,\rangle$ is an inner product in V. This inner product is also called a **dot product**.

Example 124 *Suppose we have $V = \mathbb{R}^2$. Let*

$$u = \begin{bmatrix} 3 \\ 2 \end{bmatrix}, v = \begin{bmatrix} 1 \\ 2 \end{bmatrix}.$$

Then the inner product of u, v is

$$\langle u, v \rangle = 3 \cdot 1 + 2 \cdot 2 = 3 + 4 = 7.$$

Example 125 *Suppose we have $V = \mathbb{R}^3$. Let*

$$u = \begin{bmatrix} 1 \\ 5 \\ 0 \end{bmatrix}, v = \begin{bmatrix} 1 \\ 0 \\ 3 \end{bmatrix}.$$

Then the inner product of u, v is

$$\langle u, v \rangle = 1 \cdot 1 + 5 \cdot 0 + 0 \cdot 3 = 1 + 0 + 0 = 1.$$

Example 126 *Suppose we have $V = \mathbb{R}^3$. Then we let*

$$\langle u, v \rangle = 3 \cdot u_1 \cdot v_1 + 2\dot{u}_2 \cdot v_2 + u_3 \cdot v_3,$$

where

$$u = \begin{bmatrix} u_1 \\ u_2 \\ u_3 \end{bmatrix}, v = \begin{bmatrix} v_1 \\ v_2 \\ v_3 \end{bmatrix}$$

are vectors in V. Then $\langle , \rangle$ is an inner product in V.

To compute the inner product of two vectors in $\mathbb{R}^d$ in R, we can use the sum() function. From Example 125, suppose we have $V = \mathbb{R}^3$. Let

$$u = \begin{bmatrix} 1 \\ 5 \\ 0 \end{bmatrix}, v = \begin{bmatrix} 1 \\ 0 \\ 3 \end{bmatrix}.$$

First we define two vectors:

```
u <- c(1, 5, 0)
v <- c(1, 0, 3)
```

Then we use the sum() function to find the product of two vectors:

```
sum(u * v)
```

This gives the inner product of these vectors:

```
> sum(u * v)
[1] 1
```

To analyze data sets in $\mathbb{R}^d$ or to measure the strength of a force in mechanical physics, it is very useful to measure the length of a vector in $\mathbb{R}^d$. Here we can define the length of a vector in $\mathbb{R}^d$ using an inner product.

Definition 41 *The length of a vector v in a vector space $V = \mathbb{R}^d$ is defined as*

$$||v|| = \sqrt{\langle v, v \rangle}.$$

Example 127 *Suppose we have $V = \mathbb{R}^2$. Let*

$$v = \begin{bmatrix} 1 \\ 2 \end{bmatrix}.$$

Then the length of v is

$$||v|| = \sqrt{1^2 + 2^2} = \sqrt{5}.$$

Example 128 *Suppose we have $V = \mathbb{R}^3$. Let*

$$v = \begin{bmatrix} 1 \\ 2 \\ 3 \end{bmatrix}.$$

Then the length of v is

$$||v|| = \sqrt{1^2 + 2^2 + 3^2} = \sqrt{1 + 4 + 9} = \sqrt{14}.$$

To compute the inner product of two vectors in $\mathbb{R}^d$ in R, we can use the sum() function. From Example 128, suppose we have $V = \mathbb{R}^3$. Let

$$v = \begin{bmatrix} 1 \\ 2 \\ 3 \end{bmatrix}.$$

First we define a vector:

```
v <- c(1, 2, 3)
```

Then we use the sum() function and the sqrt() to compute the inner product of two vectors:

```
sqrt(sum(v * v))
```

This gives the length of a vector:

```
> sqrt(sum(v * v))
[1] 3.741657
```

Measuring a distance between two vectors in $\mathbb{R}^d$ is used in many problems in many areas. For example, in statistics, to compute the sample standard deviation from a sample in $\mathbb{R}^d$ we need to measure distances between each data point and its sample mean. With the definition of the length of a vector in $\mathbb{R}^d$, we can define a distance between two vectors in $\mathbb{R}^d$.

Definition 42 *Suppose $V = \mathbb{R}^d$. Let v, u be vectors in $\mathbb{R}^d$. Then the distance between v and u is defined as*

$$||v - u||.$$

Example 129 *Suppose we have $V = \mathbb{R}^3$. Let*

$$v = \begin{bmatrix} 1 \\ 2 \\ 3 \end{bmatrix}, u = \begin{bmatrix} 2 \\ 2 \\ 2 \end{bmatrix}$$

Then the length of v is

$$||v - u|| = \sqrt{(1-2)^2 + (2-2)^2 + (3-2)^2} - \sqrt{1 + 0 + 1} = \sqrt{2}.$$

To compute the inner product of two vectors in $\mathbb{R}^d$ in R, we can use the sum() function. From Example 129 suppose we have $V = \mathbb{R}^3$. Let

$$v = \begin{bmatrix} 1 \\ 2 \\ 3 \end{bmatrix}, u = \begin{bmatrix} 2 \\ 2 \\ 2 \end{bmatrix}$$

First we define vectors:

```
v <- c(1, 2, 3)
u <- c(2, 2, 2)
```

Then we use the sum() function and the sqrt() functions:

```
sqrt(sum((v - u)^2))
```

This gives the length of a vector:

```
> sqrt(sum((v - u)^2))
[1] 1.414214
```

The following theorem is useful when we simplify computations including inner products.

Theorem 5.1 *Suppose u, v, w are vectors in a vector space V with an inner product and c is a scalar. Then we have*

- $\langle u, \mathbf{0} \rangle = \langle \mathbf{0}, u \rangle = \mathbf{0}$,

- $\langle u, v + w \rangle = \langle u, v \rangle + \langle u, w \rangle$,

- $\langle u, c \cdot v \rangle = c \cdot \langle u, v \rangle$,

- $\langle u - v, w \rangle = \langle u, w \rangle - \langle v, w \rangle$, *and*

- $\langle u, v - w \rangle = \langle u, v \rangle - \langle u, w \rangle$.

Example 130 *Suppose we have* $V = \mathbb{R}^3$. *Let*

$$ u = \begin{bmatrix} 1 \\ 0 \\ 2 \end{bmatrix}, \, v = \begin{bmatrix} 1 \\ -5 \\ 3 \end{bmatrix}, \, w = \begin{bmatrix} -1 \\ 5 \\ 0 \end{bmatrix}. $$

Then we want to compute $\langle u - 2 \cdot v, 3 \cdot u + w \rangle$.

$$
\begin{aligned}
\langle u - 2 \cdot v, 3 \cdot u + w \rangle &= \langle u, 3 \cdot u + w \rangle - \langle 2 \cdot v, 3 \cdot u + w \rangle \\
&= \langle u, 3 \cdot u \rangle + \langle u, w \rangle - (\langle 2 \cdot v, 3 \cdot u \rangle + \langle 2 \cdot v, w \rangle) \\
&= 3 \cdot \langle u, u \rangle + \langle u, w \rangle - 2 \cdot (3 \cdot \langle v, u \rangle + \langle v, w \rangle) \\
&= 3 \cdot ||u||^2 + \langle u, w \rangle - 2 \cdot (3 \cdot \langle u, v \rangle + \langle v, w \rangle).
\end{aligned}
$$

Thus we can compute only $||u||^2$, $\langle u, w \rangle \, \langle u, v \rangle$, $\langle v, w \rangle$. *We have*

$$
\begin{aligned}
||u||^2 &= 1^2 + 0^2 + 2^2 = 5, \\
\langle u, w \rangle &= 1 \cdot (-1) + 0 \cdot 5 + 2 \cdot 0 = -1, \\
\langle u, v \rangle &= 1 \cdot 1 + 0 \cdot (-5) + 2 \cdot 3 = 1 + 0 + 6 = 7, \\
\langle v, w \rangle &= 1 \cdot (-1) + 5 \cdot (-5) + 3 \cdot 0 = -1 - 25 + 0 = -26.
\end{aligned}
$$

Thus we have

$$
\begin{aligned}
&\langle u - 2 \cdot v, 3 \cdot u + w \rangle \\
&= 3 \cdot ||u||^2 + \langle u, w \rangle - 2 \cdot (3 \cdot \langle u, v \rangle + \langle v, w \rangle) \\
&= 3 \cdot 5 + (-1) - 2 \cdot (3 \cdot 7 + (-26)) \\
&= 24.
\end{aligned}
$$

5.2.4 Checkmarks

1. The definition of an inner product in a vector space.

2. The definition of the length of a vector in a vector space.

3. The definition of the distance between two vectors in a vector space.

4. You can give an example of an inner product in a vector space.

5. You can compute the length of a vector in a vector space.

6. You can compute the distance between two vectors in a vector space.

7. With R, you can perform computational examples in this section.

5.2.5   Conceptual Quizzes

Quiz 151 True or False: *Suppose $V = \mathbb{R}^2$. For any vectors u, v in V,*

$$\langle u, v \rangle = \min\{u_1 - v_1, u_2 - v_2\}$$

for

$$u = \begin{bmatrix} u_1 \\ u_2 \end{bmatrix}, v = \begin{bmatrix} v_1 \\ v_2 \end{bmatrix}$$

is an inner product.

Quiz 152 True or False: *Suppose $V = \mathbb{R}^2$. For any vectors u, v in V,*

$$\langle u, v \rangle = \max\{u_2, v_2\} - \min\{u_1, v_1\}$$

for

$$u = \begin{bmatrix} u_1 \\ u_2 \end{bmatrix}, v = \begin{bmatrix} v_1 \\ v_2 \end{bmatrix}$$

is an inner product.

Quiz 153 True or False: *Suppose $V = \mathbb{R}^2$. For any vectors u, v in V,*

$$\langle u, v \rangle = 2 \cdot u_1 \cdot v_1 - 5 \cdot u_2 \cdot v_2$$

for

$$u = \begin{bmatrix} u_1 \\ u_2 \end{bmatrix}, v = \begin{bmatrix} v_1 \\ v_2 \end{bmatrix}$$

is an inner product.

Quiz 154 True or False: *Suppose $V = \mathbb{R}^2$. For any vectors u, v in V,*

$$\langle u, v \rangle = u_1^2 + u_2^2$$

for

$$u = \begin{bmatrix} u_1 \\ u_2 \end{bmatrix}, v = \begin{bmatrix} v_1 \\ v_2 \end{bmatrix}$$

is an inner product.

Quiz 155 True or False: *Suppose $V = \mathbb{R}^2$. Let*

$$u = \begin{bmatrix} 1 \\ 2 \end{bmatrix}.$$

Then the length of u is $\sqrt{5}$.

Quiz 156 True or False: *Suppose $V = \mathbb{R}^3$. Let*

$$u = \begin{bmatrix} 2 \\ 3 \\ 0 \end{bmatrix}.$$

Then the length of u is $\sqrt{3}$.

Quiz 157 True or False: *Suppose $V = \mathbb{R}^4$. Let*

$$u = \begin{bmatrix} 1.1 \\ 0 \\ 2.1 \\ 1 \end{bmatrix}.$$

Then the length of u is $\sqrt{5.5}$.

Quiz 158 True or False: *Suppose $V = \mathbb{R}^2$. Let*

$$u = \begin{bmatrix} 1 \\ 2 \end{bmatrix}, v = \begin{bmatrix} 3 \\ 2 \end{bmatrix}.$$

The distance between u and v is 1.

Quiz 159 True or False: *Suppose $V = \mathbb{R}^3$. Let*

$$u = \begin{bmatrix} 1.1 \\ 0 \\ 5 \end{bmatrix}, v = \begin{bmatrix} 2.1 \\ -2 \\ -1 \end{bmatrix}.$$

The distance between u and v is 3.

Quiz 160 True or False: *Suppose $V = \mathbb{R}^4$. Let*

$$u = \begin{bmatrix} 1.1 \\ -2 \\ 0 \\ 2.1 \end{bmatrix}, v = \begin{bmatrix} -1.9 \\ 2 \\ 1 \\ 2.1 \end{bmatrix}.$$

The distance between u and v is 5.

Quiz 161 True or False: *Suppose $V = \mathbb{R}^3$. Let u, v be vectors in V. Then*

$$\langle u - v, u - v \rangle = ||u||^2 - 2 \cdot \langle u, v \rangle + ||v||^2.$$

Quiz 162 True or False: *Suppose $V = \mathbb{R}^3$. Let u, v be vectors in V. Then*

$$\langle u - v, u + v \rangle = ||u||^2 - ||v||^2.$$

5.2.6   Regular Exercises

Exercise 5.1 *Let $V = \mathbb{R}^2$. Compute the lengths for the following vectors:*

1.
$$\begin{bmatrix} 2 \\ 3 \end{bmatrix}$$

2.
$$\begin{bmatrix} -4 \\ 0 \end{bmatrix}$$

3.
$$\begin{bmatrix} -1 \\ 1 \end{bmatrix}$$

4.
$$\begin{bmatrix} 1/2 \\ -1/2 \end{bmatrix}$$

Exercise 5.2 *Let $V = \mathbb{R}^3$. Compute the lengths for the following vectors:*

1.
$$\begin{bmatrix} 1 \\ -3 \\ -1 \end{bmatrix}$$

2.
$$\begin{bmatrix} -1 \\ 1 \\ 1 \end{bmatrix}$$

3.
$$\begin{bmatrix} -0.1 \\ 0.2 \\ -2 \end{bmatrix}$$

4.
$$\begin{bmatrix} 2 \\ -1/2 \\ 2/3 \end{bmatrix}$$

Exercise 5.3 *Let $V = \mathbb{R}^3$. Compute the distances for the following vectors:*

1.
$$\begin{bmatrix} 1 \\ -3 \\ -1 \end{bmatrix}, \begin{bmatrix} 1 \\ 0 \\ -1 \end{bmatrix}$$

2.

$$\begin{bmatrix} -1 \\ 1 \\ 1 \end{bmatrix}, \begin{bmatrix} 1 \\ 1 \\ 1 \end{bmatrix}$$

3.

$$\begin{bmatrix} -0.1 \\ 0.2 \\ -1 \end{bmatrix}, \begin{bmatrix} -0.4 \\ -1 \\ -1.5 \end{bmatrix}$$

4.

$$\begin{bmatrix} 5 \\ 2 \\ -2/3 \end{bmatrix}, \begin{bmatrix} 4 \\ -2 \\ -1/3 \end{bmatrix}$$

Exercise 5.4 *Verify the operation in Example 126 is an inner product.*

Exercise 5.5 *Suppose we have the following vectors in $\mathbb{R}^3$:*

$$u = \begin{bmatrix} 1 \\ -1 \\ -2 \end{bmatrix}, v = \begin{bmatrix} 2 \\ -2 \\ -1 \end{bmatrix}, w = \begin{bmatrix} 1 \\ 0 \\ -1 \end{bmatrix}, z = \begin{bmatrix} 0 \\ 2 \\ -1 \end{bmatrix}.$$

Then compute the following:

1. $\langle 2 \cdot u, 3 \cdot v \rangle$,

2. $\langle u, v + w \rangle$,

3. $\langle 3 \cdot u - 2 \cdot w, v \rangle$,

4. $\langle z + y, v - w \rangle$,

5. $\langle 2 \cdot u + 5 \cdot z, v - z \rangle - \langle u, 2 \cdot v \rangle$, and

6. $\langle \frac{1}{2} \cdot u + \frac{5}{2} \cdot z, z \rangle - \langle \frac{1}{2} \cdot u - \frac{3}{2} \cdot 2 \cdot v, z \rangle$.

5.2.7 Lab Exercises

Lab Exercise 166 *Using the sum() function or/and the sqrt() function in* R, *repeat Exercise 5.2.*

Lab Exercise 167 *Using the sum() function or/and the sqrt() function in* R, *repeat Exercise 5.3.*

Lab Exercise 168 *Using the sum() function or/and the sqrt() function in* R, *repeat Exercise 5.5.*

Lab Exercise 169 *Suppose we have a matrix*

$$A = \begin{bmatrix} -2 & 0 & 0 & 0 \\ 3 & 0 & 1 & 1 \\ 2 & 0 & -2 & -1 \end{bmatrix}.$$

1. *Compute a basis for the null space of A.*

2. *Compute a basis for the row space of A.*

3. *For each vector in a basis for the null space of A and for each vector in a basis for the row space of A, take an inner product. What do you obtain?*

4. *Repeat (1), (2), (3) for the matrix*

$$A = \begin{bmatrix} -3 & -1 & -3 & -3 & 2 \\ -2 & 2 & 1 & -3 & 1 \\ 3 & 0 & -2 & 3 & -3 \end{bmatrix}.$$

Is there any pattern of inner products of a vector in a basis for the null space of A and a vector in a basis of row space of A? If so, what did you find?

5.2.8 Practical Applications

The algorithm of the k-means clustering method is to repeatedly compute a mean and assign new members to each cluster by computing distances between the mean and observations. A distance between the mean and an observation is defined as

$$||\mu_i - x_j||,$$

where μ_i is the mean of the cluster i and x_j is the jth observation.

More specifically, the outline of the algorithm is the following:

1. Initialize means μ_i for $i = 1, \ldots, k$ by picking random observations in the input data set.

2. Assign a member x_j for $j = 1, \ldots n$ to the ith cluster from the data set if

$$(||\mu_i - x_1|| + \ldots + ||\mu_i - x_n||)$$

   is minimized.

3. Recompute the mean μ_i for the cluster i and repeat the process.

If you are interested in how the kmeans() function is written in R, we can use the print() function.

```
print(kmeans)
```

The kmeans.ani() function from the **animation** package in R [47] shows how the k-means clustering method works on an input data set. First, we load the **animation** package:

```
library(animation)
```

Then we use the kmeans.ani() function. The arguments of the kmeans.ani() function are the same as the kmeans() function:

```
kmeans.ani(df, 4)
```

Then this will produce 14 figures, which are shown in Figures 5.3 and 5.4.

As we discussed earlier, one disadvantage of k-means clustering is that we have to know how many clusters are in the data set before running it. For finding the optimal number of clusters, k, we can use the **elbow method**. This uses the sum of squares, which uses distances between two vectors in a vector space. We can conduct this procedure using the fviz_nbclust() function from the **factoextra** package:

```
fviz_nbclust(df, kmeans, method = "wss")
```

The first argument of the function is the name of the data set, the second argument is the name of the function which we used, and we choose the method "wss" for "within sum of squares". The output is shown in Figure 5.5: From Figure 5.5 after $k = 4$ in the x-axis, the total sum of squares would not be much smaller. In the elbow method, we pick the smallest number which would not significantly improve the total sum of squares.

5.2.9 Supplements with python Code

In python, we can use the np.dot() function from the numpy package to compute a dot product of two vectors in $\mathbb{R}^d$. For example, if we want to compute the dot product of the vectors:

$$\begin{bmatrix} 5 \\ 2 \\ -2/3 \end{bmatrix}, \begin{bmatrix} 4 \\ -2 \\ -1/3 \end{bmatrix},$$

then we can do:

```
import numpy.matlib
import numpy as np
u = np.array([5, 2, -2/3])
v = np.array([4, -2, -1/3])
np.dot(u,v)
```

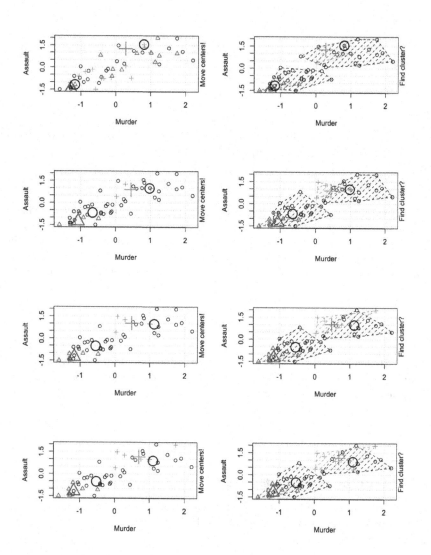

FIGURE 5.3
The figures in the left column show means' locations and the figures in the
right column show the clusters at the ith row. The first row shows the first
iteration, the second row shows the second iteration, the third row shows the
third iteration, and the fourth row shows the fourth iteration.

It will output:

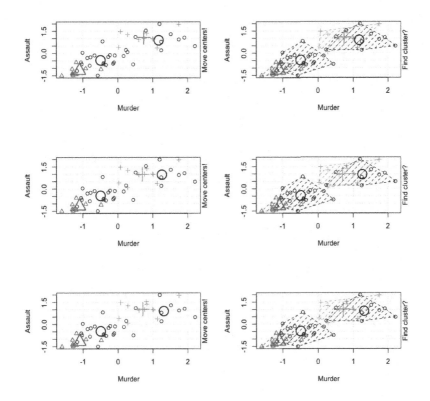

FIGURE 5.4
The figures in the left column show means' locations and the figures in the
right column show the clusters at the ith row. The first row shows the fifth
iteration, the second row shows the sixth iteration, and the third row shows
the seventh iteration.

```
>>> np.dot(u,v)
16.22222222222222
```

To compute the root square, we can use **, which is for the exponent. For
example, if we want to compute the length of a vector:

$$\begin{bmatrix} 5 \\ 2 \\ -2/3 \end{bmatrix},$$

then we can do

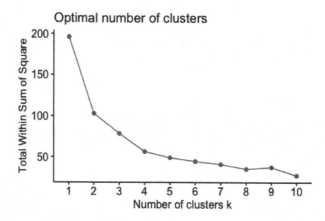

FIGURE 5.5
The output from the elbow method for the k-means clustering method on the "USArrests" data set.

```
import numpy.matlib
import numpy as np
u = np.array([5, 2, -2/3])
np.dot(u,u) ** (0.5)
```

Then it will return

```
>>> np.dot(u,u) ** (0.5)
5.426273532033235
```

5.3 Angles and Orthogonality

There are infinitely many ways to form a basis for a vector space. However, we might want to have a basis where each vector is orthogonal to the others in many cases. For example, basis V is spanned by them as shown Figure 4.9. If they do not, then V is spanned by them as shown in Figure 4.13. Therefore, in this section we will discuss how we can measure the angle between two vectors in an inner product space.

5.3.1 Task Completion Checklist

- During the Lecture:

    1. Read the definition of orthogonality of an inner product.

2. Read the definition of a projection of a vector onto another vector.

3. Read the definition of an orthogonal basis for a vector space.

4. Read the definition of an orthonormal basis for a vector space.

5. Read the definition of an orthogonal complement of a vector subspace.

6. Learn how to use the orth() function from the **pracma** package to compute an orthonormal basis for a vector space.

7. Learn how to compute an angle between two vectors in a vector space.

8. Learn how to give an example of an orthogonal basis of a vector space.

9. With R, learn how to perform computational examples in this section.

- After the Lecture:

  1. Take conceptual quizzes to make sure you understand the materials in this section.

  2. Do some regular exercises.

  3. Conduct lab exercises with R.

  4. Conduct practical applications with R.

  5. If you are interested in **python**, read the supplement in this section and conduct lab exercises and practical applications with **python**.

5.3.2 Working Examples

In PCA, principal components form a basis of a vector space $\mathbb{R}^d$, where d is the number of variables in an input data set, and they also form an orthogonal basis. When we draw pictures of clusters on the "USArrests" data set which we computed via the k-means clustering method in the previous section and also via the hierarchical clustering method in Chapter 4, we used the first principal component and the second principal component to visualize the data. PCA is often used to visualize an input data set in a lower-dimensional vector space. In the Working Example in this section we will apply PCA to the "USArrests" data set so we can closely take a look at all principal components.

First, let us plot the data set on the first and second principal components. To do so, we need to use the **ggplot2**, **ggfortify**, and **gridExtra** packages.

```
library(ggplot2)
library(ggfortify)
library(gridExtra)
```

First, we upload and clean the data:

```
df <- USArrests
df <- na.omit(df)
df <- scale(df)
```

To visualize the data set via PCA, we use the autoplot() function:

```
pc <- prcomp(df, scale=TRUE)
autoplot(pc, data=df)
```

It will produce the plot shown in Figure 5.6.

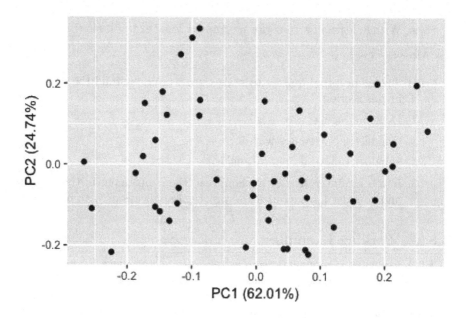

FIGURE 5.6
This plot shows the "USArrests" data via PCA. The x-axis shows the first principal component and the y-axis shows the second principal component. We created this plot using the autoplot() function from the **ggplot2** package.

5.3.3   Angles and Orthogonality

In many situations, we have to measure the angle between two vectors. For example, in mechanical physics, we have to measure the angle of the projectile motion in order to predict the distance from traveled. In probability, this relates to a correlation between two random variables. Therefore, first, we show how to compute the angle between vectors using inner products in a vector space.

Theorem 5.2 *Suppose u, v are vectors in a vector space $V = \mathbb{R}^d$. Then the angle θ between u and v is*

$$\cos\theta = \frac{\langle u, v \rangle}{||u|| \cdot ||v||}.$$

Example 131 *Suppose $V = \mathbb{R}^3$ and suppose*

$$u = \begin{bmatrix} 1 \\ 0 \\ 2 \end{bmatrix}, v = \begin{bmatrix} 1 \\ -5 \\ 3 \end{bmatrix}.$$

Then we have

$$\begin{array}{rcl} \langle u, v \rangle & = & 7, \\ ||u|| & = & \sqrt{5} \\ ||v|| & = & \sqrt{35}. \end{array}$$

Thus,

$$\cos\theta = \frac{\langle u, v \rangle}{||u|| \cdot ||v||} = \frac{7}{\sqrt{5} \cdot \sqrt{35}} = \frac{7}{\sqrt{175}}.$$

Therefore

$$\theta = \arccos\left(\frac{7}{\sqrt{175}}\right),$$

where $\arccos \cdot$ *is the arc cosine.*

Remark 5.3 *Suppose u, v are vectors in a vector space $V = \mathbb{R}^d$. Then the angle θ between u and v is between 0 and π.*

Suppose u, v are vectors in a vector space $V = \mathbb{R}^d$. Then the angle θ between u and v is

$$\cos\theta = \frac{\langle u, v \rangle}{||u|| \cdot ||v||}$$

by Theorem 5.2. Thus we have

$$\langle u, v \rangle = ||u|| \cdot ||v|| \cdot \cos\theta.$$

Then, when the angle between two vectors in $\mathbb{R}^d$ is between 0 and π, we have

$$\langle u, v \rangle = ||u|| \cdot ||v|| \cdot \cos\frac{\pi}{2} = 0.$$

This leads to the definition of the **orthogonality**:

Definition 43 *Suppose u, v are vectors in a vector space $V = \mathbb{R}^d$. Then, if*

$$\langle u, v \rangle = 0,$$

we say u and v are **orthogonal** *to each other.*

Example 132 *Suppose $V = \mathbb{R}^3$ and suppose*

$$u = \begin{bmatrix} 1 \\ 0 \\ 2 \end{bmatrix}, v = \begin{bmatrix} 0 \\ -5 \\ 0 \end{bmatrix}.$$

Since

$$\langle u, v \rangle = 1 \cdot 0 + 0 \cdot (-5) + 2 \cdot 0 = 0,$$

u and v are orthogonal to each other.

A **projection** of a vector onto another vector in $\mathbb{R}$ is an important operation in data analysis, especially fitting a data set to a linear plane. Here we define a projection of a vector onto another vector.

Definition 44 *Suppose $V = \mathbb{R}^d$ and suppose u and v are vectors in V such that $u \neq v$. Then the **projection** of v onto u is another vector w which is a scalar multiplication of u such that*

$$w = \frac{\langle u, v \rangle}{||u||} \cdot \frac{u}{||u||}.$$

The projection of a vector v onto another vector u in a vector space V is illustrated in Figure 5.7.

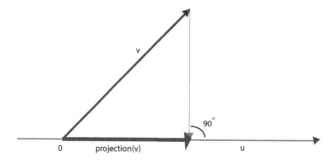

FIGURE 5.7
A vector v is projected onto another vector u in a vector space V.

Example 133 *Suppose $V = \mathbb{R}^3$ and suppose*

$$u = \begin{bmatrix} 1 \\ 1 \\ 0 \end{bmatrix}, v = \begin{bmatrix} 1 \\ 1 \\ 1 \end{bmatrix}.$$

Then the projection of v onto u is

$$w = \frac{\langle u, v \rangle}{||u||} \cdot \frac{u}{||u||} = \frac{2}{\sqrt{2}} \cdot \frac{u}{\sqrt{2}} = u = \begin{bmatrix} 1 \\ 1 \\ 0 \end{bmatrix}.$$

If two vectors are orthogonal to each other, we can simplify some computations. For example, we have the following theorem:

Theorem 5.4 *Suppose $V = \mathbb{R}^d$ and suppose u and v are vectors in V. If u and v are orthogonal, then*

$$||u + v||^2 = ||u||^2 + ||v||^2.$$

There are several nice properties for vectors which are orthogonal to each other, so orthogonality is important in many areas. Especially to describe a vector space, we want to choose a basis whose elements are orthogonal to each other. We have a special name for such bases for a vector space.

Definition 45 *Suppose $V = \mathbb{R}^d$ and*

$$\mathcal{B} = \{b_1, b_2, \ldots b_d\}$$

forms a basis for V. Then if

$$\langle b_i, b_j \rangle = 0$$

for all i and j with $i \neq j$, then we call $\mathcal{B}$ an **orthogonal basis** *for V.*

Example 134 *Suppose $V = \mathbb{R}^3$ and suppose we have a basis*

$$\mathcal{B} = \left\{ e_1 = \begin{bmatrix} 1 \\ 0 \\ 0 \end{bmatrix}, e_2 = \begin{bmatrix} 0 \\ 1 \\ 0 \end{bmatrix}, e_3 = \begin{bmatrix} 0 \\ 0 \\ 1 \end{bmatrix} \right\}.$$

Then $\mathcal{B}$ is an orthogonal basis for $\mathbb{R}^3$.

Example 135 *Suppose $V = \mathbb{R}^3$ and suppose we have a basis*

$$\mathcal{B} = \left\{ b_1 = \begin{bmatrix} 1 \\ 0 \\ 1 \end{bmatrix}, b_2 = \begin{bmatrix} 1 \\ 0 \\ -1 \end{bmatrix}, b_3 = \begin{bmatrix} 0 \\ 1 \\ 0 \end{bmatrix} \right\}.$$

Then $\mathcal{B}$ is an orthogonal basis for $\mathbb{R}^3$.

If each element in an orthogonal basis for a vector space has length 1, describing any vectors in the vector space as a linear combination of elements in the basis can be simplified. Thus, if it is possible, we want to have an orthogonal basis whose elements have the length 1. Such an orthogonal basis is called an **orthonormal basis**.

Definition 46 *Suppose $V = \mathbb{R}^d$ and*

$$\mathcal{B} = \{b_1, b_2, \ldots b_d\}$$

forms a basis for V. Then if

$$\langle b_i, b_j \rangle = 0$$

for all i and j with $i \neq j$ and if

$$||b_i|| = 1$$

for all i, then we call $\mathcal{B}$ an **orthonormal basis** *for V.*

Example 136 *Suppose $V = \mathbb{R}^3$ and suppose we have a basis*

$$\mathcal{B} = \left\{ e_1 = \begin{bmatrix} 1 \\ 0 \\ 0 \end{bmatrix}, e_2 = \begin{bmatrix} 0 \\ 1 \\ 0 \end{bmatrix}, e_3 = \begin{bmatrix} 0 \\ 0 \\ 1 \end{bmatrix} \right\}.$$

Then $\mathcal{B}$ is an orthonormal basis for $\mathbb{R}^3$.

Example 137 *Suppose $V = \mathbb{R}^3$ and suppose we have a basis*

$$\mathcal{B} = \left\{ b_1 = \begin{bmatrix} 1/\sqrt{2} \\ 0 \\ 1/\sqrt{2} \end{bmatrix}, b_2 = \begin{bmatrix} 1/\sqrt{2} \\ 0 \\ -1/\sqrt{2} \end{bmatrix}, b_3 = \begin{bmatrix} 0 \\ 1 \\ 0 \end{bmatrix} \right\}.$$

Then $\mathcal{B}$ is an orthonormal basis for $\mathbb{R}^3$.

With the orth() function from the **pracma** package, we can compute an orthonormal basis for a vector space. For example, suppose

$$A = \begin{bmatrix} 1 & 1 & 1 \\ 1 & 1 & 0 \\ 0 & 0 & 1 \end{bmatrix}.$$

Now we want to compute an orthonormal basis for the row space of A. Then we have

```
library(pracma)
A <- matrix(c(1, 1, 1, 1, 1, 0, 0, 0, 1), 3, 3, byrow = TRUE)
B <- orth(A)
```

Then R outputs:

```
> B
           [,1]        [,2]
[1,] -0.7886751  0.2113249
[2,] -0.5773503 -0.5773503
[3,] -0.2113249  0.7886751
```

Each column of B from the orth() function is an element in the orthonormal basis of the row space of A.

Remark 5.5 *There are several nice algorithms to compute an orthonormal basis for a vector space, such as the Gram-Schmidt process. In this book we are not going to discuss details of the procedure, since we will use the orth() function to compute an orthonormal basis for a vector space.*

Remark 5.6 *For the null space of A, the output from the Null() function from the* MASS *package forms an orthogonal basis for the null space of A.*

Suppose we have a matrix A. We have discussed the null space, the row space and the column space of A in Section 4.3. There are some special properties between these spaces, and in order to describe them, we need the following definition:

Definition 47 *Suppose V is an inner product space and W is a vector subspace of V. A set of vectors in V, which are orthogonal to any vectors in W is called the* **orthogonal complement** *of W.*

Remark 5.7 *We denote the orthogonal complement of W as $W^\perp$.*

There are some properties of the orthogonal complement of W, which is a vector subspace of a vector space V. These properties are useful when we want to compute the orthogonal complement.

Theorem 5.8 *Suppose V is an inner product space and W is a vector subspace of V. Then $W^\perp$ is a vector subspace of V.*

Theorem 5.9 *Suppose V is an inner product space and W is a vector subspace of V. Then the intersection of W and $W^\perp$ is $\{\mathbf{0}\}$.*

Theorem 5.10 *Suppose V is an inner product space and W is a vector subspace of V. Then the orthogonal complement of $W^\perp$ is W.*

Example 138 *Suppose $V = \mathbb{R}^3$ and suppose W is a vector subspace of V whose basis is*

$$\left\{ e_1 = \begin{bmatrix} 1 \\ 0 \\ 0 \end{bmatrix}, e_2 = \begin{bmatrix} 0 \\ 1 \\ 0 \end{bmatrix} \right\}.$$

Then $W^\perp$ is a vector subspace spanned by a vector

$$e_3 = \begin{bmatrix} 0 \\ 0 \\ 1 \end{bmatrix}.$$

Example 139 *Suppose $V = \mathbb{R}^3$ and suppose W is a vector subspace whose basis is*

$$\left\{ b_1 = \begin{bmatrix} 1/\sqrt{2} \\ 0 \\ 1/\sqrt{2} \end{bmatrix}, b_2 = \begin{bmatrix} 1/\sqrt{2} \\ 0 \\ -1/\sqrt{2} \end{bmatrix}, \right\}.$$

Then a vector subspace spanned by a vector

$$b_3 = \begin{bmatrix} 0 \\ 1 \\ 0 \end{bmatrix}$$

is $W^\perp$.

Example 140 *Suppose*

$$A = \begin{bmatrix} 1 & 1 & 1 \\ 1 & 1 & 0 \\ 0 & 0 & 1 \end{bmatrix}.$$

Then using the Null() function from the MASS *package we have:*

```
library(MASS)
A <- matrix(c(1, 1, 1, 1, 1, 0, 0, 0, 1), 3, 3, byrow = TRUE)
Null(t(A))
```

Then R *returns:*

```
> Null(t(A))
            [,1]
[1,] -7.071068e-01
[2,]  7.071068e-01
[3,] -1.110223e-16
```

Actually the null space of A is spanned by a vector

$$\begin{bmatrix} \frac{1}{\sqrt{2}} \\ -\frac{1}{\sqrt{2}} \\ 0 \end{bmatrix}.$$

Also the row space of A has a basis

$$\left\{ b_1 = \begin{bmatrix} 1 & 1 & 0 \end{bmatrix}, b_2 = \begin{bmatrix} 0 & 0 & 1 \end{bmatrix} \right\}.$$

Thus any vector in the row space of A can be written as a linear combination of elements in the basis:

$$v = \alpha_1 \cdot b_1 + \alpha_2 \cdot b_2 = \alpha \cdot \begin{bmatrix} 1 & 1 & 0 \end{bmatrix} + \alpha_2 \cdot \begin{bmatrix} 0 & 0 & 1 \end{bmatrix} = \begin{bmatrix} \alpha_1 & \alpha_1 & \alpha_2 \end{bmatrix}.$$

Any vector in the null space of A can be written as a scalar multiplication of an element in a basis of the null space of A:

$$w = \begin{bmatrix} \beta_1 \cdot \frac{1}{\sqrt{2}} \\ -\beta_1 \cdot \frac{1}{\sqrt{2}} \\ 0 \end{bmatrix}.$$

Consider an inner product of W and v:

$$\langle v, w \rangle = \alpha_1 \cdot \beta_1 \cdot \frac{1}{\sqrt{2}} + \alpha_1 \cdot (-\beta_1 \cdot \frac{1}{\sqrt{2}}) + \alpha_2 \cdot 0 = \frac{\alpha_1 \cdot \beta_1}{\sqrt{2}} - \frac{\alpha_1 \cdot \beta_1}{\sqrt{2}} = 0.$$

This example is not a special case. In fact. for any matrix A, the row space of A is the orthogonal complement to the null space of A.

Theorem 5.11 *Suppose we have a matrix A. Then the null space of A is the orthogonal complement to the row space of A.*

5.3.4 Checkmarks

1. The definition of orthogonality of an inner product.

2. Read the definition of a projection of a vector onto another vector.

3. The definition of an orthogonal basis for a vector space.

4. The definition of an orthonormal basis for a vector space.

5. The definition of an orthogonal complement of a vector subspace.

6. You can use the orth() function from the **pracma** package to compute an orthonormal basis for a vector space.

7. You can compute an angle between two vectors in a vector space.

8. You can give a example of an orthogonal basis of a vector space.

9. With R, you can perform computational examples in this section.

5.3.5 Conceptual Quizzes

Quiz 163 True or False: *Suppose $V = \mathbb{R}^2$. Let*

$$u = \begin{bmatrix} 1 \\ 2 \end{bmatrix}, v = \begin{bmatrix} -2 \\ 1 \end{bmatrix}.$$

Then u and v are orthogonal to each other.

Quiz 164 True or False: *Suppose $V = \mathbb{R}^2$. Let*

$$u = \begin{bmatrix} 1 \\ 1 \end{bmatrix}, v = \begin{bmatrix} 0 \\ 1 \end{bmatrix}.$$

Then u and v are orthogonal to each other.

Quiz 165 True or False: *Suppose $V = \mathbb{R}^3$. Let*

$$u = \begin{bmatrix} 1 \\ 0 \\ 1 \end{bmatrix}, v = \begin{bmatrix} -1 \\ 1 \\ 0 \end{bmatrix}.$$

Then u and v are orthogonal to each other.

Quiz 166 True or False: *Suppose $V = \mathbb{R}^3$. Let*

$$u = \begin{bmatrix} 1 \\ 1 \\ 1 \end{bmatrix}, v = \begin{bmatrix} -1 \\ 1 \\ 0 \end{bmatrix}.$$

Then u and v are orthogonal to each other.

Quiz 167 True or False: *Suppose* $V = \mathbb{R}^4$. *Let*

$$u = \begin{bmatrix} 1 \\ 1 \\ 1 \\ 1 \end{bmatrix}, v = \begin{bmatrix} -1 \\ 1 \\ 0 \\ 0 \end{bmatrix}.$$

Then u and v are orthogonal to each other.

Quiz 168 True or False: *Suppose* $V = \mathbb{R}^4$. *Let*

$$u = \begin{bmatrix} 1 \\ 1 \\ 1 \\ 1 \end{bmatrix}, v = \begin{bmatrix} -1 \\ 0 \\ 0 \\ 1 \end{bmatrix}.$$

Then u and v are orthogonal to each other.

Quiz 169 True or False: *Suppose* $V = \mathbb{R}^2$. *Let*

$$\mathcal{B} = \left\{ b_1 = \begin{bmatrix} 1 \\ 1 \end{bmatrix}, b_2 = \begin{bmatrix} -1 \\ 1 \end{bmatrix} \right\}$$

be a basis for V. Then $\mathcal{B}$ is an orthogonal basis for V.

Quiz 170 True or False: *Suppose* $V = \mathbb{R}^2$. *Let*

$$\mathcal{B} = \left\{ b_1 = \begin{bmatrix} 1 \\ 1 \end{bmatrix}, b_2 = \begin{bmatrix} -1 \\ 1 \end{bmatrix} \right\}$$

be a basis for V. Then $\mathcal{B}$ is an orthonormal basis for V.

Quiz 171 True or False: *Suppose* $V = \mathbb{R}^3$. *Let*

$$\mathcal{B} = \left\{ b_1 = \begin{bmatrix} 1 \\ 1 \\ 0 \end{bmatrix}, b_2 = \begin{bmatrix} -1 \\ 1 \\ 0 \end{bmatrix}, b_3 = \begin{bmatrix} 1 \\ 0 \\ 0 \end{bmatrix} \right\}$$

be a basis for V. Then $\mathcal{B}$ is an orthogonal basis for V.

Quiz 172 True or False: *Suppose* $V = \mathbb{R}^3$. *Let*

$$\mathcal{B} = \left\{ b_1 = \begin{bmatrix} 1 \\ 1 \\ 0 \end{bmatrix}, b_2 = \begin{bmatrix} -1 \\ 1 \\ 0 \end{bmatrix}, b_3 = \begin{bmatrix} 1 \\ 0 \\ 0 \end{bmatrix} \right\}$$

be a basis for V. Then $\mathcal{B}$ is an orthonormal basis for V.

Quiz 173 True or False: *Suppose* $V = \mathbb{R}^3$. *Let*

$$\mathcal{B} = \left\{ b_1 = \begin{bmatrix} \cos\theta \\ \sin\theta \\ 0 \end{bmatrix}, b_2 = \begin{bmatrix} -\cos\theta \\ \sin\theta \\ 0 \end{bmatrix}, b_3 = \begin{bmatrix} 0 \\ 0 \\ 1 \end{bmatrix} \right\}$$

be a basis for V. *Then* $\mathcal{B}$ *is an orthogonal basis for* V.

Quiz 174 True or False: *Suppose* $V = \mathbb{R}^3$. *Let*

$$\mathcal{B} = \left\{ b_1 = \begin{bmatrix} \cos\theta \\ \sin\theta \\ 0 \end{bmatrix}, b_2 = \begin{bmatrix} -\sin\theta \\ \cos\theta \\ 0 \end{bmatrix}, b_3 = \begin{bmatrix} 0 \\ 0 \\ 1 \end{bmatrix} \right\}$$

be a basis for V *for any angle* $0 < \theta < \pi$. *Then* $\mathcal{B}$ *is an orthonormal basis for* V.

5.3.6 Regular Exercises

Exercise 5.6 *Suppose* $V = \mathbb{R}^2$. *Let*

$$u = \begin{bmatrix} 1 \\ 1 \end{bmatrix}, v = \begin{bmatrix} 0 \\ 1 \end{bmatrix}.$$

Compute the angle θ *between* u *and* v.

Exercise 5.7 *Suppose* $V = \mathbb{R}^2$. *Let*

$$u = \begin{bmatrix} -1 \\ 1 \end{bmatrix}, v = \begin{bmatrix} 1 \\ -1 \end{bmatrix}.$$

Compute the angle θ *between* u *and* v.

Exercise 5.8 *Suppose* $V = \mathbb{R}^3$. *Let*

$$u = \begin{bmatrix} 1 \\ 1 \\ 3 \end{bmatrix}, v = \begin{bmatrix} 1 \\ 0 \\ 1 \end{bmatrix}.$$

Compute the angle θ *between* u *and* v.

Exercise 5.9 *Suppose* $V = \mathbb{R}^3$. *Let*

$$u = \begin{bmatrix} -1 \\ 2 \\ -2 \end{bmatrix}, v = \begin{bmatrix} 1 \\ 0 \\ 1 \end{bmatrix}.$$

Compute the angle θ *between* u *and* v.

Exercise 5.10 *Suppose $V = \mathbb{R}^4$. Let*

$$u = \begin{bmatrix} -1 \\ 1 \\ 1 \\ -1 \end{bmatrix}, v = \begin{bmatrix} 1 \\ 0 \\ 1 \\ 0 \end{bmatrix}.$$

Compute the angle θ between u and v.

Exercise 5.11 *Suppose $V = \mathbb{R}^4$. Let*

$$u = \begin{bmatrix} 1 \\ 1 \\ 1 \\ 1 \end{bmatrix}, v = \begin{bmatrix} 1 \\ 0 \\ 1 \\ 1 \end{bmatrix}.$$

Compute the angle θ between u and v.

Exercise 5.12 *Suppose $V = \mathbb{R}^2$. Let*

$$u = \begin{bmatrix} 1 \\ 1 \end{bmatrix}, v = \begin{bmatrix} 0 \\ 1 \end{bmatrix}.$$

Compute the projection of v onto u.

Exercise 5.13 *Suppose $V = \mathbb{R}^2$. Let*

$$u = \begin{bmatrix} -1 \\ 1 \end{bmatrix}, v = \begin{bmatrix} 1 \\ -1 \end{bmatrix}.$$

Compute the projection of v onto u.

Exercise 5.14 *Suppose $V = \mathbb{R}^3$. Let*

$$u = \begin{bmatrix} 1 \\ 1 \\ 3 \end{bmatrix}, v = \begin{bmatrix} 1 \\ 0 \\ 1 \end{bmatrix}.$$

Compute the projection of v onto u.

Exercise 5.15 *Suppose $V = \mathbb{R}^3$. Let*

$$u = \begin{bmatrix} -1 \\ 2 \\ -2 \end{bmatrix}, v = \begin{bmatrix} 1 \\ 0 \\ 1 \end{bmatrix}.$$

Compute the projection of v onto u.

Exercise 5.16 *Suppose $V = \mathbb{R}^4$. Let*

$$u = \begin{bmatrix} -1 \\ 1 \\ 1 \\ -1 \end{bmatrix}, v = \begin{bmatrix} 1 \\ 0 \\ 1 \\ 0 \end{bmatrix}.$$

Compute the projection of v onto u.

Exercise 5.17 *Suppose $V = \mathbb{R}^4$. Let*

$$u = \begin{bmatrix} 1 \\ 1 \\ 1 \\ 1 \end{bmatrix}, v = \begin{bmatrix} 1 \\ 0 \\ 1 \\ 1 \end{bmatrix}.$$

Compute the projection of v onto u.

5.3.7 Lab Exercises

Lab Exercise 170 *Suppose we have*

$$A = \begin{bmatrix} 1 & 1 & 0 & 0 \\ 1 & 0 & 1 & 0 \\ 0 & 0 & 1 & 1 \\ 0 & 0 & 1 & 1 \end{bmatrix}.$$

1. *Compute an orthonormal basis for the row space of A using the orth() function from the* **pracma** *package.*

2. *Compute an orthonormal basis for the column space of A using the orth() function from the* **pracma** *package.*

3. *Compute an orthonormal basis for the null space of A using the Null() function from the* **MASS** *package.*

Lab Exercise 171 *Suppose we have*

$$A = \begin{bmatrix} 1 & 2 & 3 & 4 \\ 4 & 3 & 2 & 1 \end{bmatrix}.$$

1. *Compute an orthonormal basis for the row space of A using the orth() function from the* **pracma** *package.*

2. *Compute an orthonormal basis for the column space of A using the orth() function from the* **pracma** *package.*

3. *Compute an orthonormal basis for the null space of A using the Null() function from the* **MASS** *package.*

Lab Exercise 172 *Suppose we have*

$$A = \begin{bmatrix} 2 & 3 & 3 & 3 \\ -1 & -3 & 0 & 0 \\ 3 & 0 & 0 & 1 \end{bmatrix}.$$

1. *Compute an orthonormal basis for the row space of A using the orth() function from the* **pracma** *package.*

2. *Compute an orthonormal basis for the column space of A using the orth() function from the* **pracma** *package.*

3. *Compute an orthonormal basis for the null space of A using the Null() function from the* **MASS** *package.*

Lab Exercise 173 *Suppose we have*

$$A = \begin{bmatrix} 2 & 3 & 1 & 1 & -1 \\ -3 & 3 & 3 & 0 & -2 \\ -1 & 1 & 1 & 1 & 3 \end{bmatrix}.$$

1. *Compute an orthonormal basis for the row space of A using the orth() function from the* **pracma** *package.*

2. *Compute an orthonormal basis for the column space of A using the orth() function from the* **pracma** *package.*

3. *Compute an orthonormal basis for the null space of A using the Null() function from the* **MASS** *package.*

Lab Exercise 174 *Suppose we have*

$$A = \begin{bmatrix} 3 & 0 & -2 & -1 & 0 & 0 \\ 1 & 2 & -3 & -2 & -2 & 2 \\ 2 & -2 & -1 & -3 & 3 & -3 \\ -3 & -1 & -1 & -2 & -3 & 0 \end{bmatrix}.$$

1. *Compute an orthonormal basis for the row space of A using the orth() function from the* **pracma** *package.*

2. *Compute an orthonormal basis for the column space of A using the orth() function from the* **pracma** *package.*

3. *Compute an orthonormal basis for the null space of A using the Null() function from the* **MASS** *package.*

5.3.8   Practical Applications

Going back to the "USArrests" data set, note that since we have four variables, the data set is in a vector space $V = \mathbb{R}^4$. In order to compute the principal components, which form an orthonormal basis for the vector space $V = \mathbb{R}^4$, we will use the princomp() function in R. The "loadings" is an output from the princomp() function and it is a set of principal components, which is an orthonormal basis for $V = \mathbb{R}^4$:

```
pca<- princomp(data.frame(scale(df)))
pca$loadings
```

Then R outputs:

```
> pca$loadings

Loadings:
         Comp.1 Comp.2 Comp.3 Comp.4
Murder   -0.536  0.418  0.341  0.649
Assault  -0.583  0.188  0.268 -0.743
UrbanPop -0.278 -0.873  0.378  0.134
Rape     -0.543 -0.167 -0.818

                Comp.1 Comp.2 Comp.3 Comp.4
SS loadings       1.00   1.00   1.00   1.00
Proportion Var    0.25   0.25   0.25   0.25
Cumulative Var    0.25   0.50   0.75   1.00
```

This means that the first principal component is

$$\begin{bmatrix} -0.536 \\ -0.583 \\ -0.278 \\ -0.543 \end{bmatrix},$$

the second principal component is

$$\begin{bmatrix} 0.418 \\ 0.188 \\ -0.873 \\ -0.167 \end{bmatrix},$$

the third principal component is

$$\begin{bmatrix} 0.341 \\ 0.268 \\ 0.378 \\ -0.818 \end{bmatrix},$$

and the fourth principal component is

$$\begin{bmatrix} 0.649 \\ -0.743 \\ 0.134 \\ 0 \end{bmatrix}.$$

5.3.9   Supplements with python Code

In python, we can use the orth() function from the scipy.linalg package [43] to compute an orthonormal basis for $\mathbb{R}^d$. For example, suppose

$$A = \begin{bmatrix} 1 & 1 & 1 \\ 1 & 1 & 0 \\ 0 & 0 & 1 \end{bmatrix}.$$

Now we want to compute an orthonormal basis for the row space of A. Then we have

```
import numpy.matlib
import numpy as np
from scipy.linalg import orth
A = np.array([[1, 1, 1], [1, 1, 0], [0, 0, 1]])
orth(A)
```

Then python outputs:

```
>>> orth(A)
array([[-0.78867513,  0.21132487],
       [-0.57735027, -0.57735027],
       [-0.21132487,  0.78867513]])
```

5.4   Discussion

We apply the Orthogonal Distance Regression (ODR) to find the "best" fitted linear plane in $\mathbb{R}^2$ to predict the "Salary" from the "Hits" and "AtBat" of baseball players based on the "Hitters" data set from the ISLR package [24].

The ODR is a regression analysis for a data set in $\mathbb{R}^d$ which tries to find a linear plane such that the sum of the squares of distances between each data point and its projection on the linear plane is minimized over all possible linear planes in $\mathbb{R}^d$.

First we upload the ISLR package and the data set. Then, since there are some missing values we use the na/omit() function to remove all missing values:

```
library(ISLR)
data(Hitters)
D <- na.omit(Hitters)
```

Then we will use the odregress() function from the **pracma** package:

```
library(pracma)
res <- odregress(as.matrix(D[,1:2]), as.matrix(D$Salary))
```

The as.matrix() function makes the data set into a numerical matrix. The first argument of the odregress() function is a set of independent variables. In this case they are the "AtBat" (the first column of the data) and "Hits" (the second column of the data). The second argument of the function is the dependent variable. In this case it is the "Salary".

Then if you type "res$coeff" which outputs the coefficient of the linear plane:

```
> res$coeff
[1]   -43.53176   150.65796 1861.91571
```

This means that the coefficient for the "AtBat" is -43.53176 and the coefficient for the "Hits" is 150.65796. The coefficient for the intercept is 1861.91571. This means that we can estimate the "Salary" as

$$\text{Salary} = -43.53176 \cdot \text{AtBat} + 150.65796 \cdot \text{Hits} + 1861.91571.$$

6

Eigen Values and Eigen Vectors

Eigen values and eigen vectors of a matrix are very useful tool in many areas, such as data science, statistics, probability, physics, and graph theory etc. For the Introductory Example in this chapter we will discuss image compression using the principal component analysis (PCA) in R.

6.1 Introductory Example from Data Science: Image Compression

In order to process images for video games, analyzing data, security cameras, etc., it is important to be able to compress an image efficiently, since images can take large amount of memory. If we do not compress an image efficiently then we might lose too much information so that we cannot recover the original image, or if we do not compress enough, then we might not be able to process the image. For this example we will discuss how we compress an image using PCA. In the process of PCA, we compute eigen values and eigen vectors from the matrix computed from an input image.

We will use the cat picture shown in Figure 6.1:

This image is saved as a jpeg file and it has about 1.8MB of the size. In order to read the jpeg file we will use the readJPEG() function from the jpeg package in R [41]:

```
## library need to open a jpg file
library(jpeg)
## Download from the author's website
myurl <- "http://polytopes.net/Tora_Sleeping.JPG"
z <- tempfile()
download.file(myurl,z,mode="wb")
Kitty <- readJPEG(z)
## dimension
d <- dim(Kitty)
```

The argument of the readJPEG() function is the file name. The dim() function will return the dimension of matrix "Kitty".

DOI: 10.1201/9781003042259-6

FIGURE 6.1
A cat picture for the Introductory example. This is a jpeg file with 1.8MB size. We will compress the image via PCA.

```
> d
[1] 2340 4160    3
```

It has the dimension $2340 \times 4160 \times 3$. The first layer of the 2340×4160 matrix of "Kitty" stores all scores of red color. The second layer of the 2340×4160 matrix of "Kitty" stores all scores of green color, and the third layer of the 2340×4160 matrix of "Kitty" stores all scores of blue color. So we separate them as three matrices:

```
r <- Kitty[,,1]
g <- Kitty[,,2]
b <- Kitty[,,3]
```

In the Discussion, we will show how to compress this image to smaller sizes. We will use some of the R code from [36].

6.2   Eigen Values and Eigen Vectors

In Section 2.2, we briefly covered this subject in the working example. Eigen values and eigen vectors of a matrix have applications in many areas because they capture information on the input matrix. Here we will discuss details on eigen values and eigen vectors of a matrix including how to compute them as well as properties of eigen values and eigen vectors of a matrix.

6.2.1 Task Completion Checklist

- During the Lecture:

  1. Read the definition of eigen values and eigen vectors of a matrix.
  2. Read the definition of a characteristic polynomial of a matrix.
  3. Learn how to compute eigen values and eigen vectors of a matrix using the characteristic polynomial of the matrix.
  4. Use the eigen() function to compute eigen values and eigen vectors of a matrix.
  5. Learn how to compute an angle between two vectors in a vector space.
  6. Learn how to compute eigen values and eigen vectors of a matrix using theorems in this section.
  7. With R, learn how to perform computational examples in this section.

- After the Lecture:

  1. Take conceptual quizzes to make sure you understand the materials in this section.
  2. Do some regular exercises.
  3. Conduct lab exercises with R.
  4. Conduct practical applications with R.
  5. If you are interested in python, read the supplement in this section and conduct lab exercises and practical applications with python.

6.2.2 Working Examples

We are using the working example in Section 2.2. As in Section 2.2, we generate simulated data points under the bivariate normal distribution. Recall that a bivariate normal distribution is defined by a mean and standard deviation, but unlike a univariate normal distribution, its mean is defined by a two-dimensional vector and its standard deviation is defined by a 2×2 matrix. Again, we use the R package mvtnorm [18] and we use the matlib package for matrix operations. First we upload all packages needed:

```
library(mvtnorm)
library(ggplot2)
library(matlib)
```

The ggplot2 package is for plotting.
Then, we define the mean and the standard deviation as:

```
## Standard deviation
sigma <- matrix(c(4,3,3,4), ncol = 2, nrow = 2)
## Mean
mu <- c(5, 5)
```

This gives that the mean is defined by a vector

$$\mu = \begin{bmatrix} 5 \\ 5 \end{bmatrix}.$$

This means that the center of this distribution is at μ in $\mathbb{R}^2$. Also, for the standard deviation, it is defined by a 2×2 matrix:

$$\sigma = \begin{bmatrix} 4 & 3 \\ 3 & 4 \end{bmatrix}.$$

For this data, we have the sample size $n = 10000$.

```
n <- 1000
```

The function set.seed() sets the seed, which is a value to start generating random numbers:

```
set.seed(123)
```

Finally, we generate the data points by

```
x <- rmvnorm(n = n, mean = mu, sigma = sigma)
```

To plot the data we set the points in a "data frame":

```
d <- data.frame(x)
```

Using the **ggplot2** package, we plot the data points:

```
p2 <- ggplot(d, aes(x = X1, y = X2)) +
  geom_point(alpha = .5) +
  geom_density_2d()

p2
```

This creates the plot shown in Figure 6.2.

You might notice that if we draw a green line in Figure 6.3 in $\mathbb{R}^2$, most of the data points are very close to the line. So how can we compute this green line? If there is an easy way to find it, we can model these data points by a linear line in $\mathbb{R}^2$. Using the eigen vectors of the matrix computed from the data set, we can compute this line. We will discuss how to compute them in the Discussion for this section.

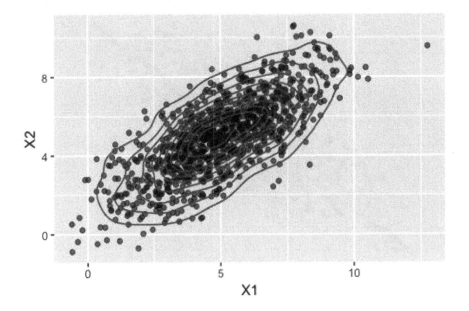

FIGURE 6.2
Bivariate normal distribution defined by μ and σ with the sample size equal
to 1000.

6.2.3 Eigen Values and Eigen Vectors

Suppose we have a 2×2 matrix:

$$A = \begin{bmatrix} 2 & 1 \\ 1 & 2 \end{bmatrix}.$$

Then notice that if I have a vector

$$v = \begin{bmatrix} -1 \\ 1 \end{bmatrix},$$

then we have

$$A \cdot v = \begin{bmatrix} 2 & 1 \\ 1 & 2 \end{bmatrix} \cdot \begin{bmatrix} -1 \\ 1 \end{bmatrix} = \begin{bmatrix} -1 \\ 1 \end{bmatrix}.$$

Therefore, we have a relation:

$$A \cdot v = v.$$

In addition, suppose we have another vector:

$$u = \begin{bmatrix} 1 \\ 1 \end{bmatrix},$$

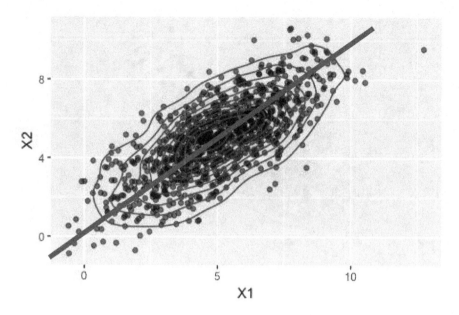

FIGURE 6.3
Bivariate normal distribution defined by μ and σ with the sample size equal
to 1000.

then we have

$$A \cdot u = \begin{bmatrix} 2 & 1 \\ 1 & 2 \end{bmatrix} \cdot \begin{bmatrix} 1 \\ 1 \end{bmatrix} = \begin{bmatrix} 3 \\ 3 \end{bmatrix}.$$

Therefore, we have another relation:

$$A \cdot u = 3 \cdot u.$$

For each matrix there are special vectors, such as u and v in this example,
whose images under the matrix are scalar multiplications of themselves. In
general we have the following form: For an $n \times n$ matrix A, there is a vector
v and a scalar λ such that

$$A \cdot v = \lambda \cdot v.$$

These vectors are called **eigen vectors** of a matrix and these scalars are called
eigen values.

Definition 48 *Suppose we have an $n \times n$ matrix A. Then an **eigen vector**
of A is a non-zero vector such that*

$$A \cdot v = \lambda \cdot v.$$

*A scalar λ is called an **eigen value** associate with the eigen vector v.*

In order to compute the eigen values and eigen vectors of a matrix, first we set up the matrix equation:

$$A \cdot v = \lambda \cdot v.$$

Then we bring the right hand side to the left-hand side:

$$A \cdot v - \lambda \cdot v = 0.$$

Then we simplify this equation:

$$(A - \lambda \cdot I_n) \cdot v = 0,$$

where I_n is the identity matrix of size n. This means that an non-zero vector V is in the null space of the matrix $(A - \lambda \cdot I_n)$. Since the matrix $(A - \lambda \cdot I_n)$ has non-zero vector in the null space, the dimension of the null space is greater than or equal to 1. Thus, the determinant of the matrix $(A - \lambda \cdot I_n)$ must be equal to 0. So we set

$$\det(A - \lambda \cdot I_n) = 0,$$

which is a polynomial in terms of one variable λ. This polynomial is called the **characteristic polynomial** of a matrix A. The roots of the characteristic polynomial of A are the eigen values. From each root λ_i of the characteristic polynomial, we find the null space of the matrix $(A - \lambda \cdot I_n)$. All elements in a basis for the null space of $(A - \lambda \cdot I_n)$ are eigen vectors associated with the eigen value λ_i.

Example 141 *Suppose we have a 2×2 matrix:*

$$A = \begin{bmatrix} 2 & 1 \\ 1 & 2 \end{bmatrix}.$$

Then, a characteristic polynomial of $(A - \lambda \cdot I_n)$ is

$$\det(A - \lambda \cdot I_n) = 0,$$

which is

$$\det\left(\begin{bmatrix} 2 & 1 \\ 1 & 2 \end{bmatrix} - \lambda \cdot \begin{bmatrix} 1 & 0 \\ 0 & 1 \end{bmatrix}\right) = \det\left(\begin{bmatrix} 2 & 1 \\ 1 & 2 \end{bmatrix} - \begin{bmatrix} \lambda & 0 \\ 0 & \lambda \end{bmatrix}\right)$$

$$= \det\left(\begin{bmatrix} 2 - \lambda & 1 \\ 1 & 2 - \lambda \end{bmatrix}\right) = (2 - \lambda)^2 - 1 = \lambda^2 - 4 \cdot \lambda + 3 = 0.$$

This polynomial has two roots: $\lambda_1 = 1$, $\lambda_2 = 3$. Now we want to find eigen vectors associated with $\lambda_1 = 1$, $\lambda_2 = 3$.

For $\lambda_1 = 1$, we substitute $\lambda = 1$ in the characteristic polynomial, then we have

$$C_1 = \begin{bmatrix} 2 - 1 & 1 \\ 1 & 2 - 1 \end{bmatrix} = \begin{bmatrix} 1 & 1 \\ 1 & 1 \end{bmatrix}.$$

Since a vector

$$\begin{bmatrix} -1 \\ 1 \end{bmatrix}$$

is in the null space of the matrix C_1, a vector

$$\begin{bmatrix} -1 \\ 1 \end{bmatrix}$$

spans the null space of the matrix C_1 and is an eigen vector associated with $\lambda_1 = 1$.

For $\lambda_1 = 3$, we substitute $\lambda = 3$ to the characteristic polynomial, then we have

$$C_2 = \begin{bmatrix} 2-3 & 1 \\ 1 & 2-3 \end{bmatrix} = \begin{bmatrix} -1 & 1 \\ 1 & -1 \end{bmatrix}.$$

Since a vector

$$\begin{bmatrix} 1 \\ 1 \end{bmatrix}$$

is in the null space of the matrix C_2. Therefore a vector

$$\begin{bmatrix} 1 \\ 1 \end{bmatrix}$$

spans the null space of the matrix C_2 and is an eigen vector associated with $\lambda_2 = 3$.

In R, we can use the eigen() function to compute eigen values and eigen vectors of a matrix. For Example 141, consider a matrix

$$A = \begin{bmatrix} 2 & 1 \\ 1 & 2 \end{bmatrix}.$$

First we define the matrix in R:

```
A <- matrix(c(2, 1, 1, 2), 2, 2)
```

Then we call the eigen() function:

```
eigen(A)
```

R returns the eigen values and eigen vectors associated with them:

```
> eigen(A)
eigen() decomposition
$values
[1] 3 1

$vectors
          [,1]        [,2]
[1,] 0.7071068 -0.7071068
[2,] 0.7071068  0.7071068
```

Here, the first column of $vectors is an eigen vector associated with the eigen value equal to 3, which is the first value of $values. The second column of $vectors is an eigen vector associated with the eigen value equal to 1, which is the second value of $values. Note that these vectors are already "normalized", i.e., these length is equal to 1.

There are some theorems which help us to compute its eigen values and eigen vectors. We will discuss some nice properties of eigen values and eigen vectors of a matrix.

Theorem 6.1 *Eigen values of a triangular matrix are the entries on the diagonal of the matrix.*

Example 142 *Suppose we have a triangular matrix:*

$$A = \begin{bmatrix} 2 & 1 & 3 & -1 \\ 0 & 2 & -1 & 3 \\ 0 & 0 & 1 & 1 \\ 0 & 0 & 0 & -1 \end{bmatrix}.$$

Then the eigen values of the matrix A are $2, 1, -1$.

Example 143 *Suppose we have a triangular matrix:*

$$A = \begin{bmatrix} 3 & 0 & 0 & 0 \\ 0 & -2 & 0 & 0 \\ 0 & 0 & 0 & 0 \\ 0 & 0 & 0 & -1 \end{bmatrix}.$$

Then the eigen values of the matrix A are $3, -2, 0, -1$.

Theorem 6.2 *Suppose we have an $n \times n$ matrix A. A is invertible if and only if 0 is not an eigen value of A.*

Example 144 *From Example 141, suppose we have a 2×2 matrix:*

$$A = \begin{bmatrix} 2 & 1 \\ 1 & 2 \end{bmatrix}.$$

We computed its eigen values from the characteristic polynomial and they are $1, 3$, which are not 0, thus A is invertible.

Example 145 *From Example 142, suppose we have a triangular matrix:*

$$A = \begin{bmatrix} 2 & 1 & 3 & -1 \\ 0 & 2 & -1 & 3 \\ 0 & 0 & 1 & 1 \\ 0 & 0 & 0 & -1 \end{bmatrix}.$$

Then the eigen values of the matrix A are $2, 1, -1$ which are not equal to 0. Thus A is invertible.

322

Linear Algebra and Its Applications with R

Example 146 *From Example 143, suppose we have a triangular matrix:*

$$A = \begin{bmatrix} 3 & 0 & 0 & 0 \\ 0 & -2 & 0 & 0 \\ 0 & 0 & 0 & 0 \\ 0 & 0 & 0 & -1 \end{bmatrix}.$$

Then, the eigen values of the matrix A are $3, -2, 0, -1$. Since 0 is an eigen value, A is not invertible.

Theorem 6.3 *Suppose $v_1, v_2, \ldots, v_r$ are eigen vectors of a matrix A associated with the distinct eigen values $\lambda_1, \lambda_2, \ldots, \lambda_r$ of A. Then $v_1, v_2, \ldots, v_r$ are linearly independent.*

Example 147 *From Example 141, suppose we have a 2×2 matrix:*

$$A = \begin{bmatrix} 2 & 1 \\ 1 & 2 \end{bmatrix}.$$

Its eigen values are $1, 3$, and their corresponding eigen vectors are

$$v_1 = \begin{bmatrix} -1 \\ 1 \end{bmatrix}, v_2 = \begin{bmatrix} 1 \\ 1 \end{bmatrix},$$

which are linearly independent.

Theorem 6.4 *Suppose $v_1, v_2, \ldots, v_r$ are eigen vectors of a symmetric matrix A associated with the distinct eigen values $\lambda_1, \lambda_2, \ldots, \lambda_r$ of A. Then $v_1, v_2, \ldots, v_r$ are linearly independent and they are orthogonal to each other.*

Example 148 *From Example 141, suppose we have a 2×2 matrix:*

$$A = \begin{bmatrix} 2 & 1 \\ 1 & 2 \end{bmatrix}.$$

Its eigen values are $1, 3$, and their corresponding eigen vectors are

$$v_1 = \begin{bmatrix} -1 \\ 1 \end{bmatrix}, v_2 = \begin{bmatrix} 1 \\ 1 \end{bmatrix},$$

which are linearly independent. Also notice that

$$\langle v_1, v_2 \rangle = \langle \begin{bmatrix} -1 \\ 1 \end{bmatrix}, \begin{bmatrix} 1 \\ 1 \end{bmatrix} \rangle = (-1) \cdot 1 + 1 \cdot 1 = (-1) + 1 = 0.$$

6.2.4 Checkmarks

1. The definition of eigen values and eigen vectors of a matrix.

2. The definition of a characteristic polynomial of a matrix.

3. You can compute eigen values and eigen vectors of a matrix using the characteristic polynomial of the matrix.

4. You can use the eigen() function to compute eigen values and eigen vectors of a matrix.

5. How to compute an angle between two vectors in a vector space.

6. You can compute eigen values and eigen vectors of a matrix using theorems in this section.

7. With R, you can perform computational examples in this section.

6.2.5 Conceptual Quizzes

Quiz 175 True or False: *Suppose we have a 3×3 matrix with its eigen values $3, 0, -1$. Then A is invertible.*

Quiz 176 True or False: *Suppose we have a 3×3 matrix with its eigen values $3, 0, -1$. Then their associated eigen vectors v_1, v_2, v_3 are linearly independent.*

Quiz 177 True or False: *Suppose we have a 3×3 matrix with its eigen values $3, 0, -1$. Then their associated eigen vectors v_1, v_2, v_3 are orthogonal to each other.*

Quiz 178 True or False: *Suppose we have a 3×3 matrix with its eigen values $2, 2, 1$. Then A is invertible.*

Quiz 179 True or False: *Suppose we have a 3×3 matrix with its eigen values $2, 2, 1$. Then their associated eigen vectors v_1, v_2, v_3 are linearly independent.*

Quiz 180 True or False: *Suppose we have a 3×3 matrix with its eigen values $2, 2, 1$. Then their associated eigen vectors v_1, v_2, v_3 are orthogonal to each other.*

Quiz 181 True or False: *Suppose we have a 3×3 symmetric matrix with its eigen values $1, 2, 3$. Then A is invertible.*

Quiz 182 True or False: *Suppose we have a 3×3 symmetric matrix with its eigen values $1, 2, 3$. Then their associated eigen vectors v_1, v_2, v_3 are linearly independent.*

Quiz 183 True or False: *Suppose we have a 3×3 symmetric matrix with its eigen values $1, 2, 3$. Then their associated eigen vectors v_1, v_2, v_3 are orthogonal to each other.*

Quiz 184 Multiple Choice: *Suppose we have a 3×3 matrix:*

$$A = \begin{bmatrix} 2 & 1 & -1 \\ 0 & 0 & 5 \\ 0 & 0 & -4 \end{bmatrix}.$$

Then choose all items for its eigen values

1. *2*

2. *−1*

3. *0*

4. *−4*

Quiz 185 True or False: *Suppose we have a 3×3 matrix:*

$$A = \begin{bmatrix} 2 & 1 & -1 \\ 0 & 0 & 5 \\ 0 & 0 & -4 \end{bmatrix}.$$

Then its eigen vectors are linearly independent.

Quiz 186 True or False: *Suppose we have a 3×3 matrix:*

$$A = \begin{bmatrix} 2 & 1 & -1 \\ 0 & 0 & 5 \\ 0 & 0 & -4 \end{bmatrix}.$$

Then its eigen vectors are orthogonal to each other.

Quiz 187 Multiple Choice: *Suppose we have a 3×3 matrix:*

$$A = \begin{bmatrix} 1 & 0 & 0 \\ 0 & 0 & 0 \\ 0 & 0 & -4 \end{bmatrix}.$$

Then choose all items for its eigen values

1. *1*

2. *−1*

3. *0*

4. *−4*

Quiz 188 True or False: *Suppose we have a 3×3 matrix:*

$$A = \begin{bmatrix} 1 & 0 & 0 \\ 0 & 0 & 0 \\ 0 & 0 & -4 \end{bmatrix}.$$

Then its eigen vectors are linearly independent.

Quiz 189 True or False: *Suppose we have a* 3×3 *matrix:*

$$A = \begin{bmatrix} 1 & 0 & 0 \\ 0 & 0 & 0 \\ 0 & 0 & -4 \end{bmatrix}.$$

Then its eigen vectors are orthogonal to each other.

6.2.6 Regular Exercises

Exercise 6.1 *Compute the eigen values and eigen vectors of the following matrices without any help from a computer. Also check if they are invertible.*

1.

$$\begin{bmatrix} 1 & 3 \\ 1 & 1 \end{bmatrix}.$$

2.

$$\begin{bmatrix} -1 & 3 \\ 1 & 5 \end{bmatrix}.$$

3.

$$\begin{bmatrix} 2 & 1 \\ 4 & 2 \end{bmatrix}.$$

4.

$$\begin{bmatrix} 1 & 2 \\ 4 & 3 \end{bmatrix}.$$

5.

$$\begin{bmatrix} 1 & 3 & 0 \\ 1 & 1 & 1 \\ -1 & 1 & 0 \end{bmatrix}.$$

6.

$$\begin{bmatrix} 1 & 3 & 0 \\ 1 & 1 & 1 \\ 2 & 3 & 0 \end{bmatrix}.$$

7.

$$\begin{bmatrix} 1 & 3 & 0 \\ 0 & 1 & 1 \\ 0 & 0 & 0 \end{bmatrix}.$$

8.

$$\begin{bmatrix} 1 & 0 & 0 & 0 \\ 1 & -1 & 0 & 0 \\ -1 & 1 & 0 & 0 \\ 5 & 3 & 2 & 5 \end{bmatrix}.$$

9.
$$\begin{bmatrix} 1 & 0 & 0 & 0 \\ 0 & 15 & 0 & 0 \\ 0 & 1 & 23 & 0 \\ 5 & 3 & 2 & 5 \end{bmatrix}.$$

10.
$$\begin{bmatrix} 213 & 0 & 0 & 0 \\ 0 & 151 & 0 & 0 \\ 0 & 0 & 23 & 0 \\ 0 & 0 & 0 & 7 \end{bmatrix}.$$

11.
$$\begin{bmatrix} -3 & 0 & 0 & 0 & 0 \\ 0 & 15 & 0 & 0 & 0 \\ 0 & 0 & 3 & 0 & 0 \\ 0 & 0 & 0 & -5 & 0 \\ 0 & 0 & 0 & 0 & 7 \end{bmatrix}.$$

6.2.7 Lab Exercises

Lab Exercise 175 *Compute the eigen values and eigen vectors of the following matrices with the eigen() function.*

1.
$$\begin{bmatrix} -3 & 0 & 0 & 0 \\ 3 & -3 & 0 & 0 \\ 0 & -2 & 1 & 0 \\ 3 & -3 & -3 & -2 \end{bmatrix}$$

2.
$$\begin{bmatrix} -1 & 0 & 1 & 3 \\ 0 & 0 & 2 & 6 \\ 1 & 2 & -1 & -2 \\ 3 & 6 & -2 & -1 \end{bmatrix}$$

3.
$$\begin{bmatrix} -3 & 3 & 3 & 0 & -3 \\ 3 & 0 & 2 & -2 & 1 \\ 3 & 2 & 3 & 2 & -2 \\ 0 & -2 & 2 & -3 & 3 \\ -3 & 1 & -2 & 3 & 3 \end{bmatrix}$$

4.
$$\begin{bmatrix} 0 & 0 & 0 & 1 & 0 \\ 0 & -3 & 2 & 0 & 0 \\ 0 & -3 & -2 & 0 & -3 \\ 0 & 0 & -2 & 0 & 0 \\ 2 & -2 & 1 & 0 & 0 \end{bmatrix}$$

5.

$$\begin{bmatrix} 0 & -2 & -2 & 0 & 1 & -3 \\ 1 & 1 & -2 & 0 & 1 & 3 \\ 1 & 2 & 1 & -1 & 1 & -2 \\ 1 & -2 & 1 & 1 & 2 & -3 \\ 1 & -3 & 3 & -3 & -2 & -3 \\ -3 & 0 & -1 & 1 & -1 & 0 \end{bmatrix}$$

6.2.8 Practical Applications

For Practical Applications in this section, we will discuss how we can compute the first and second principal components of the data set. Going back to the working example for this section, we generated 1000 data points under the bivariate normal distribution with

$$\mu = \begin{bmatrix} 5 \\ 5 \end{bmatrix}, \sigma = \begin{bmatrix} 4 & 3 \\ 3 & 4 \end{bmatrix},$$

where μ is the mean and σ is the **covariance matrix** which defines the standard deviation for bivariate normal distribution. Note that this covariance matrix is symmetric. In order to define the standard deviation for a multivariate normal distribution, the covariance matrix has to be symmetric.

In real life, however, we do not observe μ and σ. We observe only the data set x which contains 1000 data points. Therefore we have to estimate them from the data points. Now we estimate a covariant matrix. Suppose x is the matrix whose rows represent each data point in the data set and whose columns represent each variable. Then we can estimate the covariance matrix $\hat{\sigma}$ by $x^T \cdot x$, where x^T is the transpose of the matrix x. $\hat{\sigma}$ is a $k \times k$ symmetric matrix where k is the number of random variables.

In R, we can estimate by:

```
y <- t(x) %*% x
```

Then we apply the eigen() function to compute the eigen values and eigen vectors of this estimated covariance matrix:

```
eigen(y)
```

This returns the following:

```
> eigen(y)
eigen() decomposition
$values
[1] 56444.411  1015.821

$vectors
            [,1]       [,2]
[1,] -0.7091532  0.7050544
[2,] -0.7050544 -0.7091532
```

The eigen values of $\hat{\sigma}$ are 6444.411, 1015.821, and their eigen vectors are

$$\begin{bmatrix} -0.7091532 \\ -0.7050544 \end{bmatrix}, \begin{bmatrix} 0.7050544 \\ -0.7091532 \end{bmatrix},$$

respectively. Since the eigen values of $\hat{\sigma}$ are distinct and $\hat{\sigma}$ is symmetric, eigen vectors of $\hat{\sigma}$ are orthogonal and they are linearly independent. In fact, $\hat{\sigma}$ is always symmetric and if these points are random enough, then the eigen values are distinct, and all eigen vectors of $\hat{\sigma}$ are orthogonal to each other. This is one of the most important properties for PCA.

Now we look into the eigen vectors outputted from the eigen() function on $\hat{\sigma}$. The first principal component for this data set x is the eigen vector for the biggest eigen value of $\hat{\sigma}$. The second principal component is the eigen vector for the second biggest eigen value of $\hat{\sigma}$. In fact, the blue line in Figure 6.4 is defined by the eigen vector of the biggest eigen value of $\hat{\sigma}$, i.e., a set of all points defined by

$$\alpha \cdot \begin{bmatrix} -0.7091532 \\ -0.7050544 \end{bmatrix}$$

where α is any real number. In 6.4, the blue line is spanned by the first principal component and the pink line is spanned by the second principal component.

To check if these vectors are in fact the first and second principal components, we use the prcomp() function to verify:

```
> prcomp(d, center = FALSE)
Standard deviations (1, .., p=2):
[1] 7.516709 1.008384

Rotation (n x k) = (2 x 2):
        PC1         PC2
X1 -0.7091532  0.7050544
X2 -0.7050544 -0.7091532
```

This shows that the first and second principal components are

$$\begin{bmatrix} -0.7091532 \\ -0.7050544 \end{bmatrix}, \begin{bmatrix} 0.7050544 \\ -0.7091532 \end{bmatrix},$$

respectively.

6.2.9   Supplements with python Code

To compute the eigen values and eigen vectors of a matrix, we can use the numpy.linalg.eig() function from the numpy package. For Example 141, consider a matrix

$$A = \begin{bmatrix} 2 & 1 \\ 1 & 2 \end{bmatrix}.$$

Then, first we upload the library and define the matrix in python:

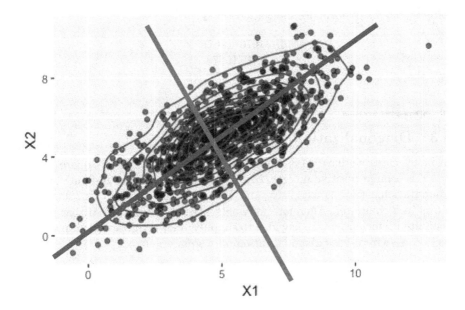

FIGURE 6.4

Bivariate normal distribution defined by μ and σ with the sample size equal to 1000. The blue line is spanned by the first principal component and the pink line is spanned by the second principal component.

```
import numpy as np
from numpy import linalg as LA
A = np.array([[2, 1], [1, 2]])
```

Then we call the numpy.linalg.eig() function from the numpy package:

```
W, V = LA.eig(A)
```

Here W stores the eigen values of A and V stores eigen vectors of A:

```
>>> W
array([3., 1.])
>>> V
array([[ 0.70710678, -0.70710678],
       [ 0.70710678,  0.70710678]])
```

Here the ith column of V is an eigen vector for the ith entry of W. Thus for $\lambda = 3$, its normalized eigen vector is

$$\begin{bmatrix} 0.70710678 \\ 0.70710678 \end{bmatrix},$$

and for $\lambda = 1$, its normalized eigen vector is

$$\begin{bmatrix} -0.70710678 \\ 0.70710678 \end{bmatrix}.$$

6.3 Diagonalization

In many cases we have to compute A^r, where A is an $n \times n$ square matrix and r is a large positive integer. In this case, if we directly compute A^r, the computational time is expensive and also we might have a large numerical errors in a computer. Therefore we would like to simplify A^r so that we can compute it efficiently. One key idea to simplify A^r is **diagonalization**. In this section we will discuss diagonalization on a square matrix.

6.3.1   Task Completion Checklist

- During the Lecture:

  1. Read the definition of a diagonalizable matrix.
  2. Read properties when a square matrix becomes diagonalizable.
  3. Learn how to perform diagonalization when A is diagonalizable.
  4. Learn how to compute A^r when A is diagonalizable.
  5. Learn how to use the eigen() function to diagonalize a diagonalizable matrix.
  6. With R, learn how to perform computational examples in this section.

- After the Lecture:

  1. Take conceptual quizzes to make sure you understand the materials in this section.
  2. Do some regular exercises.
  3. Conduct lab exercises with R.
  4. Conduct practical applications with R.
  5. If you are interested in python, read the supplement in this section and conduct lab exercises and practical applications with python.

6.3.2   Working Examples

This is from evolutionary biology. To analyze DNA sequences, such as aligning them or computing a phylogenetic tree from the alignment, we use a **continuous Markov chain model** to estimate a probability of mutations from observed DNA sequences. A continuous Markov chain model is a Markov chain with continuous time. Without discussing details, here we outline the basics of Markov chains.

A stochastic process is called a Markov process if it satisfies the **Markov property**, also called the **memoryless property**. This property provides that a sequence of random variables $\{X_t\}$ depends on time t holds such that the next state X_t depends on the current state, but does not depend on the previous state. To explain this memoryless property, we will outline a toy example: Suppose there are bars, $S = \{A, C, G, T\}$, in a town. Suppose Amy, Connie, Gail, and Tami decide to go bar hopping one night. Since they are all drunk, they decide where they are going next by rolling a four-faced die (tetrahedron). This process has the Markov property since deciding the next bar is not affected by the previous bars they visited, but is only affected by the current bar.

In this section, we consider a simplified version of analyzing DNA sequences. Instead of using a continuous Markov chain model, we will use a discrete Markov chain model. In this example, the state space is $S = \{A, C, G, T\}$, which is a set of letters on DNA sequences. Figure 6.5 shows a graphical description of a discrete time Markov chain with the state space S. In Figure 6.5, P_{ij} is a probability from a state i to a state j where i, j

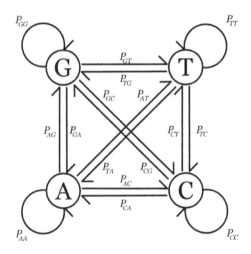

FIGURE 6.5
The graphic description of the transitions for a discrete Markov chain for analyzing DNA sequences.

are elements in $S = \{A, C, G, T\}$. We can summarize these probabilities of a matrix form:

$$P = \begin{array}{c|cccc} & A & C & G & T \\ \hline A & P_{AA} & P_{AC} & P_{AG} & P_{AT} \\ C & P_{CA} & P_{CC} & P_{CG} & P_{CT} \\ G & P_{GA} & P_{GC} & P_{GG} & P_{GT} \\ T & P_{TA} & P_{TC} & P_{TG} & P_{TT} \end{array} .$$

Note that $P_{AA} + P_{AC} + P_{AG} + P_{AT} = 1$, $P_{CA} + P_{CC} + P_{CG} + P_{CT} = 1$, $P_{GA} + P_{GC} + P_{GG} + P_{GT} = 1$, and $P_{TA} + P_{TC} + P_{TG} + P_{TT} = 1$. This matrix is called a **transition matrix** for the Markov chain.

To explain a transition matrix, we will use a toy example above. Probabilities of Amy's die, Connie's die, Gail's die, and Tami's die are listed as

	A	C	G	T
Amy's die	1/4	1/4	1/4	1/4
Connie's die	1/5	1/5	2/5	1/5
Gail's die	1/3	1/3	1/6	1/6
Tami's die T	1/6	1/3	1/3	1/6

So if they are at bar C, then Connie rolls her die to decide where they are going next. The probability of going to bar A is 1/5, the probability of going to bar C is 1/5, the probability of going to bar G is 2/5, and the probability of going to bar T is 1/5.

Going back to a DNA analysis, given the transition matrix, we would like to know after a long time what the probabilities are of mutating from A to C, from A to G and from A to T. In order to solve this problem, we need to compute $\lim_{n \to \infty} P^n$.

6.3.3   Diagonalization

Suppose we have a square matrix A. As shown in the Working Example, often we have to compute A^r for a large positive integer r. In order to compute them it would be nice to decompose A as $Q \cdot D \cdot Q^{-1}$ where D is a diagonal matrix and Q is an invertible matrix.

Suppose D is an $n \times n$ diagonal matrix such that

$$\begin{bmatrix} d_{11} & 0 & \cdots & 0 \\ 0 & d_{22} & \cdots & 0 \\ \vdots & \vdots & \vdots & \vdots \\ 0 & 0 & 0 & d_{nn} \end{bmatrix} .$$

Then

$$D^r = \begin{bmatrix} d_{11}^r & 0 & \cdots & 0 \\ 0 & d_{22}^r & \cdots & 0 \\ \vdots & \vdots & \vdots & \vdots \\ 0 & 0 & 0 & d_{nn}^r \end{bmatrix} .$$

Example 149 *Suppose we have a matrix*

$$\begin{bmatrix} 1 & 0 & 0 & 0 \\ 0 & -1 & 0 & 0 \\ 0 & 0 & 4 & 0 \\ 0 & 0 & 0 & 2 \end{bmatrix}.$$

Then we have

$$D^2 = \begin{bmatrix} 1^2 & 0 & 0 & 0 \\ 0 & (-1)^2 & 0 & 0 \\ 0 & 0 & 4^2 & 0 \\ 0 & 0 & 0 & 2^2 \end{bmatrix}, \quad D^5 = \begin{bmatrix} 1^5 & 0 & 0 & 0 \\ 0 & (-1)^5 & 0 & 0 \\ 0 & 0 & 4^5 & 0 \\ 0 & 0 & 0 & 2^5 \end{bmatrix},$$

$$D^{10} = \begin{bmatrix} 1^{10} & 0 & 0 & 0 \\ 0 & (-1)^{10} & 0 & 0 \\ 0 & 0 & 4^{10} & 0 \\ 0 & 0 & 0 & 2^{10} \end{bmatrix}, \quad D^{100} = \begin{bmatrix} 1^{100} & 0 & 0 & 0 \\ 0 & (-1)^{100} & 0 & 0 \\ 0 & 0 & 4^{100} & 0 \\ 0 & 0 & 0 & 2^{100} \end{bmatrix}.$$

Therefore if we can decompose a square matrix A as $Q \cdot D \cdot Q^{-1}$ for some diagonal matrix D and invertible matrix Q, then we can write

$$A^r = \left(Q \cdot D \cdot Q^{-1}\right)^r = Q \cdot D^r \cdot Q^{-1}.$$

Since we can compute D^r easily, we would like to decompose a matrix A first if we can. This decomposition of a square matrix A is called **diagonalization**. This gives a definition of such matrix A.

Definition 49 *A square $n \times n$ matrix A is called **diagonalizable** if it can be written as*

$$A = Q \cdot D \cdot Q^{-1},$$

where D is a diagonal matrix and Q is an invertible matrix.

Example 150 *Suppose we have a 2×2 matrix:*

$$A = \begin{bmatrix} 2 & 1 \\ 1 & 2 \end{bmatrix}.$$

Then we can decompose the matrix A as

$$\begin{bmatrix} 2 & 1 \\ 1 & 2 \end{bmatrix} = \begin{bmatrix} 1 & -1 \\ 1 & 1 \end{bmatrix} \cdot \begin{bmatrix} 3 & 0 \\ 0 & 1 \end{bmatrix} \cdot \begin{bmatrix} 1 & -1 \\ 1 & 1 \end{bmatrix}^{-1}.$$

To verify this, we use R*:*

```
Q <- matrix(c(1, 1, -1, 1), 2, 2)
D <- matrix(c(3, 0, 0, 1), 2, 2)
Q %*% D %*% inv(Q)
```

This returns:

```
> Q %*% D %*% inv(Q)
     [,1] [,2]
[1,]   2    1
[2,]   1    2
```

Now a question is, when is a square matrix A diagonalizable and how can we compute a diagonalization of A? The following theorems are useful for answering these questions:

Theorem 6.5 *Suppose A is an $n \times n$ matrix. Then A is diagonalizable if and only if A has n linearly independent eigen vectors.*

Theorem 6.6 *Suppose A is an $n \times n$ matrix and suppose A is diagonalizable. Let $\lambda_1, \lambda_2, \ldots \lambda_n$ be the eigen values of A and $v_1, v_2, \ldots, v_n$ be eigen vectors associated with the eigen values $\lambda_1, \lambda_2, \ldots \lambda_n$, respectively. Then*

$$
D = \begin{bmatrix} \lambda_1 & 0 & \ldots & 0 \\ 0 & \lambda_2 & \ldots & 0 \\ \vdots & \vdots & \vdots & \vdots \\ 0 & 0 & \ldots & \lambda_n \end{bmatrix},
$$

and P is an $n \times n$ matrix whose ith column is v_i, i.e.,

$$
P = \begin{bmatrix} v_1 & | & v_2 & | & \ldots & | & v_n \end{bmatrix}.
$$

Therefore, once we can find the eigen values and eigen vectors of a square matrix A, and if A has n many linearly independent eigen vectors, then we can compute a diagonal matrix D and an invertible matrix Q for a diagonalization of A.

Example 151 *Suppose we have a 2×2 matrix:*

$$
A = \begin{bmatrix} 2 & -1 & 3 \\ -1 & -3 & 3 \\ 1 & 1 & 3 \end{bmatrix}.
$$

Then, using the eigen() function from R, *we have*

```
A <- matrix(c(2, -1, 1, -1, -3, 1, 3, 3, 3), 3, 3)
eigen(A)
```

This returns:

```
> eigen(A)
eigen() decomposition
$values
```

```
[1]   4.397382 -3.824458  1.427076
```

```
$vectors
         [,1]       [,2]       [,3]
[1,] 0.7437078  0.2542122  0.8674476
[2,] 0.1624484  0.9508909 -0.3981322
[3,] 0.6484668 -0.1765859 -0.2983713
```

Therefore we have

$$D = \begin{bmatrix} 4.397382 & 0 & 0 \\ 0 & -3.824458 & 0 \\ 0 & 0 & 1.427076 \end{bmatrix},$$

$$P = \begin{bmatrix} 0.7437078 & 0.2542122 & 0.8674476 \\ 0.1624484 & 0.9508909 & -0.3981322 \\ 0.6484668 & -0.1765859 & -0.2983713 \end{bmatrix}.$$

By Theorem 6.3, we have the following theorem:

Theorem 6.7 *Suppose A is an n × n matrix. If A has n many distinct eigen values, then A is diagonalizable.*

Example 152 *Suppose we have a 2 × 2 matrix:*

$$A = \begin{bmatrix} 1 & -2 & 1 \\ 0 & 2 & -3 \\ 3 & -3 & -2 \end{bmatrix}.$$

Then, using the eigen() function from R, *we have*

```
A <- matrix(c(1, 0, 3, -2, 2, -3, 1, -3, -2), 3, 3)
eigen(A)
```

This returns:

```
> eigen(A)
eigen() decomposition
$values
[1]   4.50349224 -3.56576496  0.06227272

$vectors
          [,1]        [,2]        [,3]
[1,]  0.5276592 -0.01504018 -0.7715961
[2,] -0.6521968 -0.47442002 -0.5343413
[3,]  0.5442565 -0.88017012 -0.3451359
```

Since all eigen values of A are distinct, A is diagonalizable. Thus we have

$$D = \begin{bmatrix} 4.50349224 & 0 & 0 \\ 0 & -3.56576496 & 0 \\ 0 & 0 & 0.06227272 \end{bmatrix},$$

$$P = \begin{bmatrix} 0.5276592 & -0.01504018 & -0.7715961 \\ -0.6521968 & -0.47442002 & -0.5343413 \\ 0.5442565 & -0.88017012 & -0.3451359 \end{bmatrix}.$$

If you want to have matrices D, Q from the eigen values and eigen vectors of A in R, *we can do:*

```
p <- eigen(A)$vectors
D <- diag(eigen(A)$values)
p %*% D %*% solve(p)
```

Note that we use the solve() function to compute the inverse of the matrix p. Instead of the solve() function, we can use the inv() function.

6.3.4   Checkmarks

1. The definition of a diagonalizable matrix.

2. The properties when a square matrix becomes diagonalizable.

3. You can operate diagonalization when A is diagonalizable.

4. You can compute A^r when A is diagonalizable.

5. You can use the eigen() function to diagonalize a diagonalizable matrix.

6. With R, you can perform computational examples in this section.

6.3.5   Conceptual Quizzes

Quiz 190 True or False: *Suppose we have a 3×3 matrix with its eigen values $3, 0, -1$. Then A is diagonalizable.*

Quiz 191 True or False: *Suppose we have a 4×4 matrix with its eigen values $3, 0, -1, -1$. Then A is diagonalizable.*

Quiz 192 True or False: *Suppose we have a 3×3 matrix:*

$$A = \begin{bmatrix} 2 & 1 & -1 \\ 0 & 0 & 5 \\ 0 & 0 & -4 \end{bmatrix}.$$

Then A is diagonalizable.

Quiz 193 True or False: *Suppose we have a* 3×3 *matrix:*

$$A = \begin{bmatrix} 1 & 0 & 0 \\ 0 & 0 & 0 \\ 0 & 0 & -4 \end{bmatrix}.$$

Then A is diagonalizable.

Quiz 194 True or False: *Suppose a* 3×3 *matrix A has three eigen vectors such that*

$$v_1 = \begin{bmatrix} 1 \\ 0 \\ 0 \end{bmatrix}, v_2 = \begin{bmatrix} 0 \\ 1 \\ 0 \end{bmatrix}, v_3 = \begin{bmatrix} 0 \\ 0 \\ 1 \end{bmatrix}.$$

Then A is diagonalizable.

Quiz 195 True or False: *Suppose a* 3×3 *matrix A has three eigen vectors such that*

$$v_1 = \begin{bmatrix} 1 \\ 1 \\ 1 \end{bmatrix}, v_2 = \begin{bmatrix} 1 \\ 1 \\ 0 \end{bmatrix}, v_3 = \begin{bmatrix} 0 \\ 0 \\ 1 \end{bmatrix}.$$

Then A is diagonalizable.

Quiz 196 *Suppose a* 3×3 *matrix A has three eigen vectors such that*

$$v_1 = \begin{bmatrix} 1 \\ 0 \\ 0 \end{bmatrix}, v_2 = \begin{bmatrix} 0 \\ 1 \\ 0 \end{bmatrix}, v_3 = \begin{bmatrix} 0 \\ 0 \\ 1 \end{bmatrix},$$

and with its eigen values $2, -1, 1$. *Then the matrix A is*

$$A = \begin{bmatrix} 2 & 0 & 0 \\ 0 & -1 & 1 \\ 0 & 0 & 1 \end{bmatrix}.$$

Quiz 197 *Suppose a* 3×3 *matrix A has three eigen vectors such that*

$$v_1 = \begin{bmatrix} 0 \\ 1 \\ 0 \end{bmatrix}, v_2 = \begin{bmatrix} 1 \\ 0 \\ 0 \end{bmatrix}, v_3 = \begin{bmatrix} 0 \\ 0 \\ 1 \end{bmatrix},$$

and with its eigen values $2, -1, 1$. *Then the matrix A is*

$$A = \begin{bmatrix} 2 & 0 & 0 \\ 0 & -1 & 1 \\ 0 & 0 & 1 \end{bmatrix}.$$

6.3.6   Regular Exercises

Exercise 6.2 *Compute matrices D and P for diagonalization of the following matrices without any help from a computer.*

1.

$$\begin{bmatrix} -1 & -2 \\ -3 & 2 \end{bmatrix}.$$

2.

$$\begin{bmatrix} 1 & 3 \\ 1 & 1 \end{bmatrix}.$$

3.

$$\begin{bmatrix} -1 & 3 \\ 1 & 5 \end{bmatrix}.$$

4.

$$\begin{bmatrix} -2 & -2 \\ 0 & -2 \end{bmatrix}.$$

5.

$$\begin{bmatrix} 3 & 1 \\ 1 & 2 \end{bmatrix}.$$

6.

$$\begin{bmatrix} -1 & 3 & 0 \\ 1 & 5 & 1 \\ 1 & 0 & 1 \end{bmatrix}.$$

7.

$$\begin{bmatrix} -1 & 3 & 0 \\ 0 & 5 & 1 \\ 0 & 0 & 1 \end{bmatrix}.$$

8.

$$\begin{bmatrix} 7 & 0 & 0 & 0 \\ 0 & 5 & 0 & 0 \\ 0 & 0 & 1 & 0 \\ 0 & 0 & 0 & -7 \end{bmatrix}.$$

6.3.7   Lab Exercises

Lab Exercise 176 *Compute matrices D and P for diagonalization of the following matrices with the eigen() function in* R.

1.

$$\begin{bmatrix} 3 & 0 & -2 & -2 \\ -3 & 3 & 1 & 1 \\ -3 & 2 & -2 & 0 \\ -2 & 2 & 1 & 1 \end{bmatrix}.$$

2.

$$\begin{bmatrix} 2 & 2 & 0 & 0 \\ 3 & -1 & -1 & 0 \\ -2 & -2 & 1 & 0 \\ 3 & -1 & -1 & -3 \end{bmatrix}.$$

3.

$$\begin{bmatrix} -3 & 1 & 2 & -1 \\ 0 & 0 & 3 & 0 \\ 0 & 2 & -1 & 2 \\ -2 & 1 & 1 & 3 \end{bmatrix}.$$

4.

$$\begin{bmatrix} -3 & 3 & 3 & 0 & -3 \\ 3 & 0 & 2 & -2 & 1 \\ 3 & 2 & 3 & 2 & -2 \\ 0 & -2 & 2 & -3 & 3 \\ -3 & 1 & -2 & 3 & 3 \end{bmatrix}$$

5.

$$\begin{bmatrix} 0 & 0 & 0 & 1 & 0 \\ 0 & -3 & 2 & 0 & 0 \\ 0 & -3 & -2 & 0 & -3 \\ 0 & 0 & -2 & 0 & 0 \\ 2 & -2 & 1 & 0 & 0 \end{bmatrix}$$

6.

$$\begin{bmatrix} 0 & -2 & -2 & 0 & 1 & -3 \\ 1 & 1 & -2 & 0 & 1 & 3 \\ 1 & 2 & 1 & -1 & 1 & -2 \\ 1 & -2 & 1 & 1 & 2 & -3 \\ 1 & -3 & 3 & -3 & -2 & -3 \\ -3 & 0 & -1 & 1 & -1 & 0 \end{bmatrix}$$

6.3.8   Practical Applications

Suppose we have a transition probability for DNA sequences as shown in the following matrix:

$$P = \begin{bmatrix} & A & C & G & T \\ \hline A & 1/4 & 1/4 & 1/4 & 1/4 \\ C & 1/5 & 1/5 & 2/5 & 1/5 \\ G & 1/3 & 1/3 & 1/6 & 1/6 \\ TT & 1/6 & 1/3 & 1/3 & 1/6 \end{bmatrix}.$$

Also suppose we have the following observed sequences: ACTGTTC.

We can think of this sequence as a word composed by four letters, A, C, G, and T. Then we want to compute the probability of observing this sequence after a long evolutionary time. In this case, to make it very simple, we model this as a discrete time Markov chain instead of a continuous time Markov chain. Here we also assume that each site, i.e., each position in the word, is independent, so all we have to do is compute the probability of A, C, G, or T.

To compute the probability of A, C, G, or T, we need to compute the **limiting distribution**. The limiting distribution can be computed by taking:

$$\pi \cdot \lim_{n \to \infty} P^n,$$

where π is a vector:

$$\begin{bmatrix} \pi_A \\ \pi_C \\ \pi_G \\ \pi_T \end{bmatrix},$$

where π is any distribution of letters, A, C, G, T.

To compute this limit, first we need to diagonalize P. In R we will use the eigen() function to diagonalize the transition matrix P. First define the matrix:

```
A <- matrix(c(1/4, 1/5, 1/3, 1/6, 1/4, 1/5, 1/3,
1/3, 1/4, 2/5, 1/6, 1/3, 1/4, 1/5, 1/6, 1/6), 4, 4)
```

Then we use the eigen() function:

```
p <- eigen(A)$vectors
D <- diag(eigen(A)$values)
p %*% D %*% solve(p)
```

Again, note that we use the solve() function to compute the inverse of the matrix p. Instead of using the solve() function, we can use the inv() function. It returns the following:

```
> p %*% D %*% solve(p)
            [,1]           [,2]           [,3]           [,4]
[1,] 0.2500000+0i 0.2500000+0i 0.2500000+0i 0.2500000+0i
[2,] 0.2000000-0i 0.2000000+0i 0.4000000+0i 0.2000000+0i
[3,] 0.3333333+0i 0.3333333-0i 0.1666667-0i 0.1666667+0i
[4,] 0.1666667+0i 0.3333333+0i 0.3333333+0i 0.1666667+0i
```

Here, "i" means a complex number. Let's define π as a uniform distribution, i.e.:

$$\begin{bmatrix} 1/4 \\ 1/4 \\ 1/4 \\ 1/4 \end{bmatrix},$$

```
ini <- c(1/4, 1/4, 1/4, 1/4)
```

Then we take n as a large number, like 10000:

```
ini %*% p %*% D^(10000) %*% solve(p)
```

Then it will output:

```
> ini %*% p %*% D^(10000) %*% solve(p)
          [,1]        [,2]        [,3]        [,4]
[1,] 0.2435175+0i 0.276212+0i 0.2841037+0i 0.1961669+0i
```

Therefore, the probabilities of observing A, C, G, or T are

$$p = \begin{bmatrix} 0.2435175 \\ 0.276212 \\ 0.2841037 \\ 0.1961669 \end{bmatrix},$$

Therefore, the probability to observe a sequence ACTGTTC is

$$(0.2435175) \cdot (0.276212) \cdot (0.1961669) \cdot (0.2841037) \cdot (0.1961669)$$
$$\cdot (0.1961669) \cdot (0.276212) = 3.98445 \cdot 10^{-5}.$$

6.3.9 Supplements with python Code

In this section we did not use any new functions. Therefore we do not have any supplemental **python** codes in this section.

6.4 Discussion

Now we will apply PCA to the image shown in Figure 6.1. As we discussed, the first principal component catches most information, and the second principal component captures the second most, and so on. So we are taking the first principal component to the kth principal component to see how much we can compress the image as well as keeping the quality of the image.

We will use the prcomp() function to perform PCA for each layer of the matrix:

```
kitty.r.pca <- prcomp(r, center = FALSE)
kitty.g.pca <- prcomp(g, center = FALSE)
kitty.b.pca <- prcomp(b, center = FALSE)
```

The first argument of this function is the name of the data and the second argument is the logic if we want to center around the origin. In this case we do not want to shift the values around the origin, so we set as "FALSE". Recall that the Kitty variable has the dimension:

```
> d
[1] 2340 4160    3
```

This means that Kitty is the $2340 \times 4160 \times 3$ table. Also r, p, g are 2340×4160 matrices.

Recall that PCA finds a basis of the p-dimensional vector space using the eigen vectors of each matrix where $p = 2340$ in this example. Recall that all p many vectors in a basis span the vector space. So if we pick a subset of the basis, this subset spans only a subspace of the vector space, which is smaller than the original vector space. Therefore picking the "best" subset of the vector space keeps the maximum amount of information with the smallest amount of memory. Again recall that the direction of the eigen vector with the biggest eigen value contains the maximum information of the data set, and the direction of the eigen vector with the second largest eigen value contains the second largest information of the data set, and so on. Therefore we pick the eigen vectors with the first k largest eigen values for $k < p$.

Then we combine the outputs for each layer:

```
kitty.rgb.pca <- list(kitty.r.pca, kitty.g.pca, kitty.b.pca)
```

Then, we pick a set of eigen vectors with the first second largest eigen values, a set of eigen vectors with the first 50th largest eigen values, a set of eigen vectors with the first 100th largest eigen values, a set of eigen vectors with the first 1000th largest eigen values, and a set of eigen vectors with the first 2000th largest eigen values.

```
num <- c(2, 50, 100, 1000, 2000)
```

Then now we create the compressed images using the set of eigen vectors with the first kth largest eigen values for $k = 2, 50, 100, 1000, 2000$:

```
for (i in num)
    pca.img <- sapply(kitty.rgb.pca, function(j)
        compressed.img <- j$x[,1:i] %*% t(j$rotation[,1:i])
    , simplify = 'array')
    writeJPEG(pca.img, paste('/tora_compressed_',
    round(i,0), '_components.jpg', sep = ''))
```

This for loop creates jpg files with the name "tora_compressed_i" for $i = 2, 50, 100, 1000, 2000$ in the working directory. These pictures are shown in Figure 6.6. Basically, the eigen vectors, which form a subset of a basis to span the vector space, correspond to these eigen values to span the images instead of using all vectors in a basis to span the vector space. By reducing the number of vectors to span the image, we reduce the size of memory. Therefore, we can see that indeed, reducing the number of vectors to span the image reduces the size of images. Specifically, the top left image shown in Figure 6.6 has about 230KB. The top right one in Figure 6.6 has about 412KB. The middle left image shown in Figure 6.6 has about 472KB. The middle right image shown in Figure 6.6 has about 230KB. The bottom image shown in Figure 6.6 has about 412KB.

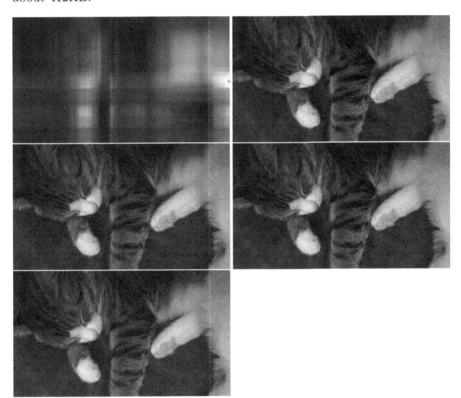

FIGURE 6.6
The top left image is spanned by the eigen vectors with the first two largest eigen values. The top right image is spanned by the eigen vectors with the first 50 largest eigen values. The middle left image is spanned by the eigen vectors with the first 100 largest eigen values. The middle right image is spanned by the eigen vectors with the first 1000 largest eigen values. The bottom left image is spanned by the eigen vectors with the first 2000 largest eigen values.

Note that the second image, with the first principal component to the 50th principal component is clear enough. Also it has only 412KB instead of 1.8MB, which the original jpeg file has. We can apply the elbow method to find the best k for the number of principal components using the plot() function. If a reader is interested in this topic, we recommend this website `https://theanlim.rbind.io/project/image-compression-with-principal-component-analysis/`.

7

Linear Regression

Linear regression is one of the oldest statistical models and a foundation for other models. Linear regressions use tools from linear algebra to fit an input data set with a linear plane in $\mathbb{R}^d$. In many situations, we would like to predict some values based on other information. For example, if we want to buy a house and we send an offer to the seller, we try to obtain the "best guess" for the best price to offer based on other information, like the square footage of the house, the location of the house, etc. Linear regression is the simplest model to predict or guess the price of the house for this situation. In machine learning, linear regression is a supervised learning model and in statistics, linear regression is an inferential statistical model. In this chapter, we will discuss the basics of linear regression and how we apply tools from linear algebra to fit the input data to a linear regression model.

7.1 Introductory Example from Statistics

We are going to use the "Boston" data set from the MASS package in R. The "Boston" data set contains 506 observations of the houses in Boston area and it has 14 variables as shown in Table 7.1. This data set is from Harrison and Rubinfeld [20].

In R, to upload this data set, you need to upload the MASS package:

```
library(MASS)
data(Boston)
```

In this introductory example, we are predicting the median of the house price based on other variables using linear regressions. To make it simple and to avoid confusion, through this chapter we will use this example as the working example for each section.

Variable Name	Information
crim	Crime rate per town
zn	Proportion of residential land zoned for lots over 25,000 sq.ft.
indus	Proportion of non-retail business acres per town.
chas	Charles River dummy variable (= 1 if tract bounds river; 0 otherwise).
nox	Nitrogen oxides concentration (parts per 10 million).
rm	Average number of rooms per dwelling.
age	Proportion of owner-occupied units built prior to 1940.
dis	Weighted mean of distances to five Boston employment centres.
rad	Index of accessibility to radial highways.
tax	Full-value property-tax rate per $10,000.
ptratio	Pupil-teacher ratio by town.
black	$1000(Bk - 0.63)^2$ where Bk is the proportion of blacks by town.
lstat	Lower status of the population (percent).
medv	Median value of owner-occupied homes in $1000s.

TABLE 7.1
Information of the variables in the "Boston" data set from the MASS library.

7.2   Simple Linear Regression

Simple linear regression is a linear regression with one explanatory variable. In this section we will define basic terminology and focus on simple linear regression.

7.2.1   Task Completion Checklist

- During the Lecture:

  1. Read the definition of an explanatory variable.
  2. Read the definitions of basic terminology from statistics.
  3. Read the definition of a response variable.
  4. Read the definition of simple linear regression.
  5. Learn how to fit a data set to a simple linear regression.
  6. With R, learn how to perform computational examples in this section.

- After the Lecture:

  1. Conduct practical applications with R.
  2. If you are interested in python, read the supplement in this section and conduct lab exercises and practical applications with python.

7.2.2 Basic Terminology

With the "Boston" data set, suppose we want to predict the median value of owner-occupied homes in the Boston area based on the average number of rooms. Then the variable "medv" (median value of the house in thousands of dollars) is a dependent variable and the variable "rm" (the average number of rooms in a house) is an independent variable. In data science and statistics, we call a dependent variable a **response variable** and we call an independent variable an **explanatory variable**. In this example, the variable "medv" is a response variable and the variable "rm" is an explanatory variable. For linear regression, our goal is to predict the value of the response variable based on values of explanatory variables using a linear model.

There are basically two different types of scales for a variable: **qualitative** and **quantitative scales**.

A qualitative variable is also called a categorical variable since it is measured by categories. For example, it can be "yes" or "no". Another example is gender, female or male. It can be grades for a linear algebra course. A quantitative variable is measured numerically. For example, it can be weight, height, distance between towns, etc.

Among qualitative variables, there are two types of scales: **nominal** and **ordinal** scales. A nominal scale is a qualitative variable such that there is no natural order among its categories. For example, it can be gender, nationality or yes/no answer. An ordinal scale is also a categorical variable such that there is a natural ordering among its categories.

7.2.3 Simple Linear Regression

For simple linear regression, we fit the data as a linear equation such that

$$Y = \beta_1 \cdot X + \beta_0 + \epsilon, \tag{7.1}$$

where β_1 is a coefficient for the explanatory variable X, β_0 is an intercept, and ϵ is an error term. In this model, the response variable is written as a linear combination of the explanatory variable X and the error term ϵ. Because of the error terms, the value of the response variable for each data point in the input data set might not be exactly on the line defined by the above equation in (7.1). However, we try to find the "best" line to fit the data to the equation with the smallest errors, i.e., find the coefficient β_1 and intercept β_0 with the "smallest" errors.

There are several assumptions for this model: If you want to apply a statistical model to fit the data, you have to make sure you check its assumptions. These are very important: Linear regression assumes:

- For each data point

$$\begin{bmatrix} x \\ y \end{bmatrix},$$

  there is a linear line such that

$$y = \beta_1 \cdot x + \beta_0 + \epsilon.$$

- The response variable is numerical.

- An explanatory variable is numerical.

- The error term is generated under the normal distribution with the mean equal to zero and with a fixed standard deviation σ.

- σ is independent of any value $Y = y$.

So if your response variable or explanatory variable is categorical, such as "yes" or "no", then we should not use linear regressions.

How do we find the "best" line for simple regression? How do we measure the "smallest errors"? In order to demonstrate, we take the first 100 data points from the "Boston" data set from the **MASS** package.

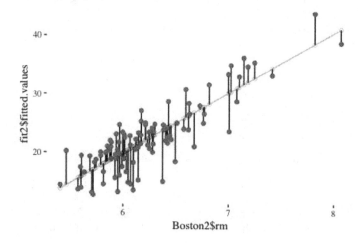

FIGURE 7.1
Simple linear regression line is written in blue and red points are data points from the "Boston" data set.

Let

$$\hat{y} = \hat{\beta}_1 \cdot x + \hat{\beta}_0$$

where $\hat{y}$ is an estimated value for Y, $\hat{\beta}_1$ is an estimate of β_1, $\hat{\beta}_0$ is an estimate of β_0, and x is the value of the explanatory variable X from an observation. Then, the end of the black line on the blue line in Figure 7.1 is

$$\begin{bmatrix} x \\ \hat{y} \end{bmatrix}$$

and a red circle is a vector

$$\begin{bmatrix} x \\ y \end{bmatrix}$$

for each data point. Suppose we have a set of data points

$$\left\{ \begin{bmatrix} x_1 \\ y_1 \end{bmatrix}, \begin{bmatrix} x_2 \\ y_2 \end{bmatrix}, \dots \begin{bmatrix} x_n \\ y_n \end{bmatrix} \right\}.$$

Then we want to find the blue line in Figure 7.1 which minimizes

$$\sum_{i=1}^{n} (\hat{y}_i - y_i)^2. \tag{7.2}$$

Intuitively, this sum is the sum of the squared lengths of black lines for each observation. If this value is small, then it means that the total error is small to fit data by a linear line.

Since we do not know the true β_1 and β_0, we are going to estimate β_1 and β_0 by finding the "best" line. Let $\hat{\beta}_1$ be an estimate of β_1 and let $\hat{\beta}_0$ be an estimate of β_0. Then we obtain $\hat{\beta}_1$ and $\hat{\beta}_0$ by minimizing Equation (7.2). This method is also called the **least squares method**.

In R we can use the lm() function. This "lm" stands for "Linear Models". Let us look at the first 100 observations in the "Boston" data set:

```
Boston2 <- Boston[1:100,]
```

Then we use the lm() function to fit the data to a linear regression line:

```
fit2 <- lm(medv ~ rm, data=Boston2)
```

The response variable should be in the left side of $\sim$ mark and the explanatory variable should go to the right side in the lm() function. Now the variable "fit2" contains all results from the lm() function. Use the summary() function to see what we have:

```
summary(fit2)
```

Then it returns:

```
> summary(fit2)

Call:
lm(formula = medv ~ rm, data = Boston2)

Residuals:
    Min      1Q  Median      3Q     Max
-8.7668 -1.9506  0.3969  2.2360  5.8490

Coefficients:
            Estimate Std. Error t value Pr(>|t|)
(Intercept) -41.4283     4.0616  -10.20   <2e-16 ***
rm           10.2235     0.6495   15.74   <2e-16 ***
---
Signif. codes:  0 '***' 0.001 '**' 0.01 '*' 0.05 '.' 0.1 ' ' 1

Residual standard error: 3.172 on 98 degrees of freedom
Multiple R-squared:  0.7166,Adjusted R-squared:  0.7137
F-statistic: 247.8 on 1 and 98 DF,  p-value: < 2.2e-16
```

The most important part which we should focus is the estimated β_1 and β_0:

```
Coefficients:
            Estimate Std. Error t value Pr(>|t|)
(Intercept) -41.4283     4.0616  -10.20   <2e-16 ***
rm           10.2235     0.6495   15.74   <2e-16 ***
---
Signif. codes:  0 '***' 0.001 '**' 0.01 '*' 0.05 '.' 0.1 ' ' 1
```

Here we have from the lm() function that

$$\beta_1 = 10.2235, \ \beta_0 = -41.4283.$$

This means that the median of the house price can be estimated by

House price $= 10.2235 \cdot$ Number of rooms $- 41.4283.$

So, for example, if a house has six bedrooms the estimated house price will be

House price $= 10.2235 \cdot (6) - 41.4283 = 19.9127.$

Thus, the estimated house price is $19,9127.

To plot in the picture shown in Figure 7.1, we can use the ggplot2 package:

```
require(ggplot2)
require(ggthemes)
fit2 <- lm(medv ~ rm, data=Boston2)
```

```
ggplot2 <- ggplot() +
  geom_point(aes(x = Boston2$rm, y = fit2$fitted.values), shape = 1,
  alpha = 0.2) + geom_line(data = fortify(fit2),
  aes(x = rm, y = .fitted), color = "green") +
  geom_segment(aes(x = Boston2$rm, xend = Boston2$rm,
  y = fit2$fitted.values, yend = Boston2$medv)) +
  geom_point(data = Boston2, aes(x = rm, y = medv), color = "red") +
  theme_tufte()

ggplot2
```

This will output the picture shown in Figure 7.1.

7.2.4 Multiple Linear Regression

From the previous subsection, we learned that the house price in the Boston area based on the number of rooms is

$$\text{House price} = 10.2235 \cdot \text{Number of rooms} - 41.4283.$$

However, if a house has two bedrooms the estimated house price will be

$$\text{House price} = 10.2235 \cdot (2) - 41.4283 = -20.9813.$$

Thus, the estimated house price is $-\$20,9813$ which does not make sense. It seems that the estimation by fitting a simple linear regression is not so great over all. Thus, we try to apply **multiple linear regression** on this data set.

Multiple linear regression is a linear model with k many explanatory variables such that we fit the data to a linear equation:

$$Y = \beta_k \cdot X_k + \ldots + \beta_1 \cdot X_1 + \beta_0 + \epsilon, \qquad (7.3)$$

where Y is the response variable, $X_1, \ldots, X_k$ are explanatory variables, and ϵ is the error term.

For example, in the "Boston" data set, we estimate the house price based on the number of rooms, crime rate, proportion of residential land zone, proportion of non-retail business acres, age of the house, distance to the main area in Boston, accessibility to radial highways, and tax for the house. In this case, we have $k = 8$ since there are eight explanatory variables.

For multiple linear regression we have several assumptions:

- For each data point
$$\begin{bmatrix} x_1 \\ x_2 \\ \vdots \\ x_k y \end{bmatrix},$$
there is a linear line such that

$$y = \beta_k \cdot x_k + \ldots + \beta_1 \cdot x_1 + \beta_0 + \epsilon.$$

- The response variable is numerical.

- All explanatory variables are numerical.

- The error term is generated under the normal distribution with the mean equal to zero and with a fixed standard deviation σ.

- σ is independent of any value $Y = y$.

Again here we assume that all variables are qualitative variables, i.e., numerical variables.

For multiple linear regression we try to find the "best" linear plane which fits the data "well". How can we find such a linear plane? As we did for simple linear regression, we minimize the sum of squares of the lengths between each observation and its estimate.

Let $\hat{y} = \hat{\beta}_k \cdot x_k + \ldots + \hat{\beta}_1 \cdot x_1 + \hat{\beta}_0$ be an estimate from the multiple linear regression, where $\hat{\beta}_i$ is an estimate of β_i for $i = 1, \ldots, k$. Suppose we have a set of data points

$$\left\{ \begin{bmatrix} x_{11} \\ x_{12} \\ \vdots \\ x_{1k} \\ y_1 \end{bmatrix}, \begin{bmatrix} x_{21} \\ x_{22} \\ \vdots \\ x_{2k} \\ y_2 \end{bmatrix}, \ldots \begin{bmatrix} x_{n1} \\ x_{n2} \\ \vdots \\ x_{nk} \\ y_n \end{bmatrix} \right\}.$$

Then we want to find the linear plane which minimizes

$$\sum_{i=1}^{n} (\hat{y}_i - y_i)^2. \tag{7.4}$$

Intuitively, this sum is the sum of the squared lengths of black lines for each observation. If this value is small, then it means that the total error is small to fit data by a linear line.

To demonstrate what is going on visually, we take the first 50 data points in the "Boston" data set. Then we fit the data with two explanatory variables, "rm" and "age". Intuitively, Equation (7.4) is the sum of the squared lengths of red lines and blue lines. Multiple linear regression is designed to find the linear plane which minimizes the sum.

To run multiple linear regression in R, we again use the lm() function. Here we want to predict the house price "medv" as the response variable, based on the explanatory variables "crim", "zn", "indus", "rm", "age", 'dis", "tax", and "rad".

```
fit3 <- lm(medv ~ crim + zn + indus + rm + age + dis + tax + rad,
    data=Boston2)
```

The response variable should be in the left side of $\sim$ mark and the explanatory variables should go to the right side in the lm() function. Now the

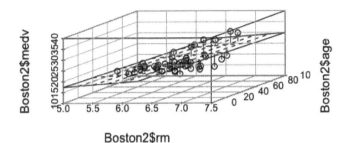

FIGURE 7.2
Multiple linear regression plane computed from the "Boston" data set with the "rm" and "age" as explanatory variables and the "medv" as the response variable. The line between $\hat{y}$ and y for each point is written in red if the data point is above the linear regression plane and it is written in blue if the data point is below the plane.

variable "fit3" contains all results from the lm() function. Use the summary() function to see what we have:

```
summary(fit3)
```

Then it returns:

```
> summary(fit3)

Call:
lm(formula = medv ~ crim + zn + indus + rm + age + dis + tax +
    rad, data = Boston2)

Residuals:
    Min      1Q  Median      3Q     Max
-5.5135 -1.3478 -0.2007  1.1933  8.1711

Coefficients:
             Estimate Std. Error t value Pr(>|t|)
(Intercept) -2.054e+01  4.515e+00  -4.549 1.66e-05 ***
crim        -2.130e+00  7.378e-01  -2.887  0.00486 **
zn          -4.792e-04  1.321e-02  -0.036  0.97113
indus       -1.893e-01  9.849e-02  -1.922  0.05770 .
```

```
rm              8.322e+00  5.424e-01  15.342  < 2e-16 ***
age            -6.504e-02  9.905e-03  -6.566 3.12e-09 ***
dis            -1.700e-01  2.378e-01  -0.715  0.47663
tax            -8.882e-03  5.834e-03  -1.522  0.13138
rad            -1.404e-02  2.022e-01  -0.069  0.94479
---
Signif. codes:  0 '***' 0.001 '**' 0.01 '*' 0.05 '.' 0.1 ' ' 1

Residual standard error: 2.204 on 91 degrees of freedom
Multiple R-squared:  0.873,Adjusted R-squared:  0.8618
F-statistic: 78.16 on 8 and 91 DF,  p-value: < 2.2e-16
```

The most important part on which we should focus is the * mark. Here, only "Intersept", "crim", "rm" and "age" have * marks below:

```
Coefficients:
              Estimate Std. Error t value Pr(>|t|)
(Intercept) -2.054e+01  4.515e+00  -4.549 1.66e-05 ***
crim        -2.130e+00  7.378e-01  -2.887  0.00486 **
zn          -4.792e-04  1.321e-02  -0.036  0.97113
indus       -1.893e-01  9.849e-02  -1.922  0.05770 .
rm           8.322e+00  5.424e-01  15.342  < 2e-16 ***
age         -6.504e-02  9.905e-03  -6.566 3.12e-09 ***
dis         -1.700e-01  2.378e-01  -0.715  0.47663
tax         -8.882e-03  5.834e-03  -1.522  0.13138
rad         -1.404e-02  2.022e-01  -0.069  0.94479
```

This means that only "Intersept", "crim", "rm" and "age" contribute to the price of the house and other explanatory variables are not necessary to fit the data. So we can remove all other variables for estimating the price of the house and we only use "Intersept", "crim", "rm" and "age" to predict the house price "medv". From the lm() function we have a multiple linear regression plane:

$$\text{medv} = -2.130 \cdot \text{crim} - 0.0004792 \cdot \text{zm} - 0.1893 \cdot \text{indus} + 8.322 \cdot \text{rm}$$
$$-0.06504 \cdot \text{age} - 0.17 \cdot \text{dis} - 0.008882 \cdot \text{tax} - 0.01404 \cdot \text{rad} - 20.54.$$

To draw the picture shown in Figure 7.2, we can use the **rgl** package: First we upload all necessary packages:

```
library(rgl)
library(ggplot2)
library(dplyr)
library(gridExtra)
library(scatterplot3d)
library(FactoClass)
library(MASS)
```

Then we take the first 50 observations in the "Boston" data set:

```
Boston2 <- Boston[1:50,]
```

Then we fit the data with multiple linear regression with the "rm" and "age" variables via the lm() function:

```
fit3 <- lm(medv ~ rm +age, data=Boston2)
```

Then we type the following so that R outputs the picture shown in Figure 7.2.

```
sp <- scatterplot3d::scatterplot3d(Boston2$rm,
                                    Boston2$age,
                                    Boston2$medv,
                                    angle = 45)
sp$plane3d(fit3, lty.box = "solid")#,
        # polygon_args = list(col = rgb(.1, .2, .7, .5)) # Fill color
orig <- sp$xyz.convert(Boston2$rm,
                       Boston2$age,
                       Boston2$medv)
plane <- sp$xyz.convert(Boston2$rm,
                        Boston2$age, fitted(fit3))
i.negpos <- 1 + (resid(fit3) > 0)
segments(orig$x, orig$y, plane$x, plane$y,
        col = c("blue", "red")[i.negpos],
        lty = 1) # (2:1)[i.negpos]
sp <- FactoClass::addgrids3d(Boston2$rm,
                             Boston2$age,
                             Boston2$medv,
                             angle = 45,
                             grid = c("xy", "xz", "yz"))
```

7.2.5   Checkmarks

1. The definition of an explanatory variable.

2. The definitions of basic terminology from statistics.

3. The definition of a response variable.

4. The definition of simple linear regression.

5. You can fit a data set to simple linear regression.

6. With R, you can perform computational examples in this section.

7.2.6   Practical Applications

We are going to use the "Advertising" data set from the website `http:` `//faculty.marshall.usc.edu/gareth-james/ISL/Advertising.csv`. This data set was used in [24]. The "TV" variable stores the amount of money spent for a TV advertisement. The "radio" variable stores the amount of money spent for a radio advertisement. The "newspaper" variable stores the amount of money spent for a newspaper advertisement. The last variable is the "sales", which stores the amount of money they solve their product. Now the question is, which advertisements are useful and which ones are not useful for their sales.

To upload the data set from the website, using the read.csv() function, you can do:

```
Ads <- read.csv("http://faculty.marshall.usc.edu/gareth-james/ISL/Advertising.csv")
```

Then we check what the data set looks like by using the summary() function:

```
> summary(Ads)
       X               TV              radio           newspaper          sales
 Min.   :  1.00   Min.   :  0.70   Min.   : 0.000   Min.   :  0.30   Min.   : 1.60
 1st Qu.: 50.75   1st Qu.: 74.38   1st Qu.: 9.975   1st Qu.: 12.75   1st Qu.:10.38
 Median :100.50   Median :149.75   Median :22.900   Median : 25.75   Median :12.90
 Mean   :100.50   Mean   :147.04   Mean   :23.264   Mean   : 30.55   Mean   :14.02
 3rd Qu.:150.25   3rd Qu.:218.82   3rd Qu.:36.525   3rd Qu.: 45.10   3rd Qu.:17.40
 Max.   :200.00   Max.   :296.40   Max.   :49.600   Max.   :114.00   Max.   :27.00
```

Here, X is just an index for each observation so we can just ignore this.

Now, we want to predict the value of the "sales" based on the budget for the TV ads "TV", the budget for the radio ads "radio", and the budget for the newspaper ads "newspaper" using the lm() function:

```
fit <- lm(sales ~ TV + radio + newspaper, data = Ads)
```

To read the output from the lm() function, we use the summary() function:

```
summary(fit)
```

Then R outputs:

```
> summary(fit)

Call:
lm(formula = sales ~ TV + radio + newspaper, data = Ads)

Residuals:
    Min      1Q  Median      3Q     Max
-8.8277 -0.8908  0.2418  1.1893  2.8292

Coefficients:
            Estimate Std. Error t value Pr(>|t|)
```

```
(Intercept)   2.938889   0.311908    9.422   <2e-16 ***
TV            0.045765   0.001395   32.809   <2e-16 ***
radio         0.188530   0.008611   21.893   <2e-16 ***
newspaper    -0.001037   0.005871   -0.177    0.86
---
Signif. codes:  0 '***' 0.001 '**' 0.01 '*' 0.05 '.' 0.1 ' ' 1

Residual standard error: 1.686 on 196 degrees of freedom
Multiple R-squared:  0.8972,Adjusted R-squared:  0.8956
F-statistic: 570.3 on 3 and 196 DF,  p-value: < 2.2e-16
```

We notice that the variable "newspaper" does not have * marks. Therefore the newspaper ads are not really effective for sales. So if you are a manager at a store, you will cut the budget for newspaper advertisement.

7.2.7   Supplements with python Code

In python we need the numpy package, pandas package [27], and sklearn package [33] for conducting linear regression analysis. For example, if we want to apply multiple linear regression on the "Advertising" data set from ISLR [24], then first we upload all packages:

```
import pandas as pd
import numpy as np
from sklearn.model_selection import train_test_split
from sklearn.linear_model import LinearRegression
from sklearn import metrics
```

Now we upload the data set from the ISLR website:

```
Ads = pd.read_csv('http://faculty.marshall.usc.edu/gareth-james/ISL/Advertising.csv')
```

The describe() function will show the summary of this data set:

```
Ads.describe()
```

Then python will output: `

```
>>> Ads.describe()
        Unnamed: 0          TV      radio   newspaper       sales
count   200.000000  200.000000  200.000000  200.000000  200.000000
mean    100.500000  147.042500   23.264000   30.554000   14.022500
std      57.879185   85.854236   14.846809   21.778621    5.217457
min       1.000000    0.700000    0.000000    0.300000    1.600000
25%      50.750000   74.375000    9.975000   12.750000   10.375000
50%     100.500000  149.750000   22.900000   25.750000   12.900000
75%     150.250000  218.825000   36.525000   45.100000   17.400000
max     200.000000  296.400000   49.600000  114.000000   27.000000
```

We first separate the explanatory variables and the response variable:

```
X = Ads[['TV', 'radio', 'newspaper']].values
Y = Ads[['sales']]
```

Then we call the LinearRegression() function:

```
regfit2 = LinearRegression()
regfit2.fit(X, Y)
```

To see the coefficients $\beta_0, \beta_1, \beta_2, \beta_3$, we do:

```
regfit2.coef_
regfit2.intercept_
```

Then **python** will output:

```
>>> regfit2.coef_
array([[ 0.04576465,  0.18853002, -0.00103749]])
>>> regfit2.intercept_
array([2.93888937])
```

This means that the estimated intercept is 2.93888937 and the estimated coefficient for "TV" is 0.04576465, the estimated coefficient for "radio" is 0.18853002, and the estimated coefficient for "newspaper" is −0.00103749.

8

Linear Programming

Linear programming is an optimization model to find an optimal solution with a linear objective function over a set of all solutions defined by a system of linear equations and linear inequalities. It is the simplest model in optimization, however, it is a useful model in real life and it is an efficient model for finding an optimal solution. In this chapter, we will discuss the basics of linear programming and we will revisit the Guinness data set discussed in Section 1.3.

8.1 Introductory Example from Optimization

Recall that a Guinness data set from [5] was collected from Guinness Ghana Ltd. and contains information on the supply of Malta Guinness from two production sites, Kaasi and Achimota, to nine key distributors geographically scattered in the regions of Ghana.

They collected data twice and the data set shown in Table 8.1 and Table 8.2 were collected July 7th 2010 to June 8th 2011. Table 8.1 shows the demands from the distributors and supplies from the production sites. Table 8.2 shows the cost of transportation from a production site to a distributor, written in thousands. The question is, how can we ship Guinness beers efficiently to distributors? Minimizing cost is an important problem for the company.

To solve this question, we formulate this problem in terms of linear programming and we will solve the problem using the `lpSolve` package [6]. For readers who are interested in solving larger linear programming problems with thousands of variables, we can use the `cplexAPI` package [35] and `Rcplex` package [8] with the commercial software `cplex` [11] from IBM to solve them.

8.2 Linear Programming

Linear programming is a simple model to find an optimal solution of a linear function over a set of all solutions defined by a system of linear equations and

DOI: 10.1201/9781003042259-8

TABLE 8.1

Demands and Supplies in thousands
for Malta Guinness

Demand		Supply	
FTA	465	Achimota	1298
RICKY	605	Kaasi	1948
OBIBAJK	451		
KADOM	338		
NAATO	260		
LESK	183		
DCEE	282		
JOEMAN	127		
KBOA	535		

TABLE 8.2

Cost to transport from
production sites to distributors
for Malta Guinness in thousands

	Achimota	Kaasi
FTA	39.99	145.36
RICKY	126.27	33.82
OBIBAJK	102.70	154.05
KADOM	81.68	64.19
NAATO	38.81	87.90
LESK	71.99	107.98
DCEE	31.21	65.45
JOEMAN	22.28	39.08
KBOA	321.04	167.38

linear inequalities. In this section we will define the basics of linear programming and discuss how to solve them using the `lpSolve` package in R.

8.2.1 Task Completion Checklist

- During the Lecture:

    1. Read the definition of a polyhedron.

    2. Read the definitions of a polytope.

    3. Visualize a polyhedron and a polytope.

    4. Learn how to solve a linear programming problem with the simplex method.

    5. With R, learn how to perform computational examples in this section.

- After the Lecture:

1. Conduct practical applications with R.

2. If you are interested in python, read the supplement in this section and conduct lab exercises and practical applications with python.

8.2.2 Basic Terminology

In this section we will define the basics for solving a linear programming problem. First we define a **polyhedron**, which is a set of all solutions to a system of linear equations and inequalities.

Definition 50 *Suppose we have a system of linear equations and inequalities such that*

$$A \cdot x = b_1, B \cdot x \leq b_2,$$

where A is an $m_1 \times n$ matrix and B is an $m_2 \times n$ matrix, b is an m_1-dimensional vector and d is an m_2-dimensional vector. Then a set of all solutions x in $\mathbb{R}^n$ satisfying

$$A \cdot x = b_1, B \cdot x \leq b_2$$

*is called a **polyhedron**.*

Example 153 *In order to draw a polyhedron in $\mathbb{R}^2$, we use the gMOIP package in R [30]. For this example, we have the following system of linear inequalities:*

$$
\begin{array}{rcrcl}
x_1 & + & x_2 & \leq & 7 \\
x_1 & + & 7 \cdot x_2 & \leq & 35 \\
-x_1 & + & x_2 & \leq & 3 \\
x_1 & & & \geq & 0 \\
& & x_2 & \geq & 0.
\end{array}
$$

The solution set for this system of linear inequalities forms a polyhedron. For example, a point

$$
\begin{bmatrix} 0 \\ 0 \end{bmatrix}
$$

is inside of the polyhedron since

$$
\begin{array}{rcrcl}
0 & + & 0 & < & 7 \\
0 & + & 7 \cdot 0 & < & 35 \\
-0 & + & 0 & < & 3 \\
0 & & & = & 0 \\
& & 0 & = & 0.
\end{array}
$$

In order to visualize the polyhedron, first we upload the gMOIP package in R:

```
library(gMOIP)
```

Then we have to have the form

$$B \cdot x \le b_2.$$

In this example, we have

$$B = \begin{bmatrix} 1 & 1 \\ 1 & 7 \\ -1 & 1 \\ -1 & 0 \\ 0 & -1 \end{bmatrix},$$

and

$$b_2 = \begin{bmatrix} 7 \\ 35 \\ 3 \\ 0 \\ 0 \end{bmatrix}.$$

Now we define the matrix B and b_2 in R:

```
A <- matrix(c(1, 1, 1, 7, -1, 1, -1, 0, 0, -1), nrow = 5,
ncol = 2, byrow = TRUE)
b <- c(7, 35, 3, 0, 0)
```

Then we call the plotPolytope() function from the gMOIP *package in* R:

```
plotPolytope(
    A,
    b,
    obj,
    type = rep("c", ncol(A)),
    crit = "max",
    faces = rep("c", ncol(A)),
    plotFaces = TRUE,
    plotFeasible = TRUE,
    plotOptimum = FALSE,
    labels = "coord"
)
```

This produce a plot shown in Figure 8.1.

Definition 51 *A* **vertex** *of a polyhedron is a corner point of the polyhedron which is an intersection of edges of the polyhedron.*

Example 154 *We consider the polyhedron in Example 153. Then there are five vertices of the polyhedron*

$$\begin{bmatrix} 0 \\ 0 \end{bmatrix}, \begin{bmatrix} 7 \\ 0 \end{bmatrix}, \begin{bmatrix} 0 \\ 3 \end{bmatrix}, \begin{bmatrix} 1.75 \\ 4.75 \end{bmatrix}, \begin{bmatrix} 2.33 \\ 4.67 \end{bmatrix},$$

shown as black points in Figure 8.1.

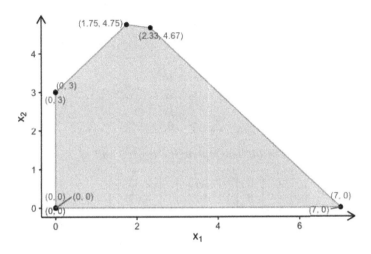

FIGURE 8.1
The polyhedron defined by a system of linear inequalities in Example 153.

Definition 52 *A **convex set** is a subset S in $\mathbb{R}^d$ such that for any points x, y in S*

$$\alpha \cdot x + (1 - \alpha) \cdot y$$

*is also in S for all $0 \leq \alpha \leq 1$. A **convex hull** of a set $\{x_1, x_2, \ldots, x_k\}$ is the smallest convex hull in $\mathbb{R}^d$ containing $\{x_1, x_2, \ldots, x_k\}$.*

Example 155 *A polyhedron in Example 153 is a convex hull of five vertices of the polyhedron.*

Definition 53 *A **polytope** is a convex hull of a set of finitely many vertices in $\mathbb{R}^d$.*

Example 156 *A polyhedron in Example 153 is a polytope of five vertices of the polyhedron.*

Theorem 8.1 *A bounded polyhedron in $\mathbb{R}^d$ is a polytope.*

By Theorem 8.1, we have two representations of a polytope. We can define a polytope by a set of a system of linear equations and inequalities. We can also define a polytope by a convex hull of a set of finitely many vertices. If we define a polytope as a system of linear equations and inequalities, then we call this a **hyperplane representation** of the polytope. If we define a polytope by a convex hull of a set of finitely many vertices, then we call this a **vertex representation** of the polytope.

364   *Linear Algebra and Its Applications with R*

Example 157 *The polytope shown in Example 153, the hyperplane represen-tation of the polytope, is the solution set of the system defined by*

$$
\begin{array}{rcrcl}
x_1 & + & x_2 & \leq & 7 \\
x_1 & + & 7 \cdot x_2 & \leq & 35 \\
-x_1 & + & x_2 & \leq & 3 \\
x_1 & & & \geq & 0 \\
& & x_2 & \geq & 0,
\end{array}
$$

and the vertex representation of the polytope is the convex hull of

$$
\left\{ \begin{bmatrix} 0 \\ 0 \end{bmatrix}, \begin{bmatrix} 7 \\ 0 \end{bmatrix}, \begin{bmatrix} 0 \\ 3 \end{bmatrix}, \begin{bmatrix} 1.75 \\ 4.75 \end{bmatrix}, \begin{bmatrix} 2.33 \\ 4.67 \end{bmatrix} \right\}.
$$

8.2.3 Linear Programming

Linear programming is a model in optimization to find an optimal solution of a linear function over a polyhedron. For example, consider the polytope in Example 153. Then suppose we want to find an optimal solution such that

$$
\begin{array}{rcrcl}
\max 2 \cdot x_1 + 7 \cdot x_2 & \text{such that} \\
x_1 & + & x_2 & \leq & 7 \\
x_1 & + & 7 \cdot x_2 & \leq & 35 \\
-x_1 & + & x_2 & \leq & 3 \\
x_1 & & & \geq & 0 \\
& & x_2 & \geq & 0.
\end{array}
$$

This is a linear programming problem. The function we want to optimize is called a **cost** function. With this example, the linear function $\max 2 \cdot x_1 + 7 \cdot x_2$ is a cost function.

With the gMOIP package in R we can visualize the linear programming problem in $\mathbb{R}^2$ and also find the optimal solution of the problem. For example, consider the polytope in Example 153 and a cost function $\max 2 \cdot x_1 + 7 \cdot x_2$.

First, we define the cost function in R:

```
obj <- c(2, 7)
```

Then we call the plotPolytope() function:

```
plotPolytope(
    A,
    b,
    obj,
    type = rep("c", ncol(A)),
    crit = "max",
    faces = rep("c", ncol(A)),
    plotFaces = TRUE,
```

```
plotFeasible = TRUE,
plotOptimum = TRUE,
labels = "coord"
)
```

Then it will output the optimal solution

$$\begin{bmatrix} 2.33 \\ 4.67 \end{bmatrix}$$

and the optimal value is 37.33. Also it produces the plot shown in Figure 8.2.

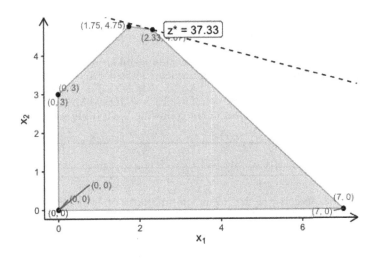

FIGURE 8.2
The polytope and cost function for the linear programming problem. The dotted line is a linear cost function for the problem and z^* is the optimal value.

There are several efficient methods to solve a linear programming problem. One of them is called the **simplex method**. This method uses the following theorem:

Theorem 8.2 *An optimal solution for a linear programming problem is always a vertex of the polyhedron.*

Due to Theorem 8.2, we know that an optimal solution of a linear programming problem is a vertex of a polyhedron defined by a system of linear equations and inequalities. Thus the simplex method goes through a vertex to another vertex of the polyhedron in a way that a value of the cost function improves. One can find details on the simplex method at http://www.phpsimplex.com/en/simplex_method_example.htm.

8.2.4 Checkmarks

1. The definition of a polyhedron.

2. The definitions of a polytope.

3. Visualize a polyhedron and a polytope.

4. You can solve a linear programming problem with the simplex method.

5. With R, you can perform computational examples in this section.

8.2.5 Practical Applications

In Practical Applications, we apply the `lpSolve` package in R to find the optimal value and an optimal solution to the linear programming problem with the Guinness data set.

First, as we discussed in Section 1.3, we set this problem as a system of linear equations and linear inequalities. For producing sites, let 1 = Achimota and 2 = Kaasi. FOr distributors, let 1 = FTA, 2 = RICKY, 3 = OBIBAJK, 4 = KADOM, 5 = NAATO, 6 = LESK, 7 = DCEE, 8 = JOEMAN, and 9 = KBOA. Then, we assign variables shown in the following 2×9 table such that

	FTA	RICKY	OBIBAJK	KADOM	NAATO	LESK	DCEE	JOEMAN	KBOA	Total
Achi-mota	x_{11}	x_{12}	x_{13}	x_{14}	x_{15}	x_{16}	x_{17}	x_{18}	x_{19}	1298
Kaasi	x_{21}	x_{22}	x_{23}	x_{24}	x_{25}	x_{26}	x_{27}	x_{28}	x_{29}	1948
Total	465	605	451	338	260	183	282	127	535	3246

where x_{ij} for $i = 1, 2$ and $j = 1, 2, \ldots 9$ is the amounts of beers from the producing site i to the distributor j. This can be written as a system of linear equations and inequalities such that:

$$
\begin{aligned}
x_{11} + x_{12} + \ldots + x_{19} &= 1298 \\
x_{21} + x_{22} + \ldots + x_{29} &= 1948 \\
x_{11} + x_{21} &= 465 \\
x_{12} + x_{22} &= 605 \\
x_{13} + x_{23} &= 451 \\
x_{14} + x_{24} &= 338 \\
x_{15} + x_{25} &= 260 \\
x_{16} + x_{26} &= 183 \\
x_{17} + x_{27} &= 282 \\
x_{18} + x_{28} &= 127 \\
x_{19} + x_{29} &= 535.
\end{aligned}
$$

for all variables $x_{ij} \geq 0$. In addition, the coefficients for the cost function can be found in Table 8.2. The cost function is

$$
\begin{aligned}
& 39.99 \cdot x_{11} + 126.27 \cdot x_{12} + 102.70 \cdot x_{13} + 81.68 \cdot x_{14} + 38.81 \cdot x_{15} \\
& + 71.99 \cdot x_{16} + 31.21 \cdot x_{17} + 22.28 \cdot x_{18} + 321.04 \cdot x_{19} + 145.36 \cdot x_{21} \\
& + 33.82 \cdot x_{22} + 154.05 \cdot x_{23} + 64.19 \cdot x_{24} + 87.90 \cdot x_{25} + 107.98 \cdot x_{26} \\
& + 65.45 \cdot x_{27} + 39.08 \cdot x_{28} + 167.38 \cdot x_{29}.
\end{aligned}
$$

For this problem we want to minimize the cost function over the polytope defined by the system of linear equations and inequalities.

In R, first we define the matrix and vectors to define the polytope using the diag() function and the rbind() function:

```
A <- matrix(c(1 , 1 , 1 , 1 , 1 , 1 , 1 , 1 , 1 , 0 ,
0 , 0 , 0 , 0 , 0 , 0 , 0 , 0 , 0 , 0 , 0 , 0 , 0 , 0 ,
0 , 0 , 0 , 1 , 1 , 1 , 1 , 1 , 1 , 1 , 1 , 1 , 1 , 0 ,
0 , 0 , 0 , 0 , 0 , 0 , 0 , 1 , 0 , 0 , 0 , 0 , 0 , 0 ,
0 , 0 , 0 , 1 , 0 , 0 , 0 , 0 , 0 , 0 , 0 , 0 , 1 , 0 ,
0 , 0 , 0 , 0 , 0 , 0 , 0 , 0 , 1 , 0 , 0 , 0 , 0 , 0 ,
0 , 0 , 0 , 1 , 0 , 0 , 0 , 0 , 0 , 0 , 0 , 0 , 0 , 1 ,
0 , 0 , 0 , 0 , 0 , 0 , 0 , 0 , 1 , 0 , 0 , 0 , 0 , 0 ,
0 , 0 , 0 , 0 , 1 , 0 , 0 , 0 , 0 , 0 , 0 , 0 , 0 , 1 ,
0 , 0 , 0 , 0 , 0 , 0 , 0 , 0 , 0 , 1 , 0 , 0 , 0 , 0 ,
0 , 0 , 0 , 0 , 1 , 0 , 0 , 0 , 0 , 0 , 0 , 0 , 0 , 0 ,
1 , 0 , 0 , 0 , 0 , 0 , 0 , 0 , 0 , 1 , 0 , 0 , 0 , 0 ,
0 , 0 , 0 , 0 , 0 , 1 , 0 , 0 , 0 , 0 , 0 , 0 , 0 , 0 ,
1 , 0 , 0 , 0 , 0 , 0 , 0 , 0 , 0 , 0 , 1 , 0 , 0 , 0 ,
0 , 0 , 0 , 0 , 0 , 1), nrow = 11, ncol = 18, byrow = TRUE)
f.con <- rbind(A, diag(18))
```

Then we define the vector for the right hand side:

```
f.rhs <- c(1298, 1948,465 ,605 ,451 ,338 ,260 ,183 ,282 ,127 ,535,
           0, 0, 0, 0, 0, 0, 0, 0, 0, 0, 0, 0, 0, 0, 0, 0, 0, 0)
```

Now we define the cost function and the equations or inequalities:

```
f.obj <- c(39.99, 126.27, 102.70, 81.68, 38.81, 71.99,
31.21, 22.28, 321.04, 145.36, 33.82, 154.05, 64.19,
87.90, 107.98, 65.45, 39.08, 167.38)
f.dir <- c("=", "=", "=", "=", "=", "=", "=", "=", "=", "=", "=",
           ">=", ">=", ">=", ">=", ">=", ">=", ">=", ">=", ">=", ">=",
           ">=", ">=", ">=", ">=", ">=", ">=", ">=", ">=")
```

Then we use the lp() function from the **lpSolve** package to find an optimal solution and the optimal value:

```
Sol <- lp ("min", f.obj, f.con, f.dir, f.rhs)
```

The first argument of the lp() function should be "min" for our problem since we are minimizing the cost function over the polytope.

To see the optimal value we type "Sol$objval". Then R returns:

```
> Sol$objval
[1] 245498.9
```

If we want to find an optimal solution, then we type "Sol$solution" in R:

```
> Sol$solution
[1] 465 0 451 0 260 122  0  0  0  0 605 0 338 0 61 282 127 535
```

8.2.6   Supplements with python Code

In python we can use the scipy.optimize package to solve a linear programming problem. As for the example, we will find an optimal solution for the following linear programming problem:

$$\max 2 \cdot x_1 + 7 \cdot x_2 \text{ such that}$$

$$
\begin{array}{rcrcl}
x_1 & + & x_2 & \leq & 7 \\
x_1 & + & 7 \cdot x_2 & \leq & 35 \\
-x_1 & + & x_2 & \leq & 3 \\
x_1 & & & \geq & 0 \\
& & x_2 & \geq & 0.
\end{array}
$$

First we upload the library and the linprog() function:

```
from scipy.optimize import linprog
```

Then we define the bounds for x_1 and x_2, which is $x_1 \geq 0$ and $x_2 \geq 0$:

```
x1_bounds = (0, None)
x2_bounds = (0, None)
```

Then we define the inequalities and the cost function:

```
A = [[1, 1], [1, 7], [-1, 1]]
b = [7, 35, 3]
c = [-2, -7]
```

Since the linprog() function finds the optimal solution by "maximizing" the cost function, and since we want to minimize the cost function, we have to multiply the cost function by -1.

Then we call the linprog() function to find an optimal solution and the optimal value:

```
res = linprog(c, A_ub=A, b_ub=b, bounds=[x1_bounds, x2_bounds])
```

To see the result we use the print() function:

```
print(res)
```

Then python prints out the results:

```
>>> print(res)
      con: array([], dtype=float64)
      fun: -37.33333333324609
  message: 'Optimization terminated successfully.'
      nit: 5
    slack: array([2.33066899e-11, 7.20632443e-11, 6.66666667e-01])
   status: 0
  success: True
        x: array([2.33333333, 4.66666667])
```

This means that the optimal value is 37.33333333324609 since we multiplied the cost function by -1 before, so we have to multiply the output from the function linprog() by -1. Thus we have the optimal value, 37.33333333324609. Since a vertex of the polytope defined by the system of linear inequalities is an optimal solution, this is a unique solution and it is

$$\begin{bmatrix} 2.333 \\ 4.667 \end{bmatrix}.$$

The linprog() function uses the **interior point** method to find the optimal value as a default method. There are many useful optimization functions to find the optimal value for a linear programming problem. Especially, Gurobi https://www.gurobi.com/ is an efficient solver for large-scale problems.

9

Network Analysis

In data science, network analysis can be a very useful tool. For example, social network analysis (SNA) can be applied to understand social connectedness between workers. Social connectedness affects stress in many ways as hypothesized by Holt-Lunstad et al. [22]. Although social connectedness usually is believed to have positive effects on health and well-being, our initial study shows that it might also have negative effects, as we linked social connectedness to increasing destructive behaviors such as illegal drug use and abusive behavior. Network analysis can be applied to neurology (for example, [29]), genetics (for example, [44]), and terrorist networks (for example, [34]). In this chapter we will show some applications of tools from linear algebra to network analysis in data science using the igraph [12] and igraphdata packages [1]. Particularly, we will use the "UKFaculty" data from the igraphdata package as the Introductory Example in this chapter.

9.1 Introductory Example from Network Analysis

The "UKFaculty" data from the igraphdata package contains a social network constructed from the personal friendship network of a faculty in a university in the United Kingdom. In this data set, there are 81 individuals (faculty members).

Basics, including definitions of a graph and network in graph theory, will be discussed in the following section. Here we would like to show the "big picture" for how we can apply networks to social network analysis in data science using this data set. In this "social network", we assign each faculty member a vertex, and we draw a "directed edge" from a vertex a to another vertex b if the faculty member (a vertex a) references the faculty member's work (a vertex b's work). In this social network there are 817 directed edges (citations among these faculty members).

In this analysis we are interested in how strong collaborations between faculty members at the university are within the community. The first thing we can do is to plot the network. The collaboration network between faculty members at a university in the UK is shown in Figure 9.1. This figure is created using the plot() function from the igraph package. As you can see in

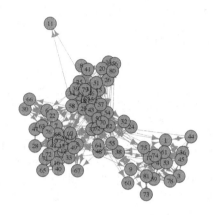

FIGURE 9.1
Visualization of a collaboration network between faculty members in a university in the UK.

Figure 9.1, it is not easy to see what is going on since this network is too big to plot. In data science this is not uncommon. Usually, in network analysis, a network has a large number of vertices and edges, so it is hard for us to visualize. Typically, a network in data science has millions of vertices and millions of edges, so it is hard for human to analyze visually. In this case, in order to visualize clearly, we take a subset of the collaboration network using the induced_subgraph() function from the `igraph` package shown in Figure 9.2.

In Figure 9.2, we see two isolated vertices, vertices 8 and 11. These are faculty members who do not collaborate among this subset of faculty members. Also for this subset of faculty members there are five components of subgroups. These subgroups represent collaborations among faculty members in the subgroup. Maybe from these collaboration networks we might be able to know what would be popular subjects among these faculty members. So now the questions are:

1. Are there any isolated vertices in the network?

2. If so how many?

3. How many components of subgroups?

To answer these questions we will use tools from linear algebra.

In the following section we will discuss some basics of graphs and networks and then we will discuss how we can answer the questions above.

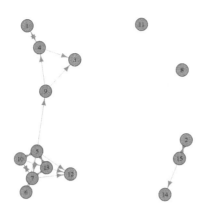

FIGURE 9.2
Visualization of a subset of a collaboration network between faculty members in a university in the UK.

9.2 Graphs and Networks

In this section, we will discuss the basics of graphs and networks including their definitions. Then we will discuss some properties from graph theory. For more details see [45], for example.

9.2.1 Task Completion Checklist

• During the Lecture:

1. Read the definition of an undirected graph.

2. Read the definition of a directed graph.

3. Read the definition of a network.

4. Read the definition of the adjacency matrix of a directed or undirected graph.

5. Read the definition of the degree matrix of a directed or undirected graph.

6. Read the definition of a Laplacian matrix of a directed or undirected graph.

7. Read the definition of a transition matrix of an undirected graph.

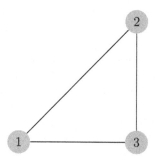

FIGURE 9.3
The triangle with a vertex set $V = \{1, 2, 3\}$ in Example 158.

8. Read the definition of a normalized Laplacian matrix of an undirected graph.

9. Read the theorem on the dimension of the null space of the Laplacian matrix of an undirected graph.

10. Read the theorem on the eigen values of the normalized Laplacian matrix of an undirected graph.

11. Learn how to visualize a graph or a network using the `igraph` package.

12. With R, learn how to perform computational examples in this section.

- After the Lecture:

  1. Conduct practical applications with R.

  2. If you are interested in `python`, read the supplement in this section and conduct lab exercises and practical applications with `python`.

9.2.2 Basic Terminology

Here we discuss some basic definitions in graph theory. First we will define a directed graph and then we will discuss directed graphs.

Definition 54 *Suppose we have a finite set of vertices (or nodes)* $V = \{1, 2, \ldots, n\}$. *Each vertex in V is labeled. If we want to pair vertices u, v in V, then we draw an edge between them. This is called an* **undirected edge.** *An* **undirected graph** *G is an object which consists of a* **vertex set** *V and a set of undirected edges E. We denote an undirected graph as $G = (V, E)$.*

Example 158 *The triangle with vertices $V = \{1, 2, 3\}$ shown in Figure 9.3 is an undirected graph.*

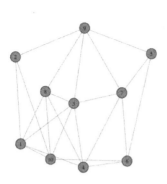

FIGURE 9.4
Random undirected graph generated under the Erdös–Rényi model with 10
vertices and a probability of connecting two vertices equal to 0.5 shown in
Example 159.

Example 159 *Suppose we have* $V = \{1, 2, \ldots, 10\}$. *Then, the undirected
graph* G *shown in Figure 9.4 has 23 undirected edges.* G *is generated by the
random graph process called the Erdös–Rényi model [45]. The* R *code that generated the undirected graph is the following:*

```
library(igraph)
G <- sample_gnp(10, 0.5)
plot(G)
```

Definition 55 *Suppose we have a finite set of vertices (or nodes)* $V =
\{1, 2, \ldots, n\}$. *Each vertex in* V *is labeled. If there is a relation from a ver-
tex* v *in* V *to a vertex* u *in* V, *then we draw an arrow. This arrow is called
a* **directed edge**. *An* **directed graph** N *is an object which consists of a*
vertex set V *and a set of directed edges* E. *We denote a directed graph as*
$N = (V, E)$. *Often a directed graph is also called a* **network**. *In this chapter,
we call a directed graph a network.*

Example 160 *Suppose* $V = \{1, 2, 3, 4, 5\}$. *A set of directed edges consists of
edges* $(1, 2)$, $(1, 3)$, $(1, 4)$, $(4, 3)$, $(3, 2)$, $(2, 5)$.

Example 161 *Suppose we have* $V = \{1, 2, \ldots, 10\}$. *Then, the network* N
shown in Figure 9.6 has 9 directed edges. N *is generated by the random graph
process called the Barabashi model [45]. We generate a random network using the barabasi.game() function from the* igraph *package. The* R *code that
generated the network is the following:*

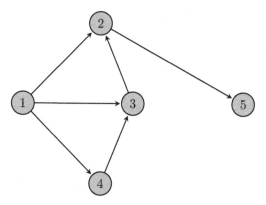

FIGURE 9.5
A directed graph with a vertex set $V = \{1, 2, \ldots, n\}$ shown in Example 160.

```
library(igraph)
N <- barabasi.game(10)
plot(N)
```

Definition 56 *Suppose we have an undirected graph $G = (V, E)$ with a vertex set $V = \{1, 2, \ldots, n\}$. Then the* **adjacency matrix** *of an undirected graph G is an $n \times n$ matrix A such that*

$$A = \begin{bmatrix} a_{11} & a_{12} & \cdots & a_{1n} \\ a_{21} & a_{22} & \cdots & a_{2n} \\ \vdots & \vdots & \vdots & \vdots \\ a_{n1} & a_{n2} & \cdots & a_{nn} \end{bmatrix}$$

where

$$a_{ij} = \begin{cases} 1 & \text{if there is an undirected edge between a vertex } i \text{ and a vertex } j \\ 0 & \text{otherwise.} \end{cases}$$

Example 162 *From Example 158, the adjacency matrix of the triangle is*

$$A = \begin{bmatrix} 0 & 1 & 1 \\ 1 & 0 & 1 \\ 1 & 1 & 0 \end{bmatrix}.$$

Remark 9.1 *The adjacency matrix of an undirected graph is a symmetric matrix.*

Example 163 *Using the as_adjacency_matrix() function from the* igraph *package, we can compute the adjacency matrix from a graph or a network. For example, we consider a random graph from Example 159. Then, if we want to compute the adjacency matrix of the graph, we type in* R:

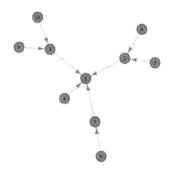

FIGURE 9.6
Random undirected graph generated under the Erdös–Rényi model with 10 vertices and a probability of connecting two vertices equal to 0.5 shown in Example 161.

```
as_adjacency_matrix(G)
```

R *outputs as follows:*

```
> as_adjacency_matrix(G)
10 x 10 sparse Matrix of class "dgCMatrix"

 [1,] . 1 1 1 . . . 1 . 1
 [2,] 1 . . . . . . . 1 1
 [3,] 1 . . 1 . . 1 1 1 1
 [4,] 1 . 1 . . 1 1 1 . .
 [5,] . . . . . 1 1 . 1 .
 [6,] . . . 1 1 . 1 . . 1
 [7,] . . 1 1 1 1 . . 1 .
 [8,] 1 . 1 1 . . . . 1 1
 [9,] . 1 1 . 1 . 1 1 . .
[10,] 1 1 1 . . 1 . 1 . .
```

Here "." means 0.

Definition 57 *Suppose we have a network* $N = (V, E)$ *with a vertex set* $V = \{1, 2, \ldots, n\}$. *Then the* **adjacency matrix** *of a network* N *is an* $n \times n$ *matrix*

A such that

$$A = \begin{bmatrix} a_{11} & a_{12} & \cdots & a_{1n} \\ a_{21} & a_{22} & \cdots & a_{2n} \\ \vdots & \vdots & \vdots & \vdots \\ a_{n1} & a_{n2} & \cdots & a_{nn} \end{bmatrix}$$

where

$$a_{ij} = \begin{cases} 1 & \text{if there is a directed edge from a vertex } i \text{ to a vertex } j \\ 0 & \text{otherwise.} \end{cases}$$

Example 164 *From Example 160, the adjacency matrix of the network is*

$$A = \begin{bmatrix} 0 & 1 & 1 & 1 & 0 \\ 0 & 0 & 0 & 0 & 1 \\ 0 & 1 & 0 & 0 & 0 \\ 0 & 0 & 1 & 0 & 0 \\ 0 & 0 & 0 & 0 & 0 \end{bmatrix}.$$

Remark 9.2 *The adjacency matrix of a network is not symmetric.*

Example 165 *Using the as_adjacency_matrix() function from the* igraph *package, we will compute the adjacency matrix of the network from Example 161.*

```
as_adjacency_matrix(N)
```

R *outputs as follows:*

```
> as_adjacency_matrix(N)
10 x 10 sparse Matrix of class "dgCMatrix"

 [1,] . . . . . . . . . .
 [2,] 1 . . . . . . . . .
 [3,] 1 . . . . . . . . .
 [4,] 1 . . . . . . . . .
 [5,] . 1 . . . . . . . .
 [6,] . 1 . . . . . . . .
 [7,] 1 . . . . . . . . .
 [8,] . . 1 . . . . . . .
 [9,] . . . . . . 1 . . .
[10,] . . 1 . . . . . . .
```

Definition 58 *The* **degree** *of a vertex v in V for a graph or a network is the number of edges adjacent to the vertex v. The* **degree matrix** *of a graph* or a network with a vertex set $V = \{1, 2, \ldots n\}$ is an $n \times n$ diagonal matrix D such that

$$
D = \begin{bmatrix}
d_1 & 0 & \cdots & 0 & 0 \\
0 & d_2 & \cdots & 0 & 0 \\
\vdots & \vdots & \ddots & \vdots & \vdots \\
0 & 0 & \cdots & d_{n-1} & 0 \\
0 & 0 & \cdots & 0 & d_n
\end{bmatrix}
$$

where d_i is the degree of a vertex i in V.

Example 166 *Consider the triangle from Example 158. The degree matrix of the triangle is*

$$
D = \begin{bmatrix}
2 & 0 & 0 \\
0 & 2 & 0 \\
0 & 0 & 2
\end{bmatrix}.
$$

Example 167 *Consider the example from Example 160. Then the degree matrix of the network is*

$$
D = \begin{bmatrix}
3 & 0 & 0 & 0 & 0 \\
0 & 3 & 0 & 0 & 0 \\
0 & 0 & 3 & 0 & 0 \\
0 & 0 & 0 & 2 & 0 \\
0 & 0 & 0 & 0 & 1
\end{bmatrix}.
$$

Definition 59 *The **Laplacian matrix** of a graph or a network with the vertex set $V = \{1, 2, \ldots, n\}$ is an $n \times n$ matrix L such that*

$$
L = D - A,
$$

where D is the degree matrix of the graph or the network and A is the adjacency matrix of the graph or the network.

Example 168 *From Example 158, the adjacency matrix of the triangle is*

$$
A = \begin{bmatrix}
0 & 1 & 1 \\
1 & 0 & 1 \\
1 & 1 & 0
\end{bmatrix},
$$

and the degree matrix of the triangle is

$$
D = \begin{bmatrix}
2 & 0 & 0 \\
0 & 2 & 0 \\
0 & 0 & 2
\end{bmatrix}.
$$

Therefore, the Laplacian matrix of the triangle is

$$
L = \begin{bmatrix}
2 & -1 & -1 \\
-1 & 2 & -1 \\
-1 & -1 & 2
\end{bmatrix}.
$$

Example 169 *From Example 160, the adjacency matrix of the network is*

$$A = \begin{bmatrix} 0 & 1 & 1 & 1 & 0 \\ 0 & 0 & 0 & 0 & 1 \\ 0 & 1 & 0 & 0 & 0 \\ 0 & 0 & 1 & 0 & 0 \\ 0 & 0 & 0 & 0 & 0 \end{bmatrix},$$

and the degree matrix of the network is

$$D = \begin{bmatrix} 3 & 0 & 0 & 0 & 0 \\ 0 & 3 & 0 & 0 & 0 \\ 0 & 0 & 3 & 0 & 0 \\ 0 & 0 & 0 & 2 & 0 \\ 0 & 0 & 0 & 0 & 1 \end{bmatrix}.$$

Therefore, the Laplacian matrix of the network is

$$L = \begin{bmatrix} 3 & -1 & -1 & -1 & 0 \\ 0 & 3 & 0 & 0 & -1 \\ 0 & -1 & 3 & 0 & 0 \\ 0 & 0 & -1 & 2 & 0 \\ 0 & 0 & 0 & 0 & 1 \end{bmatrix}.$$

Definition 60 *The* **transition matrix** *for an undirected graph $G = (V, E)$ with the vertex set $V = \{1, 2, \ldots n\}$ is an $n \times n$ matrix such that*

$$P = \begin{bmatrix} p_{11} & p_{12} & \cdots & p_{1n} \\ p_{21} & p_{22} & \cdots & p_{2n} \\ \vdots & \vdots & \vdots & \vdots \\ p_{n1} & p_{n2} & \cdots & p_{nn} \end{bmatrix}$$

where

$$p_{ij} = a_{ij}/d_i$$

with a_{ij} is the (i, j)th element of the adjacency matrix of the graph G and d_i is the degree of the vertex i in the graph G.

Example 170 *From Example 158, the transition matrix of the triangle is*

$$P = \begin{bmatrix} 0 & 1/2 & 1/2 \\ 1/2 & 0 & 1/2 \\ 1/2 & 1/2 & 0 \end{bmatrix}.$$

Definition 61 *The* **normalized Laplacian matrix** *for an undirected graph $G = (V, E)$ with the vertex set $V = \{1, 2, \ldots n\}$ is an $n \times n$ matrix L_P such that*

$$I_n - P$$

where P is the transition matrix of the graph G.

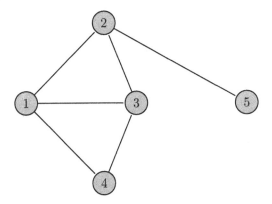

FIGURE 9.7
A undirected graph with a vertex set $V = \{1, 2, \ldots, n\}$ shown in Example 172.

Example 171 *From Example 158, the transition matrix of the triangle is*

$$P = \begin{bmatrix} 0 & 1/2 & 1/2 \\ 1/2 & 0 & 1/2 \\ 1/2 & 1/2 & 0 \end{bmatrix}.$$

Therefore, the normalized Laplacian matrix L_P is

$$L_P = \begin{bmatrix} 1 & -1/2 & -1/2 \\ -1/2 & 1 & -1/2 \\ -1/2 & -1/2 & 1 \end{bmatrix}.$$

9.2.3 Properties of Laplacian Matrices

Laplacian matrices tell us a lot about the properties about graphs. In order to show these properties we have to use tools from linear algebra. In this section we will discuss some of the well-known properties of Laplacian matrices.

Here we assume we have undirected graphs. If we have a directed graph, i.e., a network, we make a directed edge an undirected edge.

Example 172 *Consider the network from Example 160. We will make this network an undirected graph. A set of directed edges consists of undirected edges $(1,2)$, $(1,3)$, $(1,4)$, $(4,3)$, $(3,2)$, $(2,5)$.*

Then the adjacency matrix of the graph is

$$A = \begin{bmatrix} 0 & 1 & 1 & 1 & 0 \\ 1 & 0 & 1 & 0 & 1 \\ 1 & 1 & 0 & 1 & 0 \\ 1 & 0 & 1 & 0 & 0 \\ 0 & 1 & 0 & 0 & 0 \end{bmatrix},$$

and the degree matrix of the graph is

$$D = \begin{bmatrix} 3 & 0 & 0 & 0 & 0 \\ 0 & 3 & 0 & 0 & 0 \\ 0 & 0 & 3 & 0 & 0 \\ 0 & 0 & 0 & 2 & 0 \\ 0 & 0 & 0 & 0 & 1 \end{bmatrix}.$$

Therefore, the Laplacian matrix of the graph is

$$L = \begin{bmatrix} 3 & -1 & -1 & -1 & 0 \\ -1 & 3 & -1 & 0 & -1 \\ -1 & -1 & 3 & -1 & 0 \\ -1 & 0 & -1 & 2 & 0 \\ 0 & -1 & 0 & 0 & 1 \end{bmatrix}.$$

We will use the Laplacian matrix of the graph in Example 172 as a walking example in this section.

Theorem 9.3 *The Laplacian matrix L of an undirected graph G is symmetric.*

Theorem 9.4 *The dimension of the null space of the Laplacian matrix L of an undirected graph G is the number of components in the graph G.*

Using Theorem 9.4, we can see how many components are in the given graph. This can be especially useful when we are dealing with a large graph which can be seen commonly in data science.

Example 173 *From Example 172, the Laplacian matrix L is*

$$L = \begin{bmatrix} 3 & -1 & -1 & -1 & 0 \\ -1 & 3 & -1 & 0 & -1 \\ -1 & -1 & 3 & -1 & 0 \\ -1 & 0 & -1 & 2 & 0 \\ 0 & -1 & 0 & 0 & 1 \end{bmatrix}.$$

We will use the nullspace() function from the **pracma** *package from* R*:*

```
library(pracma)
L <- matrix(c(3, -1, -1, -1, 0, -1, 3, -1, 0, -1, -1, -1, 3, -1, 0, -1,
0, -1, 2, 0, 0, -1, 0, 0, 1), 5, 5)
nullspace(L)
```

R *outputs the following:*

```
> nullspace(L)
        [,1]
[1,] 0.4472136
[2,] 0.4472136
[3,] 0.4472136
[4,] 0.4472136
[5,] 0.4472136
```

Since there is only one vector to span the null space of L, the dimension of the null space of L is 1. Therefore, there is only one component in the graph which we can verify from Figure 9.7.

Theorem 9.5 ([10]) *Suppose $\lambda_0, \ldots, \lambda_{n-1}$ are distinct eigen values of the normalized Laplacian matrix L_P of an undirected graph $G = (V, E)$ with the vertex set $V = \{1, 2, \ldots, n\}$ such that*

$$\lambda_0 \leq \lambda_1 \leq \ldots \leq \lambda_{n-1}.$$

Then

$$\sum_{i=0}^{n-1} \lambda_i \leq n$$

and the equality holds if and only if there is no isolated vertices.

Example 174 *From Example 172, the normalized Laplacian matrix L_P is*

$$L = \begin{bmatrix} 1 & -1/3 & -1/3 & -1/3 & 0 \\ -1/3 & 1 & -1/3 & 0 & -1/3 \\ -1/3 & -1/3 & 1 & -1/3 & 0 \\ -1/2 & 0 & -1/2 & 1 & 0 \\ 0 & -1 & 0 & 0 & 1 \end{bmatrix}.$$

We will use the output from Example 173 and the eigen() function from R. First, we will compute the normalized Laplacian matrix of the graph:

```
LL <- L
LL[1,] <- LL[1,]/LL[1,1]
LL[2,] <- LL[2,]/LL[2,2]
LL[3,] <- LL[3,]/LL[3,3]
LL[4,] <- LL[4,]/LL[4,4]
LL[5,] <- LL[5,]/LL[5,5]
```

Then we will compute the eigen values from the eigen() function from R:

```
E <- eigen(LL)
```

This outputs eigen values and eigen vectors. To obtain eigen values, we use the $values command:

```
E$values
```

This prints all eigen values:

```
> E$values
[1] 1.767592e+00 1.333333e+00 1.333333e+00 5.657415e-01 1.340063e-16
```

Then we sum them up using the sum() function and we obtain:

```
> sum(E$values)
[1] 5
```

The sum of the eigen values of the normalized Laplacian matrix of this graph is 5, which is equal to the number of vertices. Thus there is no isolated vertex in this graph.

9.2.4   Checkmarks

1. The definition of an undirected graph.

2. The definition of a directed graph (network).

3. The definition of the adjacency matrix of a directed or undirected graph.

4. The definition of the degree matrix of a directed or undirected graph.

5. The definition of a Laplacian matrix of a directed or undirected graph.

6. The definition of a transition matrix of an undirected graph.

7. The definition of a normalized Laplacian matrix of an undirected graph.

8. The theorem on the dimension of the null space of the Laplacian matrix of an undirected graph.

9. The theorem on the Eigen values of the normalized Laplacian matrix of an undirected graph.

10. With R, you can perform computational examples in this section.

9.2.5   Practical Applications

In this practical application, we will use the macaque data set from the igraphdata package. This directed graph (network) is designed to model the visuotactile brain areas in the macaque monkey to understand how neurons in the brain fire to each other. The model consists of 45 areas (45 vertices in the network) and 463 directed connections (463 directed edges in the network). First we visualize the network shown in Figure 9.8.

In this section we ask the following questions:

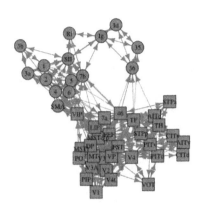

FIGURE 9.8
Network to model the visuotactile brain areas in he macaque monkey from
the macaque data set from the `igraphdata` package.

1. How many are there components in the brain network?

2. Are there any isolated vertices?

We can answer the first question using Theorem 173. We use the
graph.laplacian() function from the `igraphdata` package. First we upload libraries:

```
library(igraph)
library(igraphdata)
```

Then we upload the macaque dataset:

```
data("macaque")
```

Then we compute the Laplacian matrix of the **undirected graph** of the
macaque dataset.

```
LM <- graph.laplacian(macaque)
```

Then, in order to compute the null space of the Laplacian matrix, we have
to make a numerical matrix format in R:

```
LM <- as.numeric(LM)
LLM <- matrix(LM, 45, 45)
```

Then we use the nullspace() function to compute a basis for the null space of the Laplacian matrix of the graph:

```
nullspace(LLM)
```

This outputs one vector. Since there is only one vector in a basis for the null space of the Laplacian matrix of the graph, there is only one component in the graph.

Now we will answer the second question using Theorem 9.5. First we need to compute the normalized Laplacian matrix of the undirected graph for the macaque dataset.

```
LM2 <- graph.laplacian(macaque, normalized = TRUE)
```

Then we convert this as a numerical matrix:

```
LM2 <- as.numeric(LM2)
LLM2 <- matrix(LM2, 45, 45)
```

Now we compute the eigen values using the eigen() function:

```
EE <- eigen(LLM2)
```

Then we compute the sum of the eigen values of the normalized Laplacian matrix of the graph using the $values command and the sum() function:

```
sum(EE$values)
```

It outputs the following:

```
> sum(EE$values)
[1] 45+0i
```

Since there are 45 vertices, using Theorem 9.5, we know that there are no isolated vertices.

9.2.6   Supplements with python Code

There is the **python-igraph** package for python. In order to install it, one can type:

```
pip install python-igraph
```

To upload the package we type:

```
>>> import igraph
>>> from igraph import *
```

A nice tutorial on the `python-igraph` package for `python` can be found at `https://igraph.org/python/doc/tutorial/tutorial.html#` `starting-igraph` and we recommend that the reader consults this tutorial for more details.

9.3 Discussion

In this section we go back to the Introductory Example in this chapter. Consider the UKfactory dataset from the `igraphdata` package. Now we use Theorem 173 and Theorem 9.5 to answer the following questions:

1. How many components in the collaboration network?

2. Are there any isolated vertices in the collaboration network?

First we upload libraries:

```
library(igraph)
library(igraphdata)
```

Then we upload the UKfaculty dataset:

```
data("UKfaculty")
```

Then we compute the Laplacian matrix of the **undirected graph** of the UKfaculty dataset.

```
LF <- graph.laplacian(UKfaculty)
```

Then in order to compute the null space of the Laplacian matrix, we have to make a numerical matrix format in R:

```
LF <- as.numeric(LF)
LLF <- matrix(LF, 81, 81)
```

Then we use the nullspace() function to compute a basis for the null space of the Laplacian matrix of the graph:

```
nullspace(LLF)
```

This outputs one vector. Since there is only one vector in a basis for the null space of the Laplacian matrix of the graph, there is only one component in the graph.

Now we will answer the second question using Theorem 9.5. First we need to compute the normalized Laplacian matrix of the undirected graph for the UKfaculty dataset. However, we have trouble here. It seems that the graph.laplacian() function cannot produce the correct normalized Laplacian matrix for this example. It seems that there are some rounding errors. Therefore, we will create the normalized Laplacian matrix of the UKfaculty data from the definition. To do so, first we will compute the adjacency matrix of this undirected graph. The as.undirected() function makes a network an undirected graph. Also, the get.adjacency() function creates the ajdacency matrix of a graph:

```
AF <- get.adjacency(as.undirected(UKfaculty))
```

Then we compute the degrees of vertices using the degree() function:

```
DF<-degree(UKfaculty)
```

Then we convert the adjacency matrix as a numerical matrix:

```
AAF <- matrix(as.numeric(AF), 81, 81)
```

Now we create the transition matrix:

```
for(i in 1:81)
    AAF[i, ] <- AAF[i,]/DF[i]
```

Then we create the normalized Laplacian matrix of this graph:

```
LF3 <- eye(81) - AAF
```

Now we compute the eigen values using the eigen() function:

```
EF <- eigen(LF3)
```

Then we compute the sum of the eigen values of the normalized Laplacian matrix of the graph using the $values command and the sum() function:

```
sum(EF$values)
```

It outputs the following:

```
> sum(EF$values)
[1] 81
```

Since there are 81 vertices, using Theorem 9.5, we know that there are no isolated vertices.

Appendices

A

Introduction to RStudio via Amazon Web Services (AWS)

Amazon Web Services (AWS) is a cloud computing service by Amazon. During the epidemic episode of COVID-19, lectures given on-line have gotten much attention in order to keep social distance. However, since lectures are traditionally given in a classroom, there are new challenges for preparing and giving a lecture on-line: interactions between the instructor and students. As we see from the outline of course development shown in Table 0.1, solving problems during each lecture is an important component of the lecture. For a linear algebra course with this textbook we focus on R and RStudio to solve problems.

There are several benefits to using RStudio via AWS: (1) all students and an instructor have the same operating system, so no one has problems installing a package; (2) students do not have to install RStudio in their computer and all they need is a web browser.

If lectures are given in a classroom, then an instructor might be able to resolve such students' problems, however, it is a challenge for an instructor to help students resolve installation problems in their computers. Therefore, AWS can be a great tool for an on-line course. Therefore, in this chapter we provide a quick summary of AWS and how to create an environment for RStudio within AWS.

A.1 Setting Up AWS

Amazon Web Services (AWS) provides cloud computing services which are useful for establishing an analytical environment. Cloud computing allows a student to use services from anywhere with internet access and with a web browser. This section provides instructions for setting up an RStudio statistical software computing environment on AWS. RStudio is software that runs R.

There are several benefits for an instructor using RStudio via AWS:

1. A student does not have to install RStudio on their computer.

DOI: 10.1201/9781003042259-A

2. Since we have the same operating system (OS), we do not have any problems with dependencies for installing libraries.

3. Some R packages require compilers for C++. It can be very complicated for a student to install a compiler for C++ in their computer. Since AWS has the Linux OS, RStudio calls the C++ compiler automatically without any issues.

4. An instructor can test all code in AWS before each class, so they can prepare for any problems which students may encounter during a lab.

5. Since everyone uses the same OS with the same amount of memory, an instructor can anticipate the speed and memory requirements for each function. Therefore, the instructor can better manage and plan time for each lab. If some part of the code takes too long to compute during a lab, the instructor can provide students with an "RData" file that contains binaries before the lab.

The steps for students to set up RStudio on AWS are as follows:

1. Log in to AWS.

2. Configure and launch an RStudio Amazon Machine Image (AMI) via http://www.louisaslett.com/RStudio_AMI/.

3. Configure an RStudio Integrated Development Environment (IDE). If you successfully build an environment, then the "public DNS" is also listed at the bottom of the page. It should look something like ecX-XX-XXX-XX-X. compute-1.amazonaws.com. Log in to your RStudio IDE: The login username and password are:

> Username: rstudio
> Password: Your instance ID

4. Close down RStudio AMI. If one forgets to log out of an AMI, then it might cost them extra. It is important to log out after they finish computations.

5. Also, it is important to remember to stop the instance at AWS. If it is not stopped, then it keeps running and you might end up paying a large amount of money.

A.2   Basics in RStudio

RStudio is a great tool in data science and bioinformatics including phylogenetics. Through this course, students will be familiar with RStudio through each lab and their weekly homework. A first-week take-home lab assignment consists of some basic programming in RStudio. Before programming in RStudio, we will show basics in RStudio, such as:

- Set up a working directory. This can be done in RStudio by clicking "Session", "Set Working Directory" and then "Choose Directory."

- Save the raw data as an ".RData" file. This file is saved in the "css_data" directory in the RStudio instance on AWS. This can be done by clicking "Session", and "Save Workspace As".

- RStudio has four windows in the screen as shown in Figure A.1. You write R script in window I in Figure A.1, but it does not run until you click "Run" in the right top corner of window I to run. Running code is shown in window III in Figure A.1. Window II shows the variables you created in the session and window IV in Figure A.1 shows files, plots, packages, help, and viewer. You can find them by clicking the button in the left top of window IV.

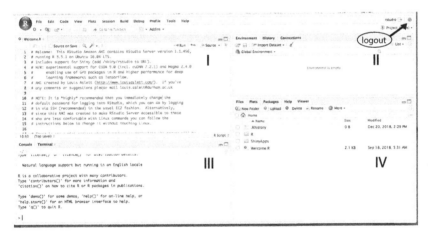

FIGURE A.1
RStudio landing page.

- To install R packages, you can click "Tools", and then click "Install Packages" or type the install.packages() command in R script in window I and click "Run". For example, if you want to install R package "ISLR" [24], then **just** type

```
install.packages("ISLR", dependencies = TRUE)
```

You should not forget the quotation marks. The second argument "dependencies" should be set as "TRUE" when you also want to install dependent packages. To call a package, use the library() function. For example

```
library(ISLR)
```

B

Introduction to R

R is a free programming language and environment for statistical analysis and data science and is especially popular in bioinformatics. The R Foundation for Statistical Computing (https://www.r-project.org/) supports and maintains the majority of R packages. All packages are freely available and we can download and install them using R. For advanced programmers, we can also integrate R with C, C++, and Fortran code. It is possible to combine R with python code for advanced computations as well.

One of the biggest advantages over python is that most R packages are maintained by the R Foundation for Statistical Computing which enforces strict rules on packages. Thus, the manuscript has the same format for all packages so that it is easy for users to read. In addition, the foundation maintains a version of each package and its dependencies so that it is easy for users to install.

B.1 Display in R

The first step to set up the R programming environment is to download R from https://www.r-project.org/. If you are using Amazon Web Services (AWS), RStudio, an integrated development environment (IDE) for R, is already set up and ready for you to use.

After setting up R or RStudio, when you open R, you can see that the R displays the screen shown in Figure B.1. If you are using RStudio, the console is the bottom left square shown in Figure A.1. Note that through this book, commands preceded with $ are at the command line, > are in an R console, and >>> are in a python console.

B.2 Setting Up the R Programming Environment

Before we begin bringing data into the R programming environment, we need to install and load the R package ISLR [24] that we will use for this lab. We will

```
(base) MacBook-Pro:ISM rurikoyoshida$ R

R version 3.6.3 (2020-02-29) -- "Holding the Windsock"
Copyright (C) 2020 The R Foundation for Statistical Computing
Platform: x86_64-apple-darwin15.6.0 (64-bit)

R is free software and comes with ABSOLUTELY NO WARRANTY.
You are welcome to redistribute it under certain conditions.
Type 'license()' or 'licence()' for distribution details.

  Natural language support but running in an English locale

R is a collaborative project with many contributors.
Type 'contributors()' for more information and
'citation()' on how to cite R or R packages in publications.

Type 'demo()' for some demos, 'help()' for on-line help, or
'help.start()' for an HTML browser interface to help.
Type 'q()' to quit R.

>
```

FIGURE B.1
Console display in R.

use the install.packages() function to install the ISLR and modeling packages.

```
install.packages("ISLR", dependencies = TRUE)
```

The first argument is the name of the package you want to install. Note that you have to add quotation marks. The second argument of the install.packages() function "dependencies = TRUE" tells the function to install any additional packages needed for the ISLR package. The library commands load the needed packages into your programming environment so that you can use the programmed functions (often called commands in R parlance) within them. You will see how to leverage these commands below.

```
library(ISLR)
```

In RStudio go to the top of your RStudio session and click "Tools" and then "Install Packages". Then type "ISLR" in Packages and make sure you check in the box to install dependencies. Then click "Install".

B.3 Getting Started with Objects in R

If you want to conduct simple arithmetic, for example 2 + 3, then just type

```
2 + 3
```

then it will output 5 as the answer. If you want to compute $e^5 = \exp(5) = 148.4132\cdots$, then just type

```
exp(5)
```

then it will output the answer.

```
> exp(5)
[1] 148.4132
```

Since R is a programming language, we have to print out "Hello World!". In order to do so, you can use the print() function.

```
print("Hello World!")
```

Do not forget to use quotation marks.

B.3.1   Assigning an Object (Variable)

Since R is a programming language, you can assign a value to a variable. For example, we are going to assign "Hello World!" to the variable "greeting".

```
greeting <- "Hello, world!"
```

Then if you type "greeting" in R, it returns:

```
> greeting
[1] "Hello, world!"
```

The object (variable) "greeting" has been assigned the expression "Hello World." This is done via the assignment operator <-. The assignment operator is made using the left-pointing arrow (the same key used to make a comma) and a dash (usually right next to the zero at the top of your keyboard). Think of the assignment operator as an arrow that maps an expression to an object (or variable).

Clearly we can assign a number to a variable as well. For example

```
seven <- 7
```

If you type "seven" in R it returns 7. Since R is case sensitive (bugs are caused by case sensitiveness), if you type

```
Seven
```

then it returns

TABLE B.1

Information on Boolean operators in R

==	The right side and the left side are equal to each other
!=	The right side and the left side are not equal to each other
>	The right side is strictly less than the left side.
>=	The right side is less than or equal to the left side.
<	The right side is strictly bigger than the left side.
<=	The right side is bigger than or equal to the left side.

```
> Seven
Error: object 'Seven' not found
```

We can use simple arithmetic with such variables. For example we can do

```
seven + 5
```

then it returns

```
> seven + 5
[1] 12
```

However, if we try to add 5 to "greeting", as

```
greeting + 5
```

then it returns an error message because the type of the variable "seven" is numeric and the type of "greeting" is character.

```
> greeting + 5
Error in greeting + 5 : non-numeric argument to binary operator
```

B.3.2   Boolean Operators

We can use Boolean operators like "==", "!=", ">", ">=", "<", and "<=". Boolean operators can be used to test whether a given statement is true or false and are used frequently in programs. The meanings of these Boolean operators are listed in Table B.1.

If you type

```
seven == 7
```

then it returns

```
> seven == 7
[1] TRUE
```

If you type

```
seven != 7
```

then it returns

```
> seven != 7
[1] FALSE
```

B.4 Saving a Session and Data

To save a session, go to "File" and click "Save". It saves a session as a plain text file. This is sometimes not enough after a long computation. In order to save all computed data, you have to save as an "RData" file. Go to "Workspace" and click "Save Workspace File..." and set up the directory and name of the workspace file with ".RDate" at the end. In order to load the workspace file, go to "Workspace" and click "Load Workspace File...".

Bibliography

[1] igraphdata, 2015. Available at https://CRAN.R-project.org/package=igraphdata.

[2] D. Adler, D. Murdoch, et al. rgl package, 2020. https://cran.r-project.org/web/packages/rgl/index.html.

[3] A. Agresti. *Categorical Data Analysis*. Wiley, New York, New York, 1990.

[4] S. A. Ambrose, M. W. Bridges, M. DiPietro, M. C. Lovett, and M. K. Norman. *How Learning Works: Seven Research-Based Principles for Smart Teaching*. John Wiley & Sons, Inc. New York, NY, USA, 2010.

[5] A. Asase. The transportation problem: Case study: Guinness Ghana Limited. Masters degree thesis of Industrial Mathematics at the department of Mathematics in University of Science and Technology, Kumasi.

[6] M. Berkelaar et al. lpsolve package, 2020. https://cran.r-project.org/web/packages/lpSolve/index.html.

[7] H.W. Borchers. pracma package, 2019. https://cran.r-project.org/web/packages/pracma/index.html.

[8] H. Corrada Bravo, K. Hornik, and S. Theussl. Rcplex package, 2016. https://cran.r-project.org/web/packages/Rcplex/index.html.

[9] Hugh Chisholm. *Encyclopædia Britannica*. Cambridge University Press, 11 edition, 1911.

[10] F.R.K. Chung. *Spectral Graph Theory*. AMS, 1997.

[11] IBM ILOG Cplex. V12. 1: User's manual for cplex, 2009. International Business Machines Corporation. Version 46. Number 53.

[12] G. Csardi and T. Nepusz. The igraph software package for complex network research. *InterJournal*, Complex Systems:1695, 2006.

[13] J. A. De Loera, R. Hemmecke, J. Tauzer, and R. Yoshida. Effective lattice point counting in rational convex polytopes. *Journal of Symbolic Computation*, 38(4):1273–1302, 2004. Symbolic Computation in Algebra and Geometry.

[14] S. Descamps, C. le Bohec, Y. le Maho, J.-P. Gendner, and M. Gauthier-Clerc. Relating demographic performance to breeding-site location in the king penguin. *Condor*, 111:81–87, 2009.

[15] B. Desgraupes. conics package, 2013. https://cran.r-project.org/web/packages/conics/index.html.

[16] P. Diaconis and B. Sturmfels. Algebraic algorithms for sampling from conditional distributions. *Annals of Statistics*, 26(1):363–397, 1998.

[17] M. Friendly. matlib package, 2020. https://cran.r-project.org/web/packages/matlib/index.html.

[18] A. Genz, F. Bretz, T. Miwa, X. Mi, F. Leisch, F. Scheipl, and T. Hothorn. *mvtnorm: Multivariate Normal and t Distributions*, 2020. R package version 1.1-1.

[19] C. R. Harris, K. J. Millman, S. J. van der Walt, R. Gommers, P. Virtanen, D. Cournapeau, E. Wieser, J. Taylor, S. Berg, N. J. Smith, R. Kern, M. Picus, S. Hoyer, M. H. van Kerkwijk, M. Brett, A. Haldane, J. Fern'andez del R'ıo, M. Wiebe, P. Peterson, P. G'erard-Marchant, K. Sheppard, T. Reddy, W. Weckesser, H. Abbasi, C. Gohlke, and T. E. Oliphant. Array programming with NumPy. *Nature*, 585(7825):357–362, September 2020.

[20] D. Harrison and D.L. Rubinfeld. Hedonic prices and the demand for clean air. *J. Environ. Economics and Management*, 5:81–102, 1978.

[21] M. Hervé. Rvaidememoire package, 2020. https://cran.r-project.org/web/packages/RVAideMemoire/index.html.

[22] J. Holt-Lunstad, T.B. Smith, and J.B. Layton. Social relationships and mortality risk: A meta-analytic review. *PLoS Med*, 7(7):e1000316. doi:10.1371/journal.pmed.1000316, 2010.

[23] J. D. Hunter. Matplotlib: A 2d graphics environment. *Computing in Science & Engineering*, 9(3):90–95, 2007.

[24] G. James, D. Witten, T. Hastie, and R. Tibshirani. *An Introduction to Statistical Learning with Applications in R*. Springer-Verlag, New York, New York, 2013.

[25] David Kahle, Luis Garcia-Puente, and Ruriko Yoshida. *latte: LattE and 4ti2 in R*, 2017. R package version 0.2.0.

[26] A. Kassambara. factoextra package, 2020. https://cran.r-project.org/web/packages/factoextra/index.html.

[27] W. McKinney et al. Data structures for statistical computing in python. In *Proceedings of the 9th Python in Science Conference*, volume 445, pages 51–56. Austin, TX, 2010.

[28] A. Meurer, C. P. Smith, M. Paprocki, O. Čertík, S. B. Kirpichev, M. Rock-lin, A. Kumar, S. Ivanov, J. K. Moore, S. Singh, T. Rathnayake, S. Vig, B. E. Granger, R. P. Muller, F. Bonazzi, H. Gupta, S. Vats, F. Johansson, F. Pedregosa, M. J. Curry, A. R. Terrel, Š. Roučka, A. Saboo, I. Fernando, S. Kulal, R. Cimrman, and A. Scopatz. Sympy: symbolic computing in python. *PeerJ Computer Science*, 3:e103, January 2017.

[29] L. Negyessy, T. Nepusz, L. Kocsis, and F. Bazso. Prediction of the main cortical areas and connections involved in the tactile function of the visual cortex by network analysis. *European Journal of Neuroscience*, 23(7):1919–1930, 2006.

[30] L.R. Nielsen. gmoip package, 2020. https://cran.r-project.org/web/packages/gMOIP/index.html.

[31] J. Ooms. magick package, 2020. https://cran.r-project.org/web/packages/magick/index.html.

[32] K. Pearson. On lines and planes of closest fit to systems of points in space. *Philosophical Magazine*, 2(11):559–572, 1901.

[33] F. Pedregosa, G. Varoquaux, A. Gramfort, V. Michel, B. Thirion, O. Grisel, M. Blondel, P. Prettenhofer, R. Weiss, V. Dubourg, J. Vander-plas, A. Passos, D. Cournapeau, M. Brucher, M. Perrot, and E. Duch-esnay. Scikit-learn: Machine learning in Python. *Journal of Machine Learning Research*, 12:2825–2830, 2011.

[34] S. Ressler. Social network analysis as an approach to combat terror-ism: Past, present, and future research. *Homeland Security Affairs 2*, 8:https://www.hsaj.org/articles/171, 2006.

[35] M. Roettger, G. Gelius-Dietrich, and C.J. Fritzemeier. cplexapi package, 2020. https://cran.r-project.org/web/packages/cplexAPI/index.html.

[36] A. Schlegel. Image compression with principal com-ponent analysis, 2016. https://rpubs.com/aaronsc32/image-compression-principal-component-analysis.

[37] C. Sievert. *Interactive Web-Based Data Visualization with R, plotly, and shiny*. Chapman and Hall/CRC, 2020.

[38] B. Smucker and A. B. Slavković. Cell bounds in two-way contingency tables based on conditional frequencies. *International Conference on Privacy in Statistical Databases: Privacy in Statistical Databases*, pages 64–76, 2008.

[39] Y. Tang, M. Horikoshi, and W. Li. ggfortify: Unified interface to visualize statistical result of popular r packages. *The R Journal*, 8, 2016.

[40] S. Urbanek. png package, 2013. `https://cran.r-project.org/web/packages/png/index.html`.

[41] S. Urbanek. jpeg package, 2019. `https://cran.r-project.org/web/packages/jpeg/index.html`.

[42] W. N. Venables and B. D. Ripley. *Modern Applied Statistics with S.* Springer, New York, fourth edition, 2002. ISBN 0-387-95457-0.

[43] P. Virtanen, R. Gommers, T. E. Oliphant, M. Haberland, T. Reddy, D. Cournapeau, E. Burovski, P. Peterson, W. Weckesser, J. Bright, S. J. van der Walt, M. Brett, J. Wilson, K. J. Millman, N. Mayorov, A. R. J. Nelson, E. Jones, R. Kern, E. Larson, C. J. Carey, İ. Polat, Y. Feng, E. W. Moore, J. VanderPlas, D. Laxalde, J. Perktold, R. Cimrman, I. Henriksen, E. A. Quintero, C. R. Harris, A. M. Archibald, A. H. Ribeiro, F. Pedregosa, P. van Mulbregt, and SciPy 1.0 Contributors. SciPy 1.0: Fundamental algorithms for scientific computing in Python. *Nature Methods*, 17:261–272, 2020.

[44] C. von Mering, R. Krause, B. Snel, M. Cornell, S.G. Oliver, S. Fields, and PeerBork. Comparative assessment of large-scale data sets of protein-protein interactions. *Nature*, 417:399–403, 2002.

[45] D.B. West. *Introduction to Graph Theory.* Prentice Hall, 1996.

[46] H. Wickham. *ggplot2: Elegant Graphics for Data Analysis.* Springer-Verlag New York, 2016.

[47] Y. Xie. animation: An R package for creating animations and demonstrating statistical methods. *Journal of Statistical Software*, 53(1):1–27, 2013.

Index

$m \times n$ matrix, 5
n-dimensional vector, 4
R, 395

adjugate matrix, 187
Amazon Web Services (AWS), 391

basis, 235

cofactor, 156

determinant, 155
diagonal entries, 12
diagonal matrix, 12
diagonalizable, 333
dimension, 238

echelon form, 60
eigen value, 318
eigen vector, 318
elementary matrix, 132
elementary row operations, 37

Gaussian Eliminations, 37

identity matrix, 10
inner product, 280
inverse, 114

linear combination, 232
linear equation, 29
linearly independent, 234
lower triangular matrix, 14

null space, 215, 220
null vector, 10

orthogonal, 297
orthogonal basis, 299

projection, 298

rank of a matrix, 239
reduced echelon form, 61
row space, 220
row vectors, 6

scalar multiplication, 91, 203
span, 232
spanning sets, 232
square matrix, 10
symmetric matrix, 13
system of linear equations, 30

transpose, 96

upper triangular matrix, 13

vector space, 204
vector subspace, 206

405

Printed in the United States
by Baker & Taylor Publisher Services